MUNICIPAL
BENCHMARKS

MUNICIPAL BENCHMARKS

*Assessing Local Performance and
Establishing Community Standards*

DAVID N. AMMONS

SAGE Publications
International Educational and Professional Publisher
Thousand Oaks   London   New Delhi

BHT 5187 - 9/3

GASB Research Reports, Service Efforts and Accomplishments Reporting: Its Time Has Come, *Fire Department Programs, Police Department Programs,* and *Mass Transit,* are copyrighted by the Governmental Accounting Standards Board, 401 Merritt 7, P.O. Box 5116, Norwalk, Connecticut 06856-5116, U.S.A. Portions are reprinted with permission. Copies of the complete documents are available from the GASB.

For information address:

SAGE Publications, Inc.
2455 Teller Road
Thousand Oaks, California 91320
E-mail: order@sagepub.com

SAGE Publications Ltd.
6 Bonhill Street
London EC2A 4PU
United Kingdom

SAGE Publications India Pvt. Ltd.
M-32 Market
Greater Kailash I
New Delhi 110 048 India

Printed in the United States of America

Library of Congress Cataloging-in-Publication Data

Ammons, David N.
  Municipal benchmarks: Assessing local performance and
establishing community standards / author, David N. Ammons.
    p. cm.
  Includes bibliographical references and index.
  ISBN 0-8039-7253-9 (cloth: acid-free paper)
  1. Municipal services—United States—Evaluation. 2. Municipal
government—United States—Evaluation. 3. Benchmarking
(Management)—United States. I. Title.
  HD4605.A667 1996
  352′.000476—dc20   96-4478

96 97 98 99 10 9 8 7 6 5 4 3 2 1

This book is printed on acid-free paper.
Sage Production Editor: Diana E. Axelsen
Sage Typesetter: Danielle Dillahunt

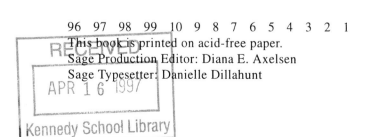

Contents

Preface

City governments need performance benchmarks, if they are serious about the efficient delivery of quality services. And their citizens need municipal benchmarks, if they are not!

Some years ago, as a local government practitioner, I fielded intermittent inquiries from council members and citizens who were concerned about our police department, our parks and recreation department, or our public works department and wondered whether local performance was up to par. Such questions were difficult to answer back then, except perhaps by anecdote—my tale of good performance balancing their bad one—or by resorting to testimonials or evidence of "busyness." Not surprisingly, however, my assurances that the police department was making as many arrests as last year or that the streets department was laying as many tons of asphalt was hardly compelling to a caller who doubted that last year's performance was too good either! Such instances begged for a standard or benchmark against which local performance could be judged.

Now, as a university faculty member with a specialty in the management of local government, I still get those questions. My perspective may have changed a little because my relationship to the caller and the municipality in question is different and the locales from which the calls come are more widespread, but the questions are basically the same. Mayors, council members, general citizens, and municipal administrators, too, want to know how to judge the service delivery performance of their local government. Often, the answers have been regrettably incomplete.

"Well, I think typical performance is about . . ." or "I have heard that professional standards may have been established and would apply there." But as one city official, contacted in the early stages of this project, said, "Unfortunately, there is no central repository of such standards."

Well, there is now.

Mayors, city council members, city managers, department heads, other municipal officials, and citizens who want a measuring rod for local government services can turn to the benchmarks presented in this book. They will find national standards, "engineered" standards, statistical norms, rules of thumb, excerpts from actual performance records, and performance targets of a collection of respected cities.

The list of benchmarks is long—and it keeps growing and becoming more ambitious. That is the nature of benchmarks, and it is also the nature of the drive for continuous improvement.

Acknowledgments

The preparation of this book turned out to be a bigger project than I imagined at its beginning. I am indebted to many persons who have graciously assisted me along the way. Jordan Davis, Jeremy Chen, Anne Lockwood, and Debra Hill, graduate students at the University of Georgia, tackled various data collection tasks and doggedly pursued authorities who could answer the recurring questions, "Are there standards on this?" and "What does this number mean?" Melanie Hardman, Joan Bertsch, Shawn Johnson, Venus Wilson, and Becky Hill of the Carl Vinson Institute of Government at the University of Georgia were immensely helpful as I pursued the tedious tasks of cataloging information, developing tables, and preparing the manuscript. This project was probably bigger than they had imagined, too.

Richard W. Campbell, my colleague at the Vinson Institute, saw value in this project from the beginning and urged me on. I thank him for that. I also wish to thank five current or former local government managers who graciously consented to review and comment on the manuscript—Harry W. Hayes, past president of the National Association of County Administrators, former county administrator of Hall County, Georgia, and now a colleague at the Vinson Institute; Arthur A. "Don" Mendonsa, former city manager of Savannah, Georgia, who is now associated with the Vinson Institute; Sylvester Murray, director of the Public Management Program at Cleveland State University and former city manager of San Diego and Cincinnati; Martin Vanacour, city manager of Glendale, Arizona; and Douglas J.

Watson, city manager of Auburn, Alabama. The final product benefited whenever I was wise enough to heed their advice.

Above all, I am indebted to the many officials who each added bits and pieces to the growing collection of material that eventually became the heart of this volume. Officials in various professional associations affiliated with, or relevant to, municipal government responded to inquiries, provided documents, and offered helpful advice when my pursuits needed to be redirected. A host of municipal officials scattered across the nation were generous with their time and with pertinent documents. Most of all, they were generous with their words of encouragement, providing reassurance time and again that this product was eagerly awaited and would be worth the effort.

1

Introduction

"How'm I doin'?" became the trademark question of Ed Koch, mayor of New York City from 1978 through 1989.

In posing his question, Mayor Koch sought reassurances regarding the depth of his political base and feedback on his stewardship of municipal operations. As most veteran mayors and city managers know, these two dimensions of municipal leadership are not independent of each other. Just as political waves can rock an administration, solid operational stewardship can influence political strength and stability.

The performance of local government operations from animal control to zoning administration can have a great deal to do with the political health of mayors and other elected executives and with the professional well-being of city managers and other appointed administrators in local government. Unless they—or trusted assistants—are "minding the store," executives' political or professional stock can quickly decline. Yet all too many officials have only a vague sense of how their own municipal operations are faring. "How are we doing?" is a question that should be asked—and deserves to be answered.

1

Gauging Municipal Performance

Surveys of local government officials indicate a broad awareness of the importance of performance measurement and monitoring. Tangible evidence, however, reveals the back-burner status of performance measurement in most communities relative to the hotter issues cooking at any given time.

By most counts, more than half of all U.S. cities collect performance measures of some type. In most cases, however, the vast majority of the measures collected by a given municipality merely reflect workload—for example, the number of applications processed, arrests made, recreation class participants, or inspections completed.

In essence, workload measures are a form of "bean counting." Such a count is important. To anyone wanting to get ahead in the bean business, however, it is also important to know the *quality* of the beans and the *efficiency* with which they are grown and harvested.

Relatively few local governments systematically assess quality and efficiency, meticulously gauging and reporting the effectiveness and unit costs of even a single one of their operations. An even smaller cadre measures effectiveness and efficiency for more than a handful of departments. Despite the existence of how-to manuals for local government performance measurement and encouragement from a host of professional associations, relatively few local governments have invested the resources and energy needed to advance beyond the bean-counting stage.

Advanced Performance Measures

Measures of efficiency, effectiveness, and productivity have been developed for local government services. Each reveals much more about municipal operations than does a simple workload figure. Efficiency measures—often expressed as unit costs or as units produced per employee-hour—depict the relationship between services or products and the resources required to produce them. Effectiveness measures depict the quality of municipal performance or indicate the extent to which a department's objectives are achieved. Measures of timeliness and citizen satisfaction are among the common forms of effectiveness measures.

Productivity measures, although extremely rare in municipal performance monitoring systems, combine efficiency and effectiveness components in a single indicator. "Utility meters read *without error* per employee-hour" would be an example.

Performance Compared With What?

For decades, municipal officials have been urged to measure performance and have been given advice on how to get started. Those who have followed the prescriptions have soon discovered that a performance measure is virtually value-less if it appears in the form of an isolated, abstract number. Its value comes in comparison with a relevant peg.

Several options exist. Authorities have long suggested that current marks could be compared with those from earlier periods, with other units in the same organi-zation, with relevant outside organizations, with preestablished targets, or with existing standards (Government Accounting Standards Board [GASB], 1992; Hatry, 1980b; Hatry, Fountain, & Sullivan, 1990). In theory, such comparisons would provide municipal officials both an internal gauge, marking year-to-year progress of a work unit or highlighting unit-to-unit performance differences, and an external gauge, showing how a municipality's operations stack up against other jurisdictions or professional ideals. In reality, only the internal gauge has proved to be sufficiently practical to receive widespread application. Many local govern-ments report year-to-year comparisons of performance indicators—although often they are merely workload measures. Few report external comparisons with stan-dards or with performance indicators from other jurisdictions.

Why So Few External Comparisons?

Undoubtedly, some local government officials are happy with the status quo, preferring not to have the performance of their organization compared with others. After all, only in Garrison Keillor's fictitious village of Lake Wobegon can *everyone* be "above average."[1] In real life, only about half can achieve that status, and, for the rest, the desire to know one's ranking must be motivated not by a yen for publicity and praise but instead by a yearning to improve. Sadly, many of those jurisdictions mired at the bottom of the performance scale—the ones, they might say, that make the upper 90% possible—may prefer to remain oblivious to their status.

In contrast, some local government officials not only are willing to engage in external comparisons but are eager to do so. Driven by a desire to climb into the ranks of outstanding municipal performers or to be recognized for already being there, these officials desire external benchmarks but often are disappointed by finding no more than a handful having any usefulness. Those hoping to spur greater accomplishments by urging their organization to keep up with the "municipal Joneses" often are disappointed to find that few relevant statistics on the Jones family even exist—at least in a form usable for this purpose. When the most that the city manager can discover about the performance of other cities is a set of workload measures and expenditure levels, few useful comparisons are possible.

If good comparison statistics are so hard to find, why not rely on municipal performance standards set by professional associations and others? It sounds simple—and in some cases, it is possible to make such comparisons against standards (Ammons, 1994, 1996). Unfortunately, however, many so-called stan-dards are vague or ambiguous, have been developed from limited data or by questionable methods, or may be self-serving. Furthermore, most standards are not widely known and often are difficult to track down. There has been no single repository of standards relevant to municipal operations—until now. Mayors, council members, city managers, department heads, other officials, and citizens wishing to sort well-developed and usable standards for a particular operation from the rest—or even to *find* those standards—faced the prospect of an often difficult and time-consuming search. Understandably, few persevered.

Meeting the Information Needs of Local Government Officials and Citizens

The practical reality of competing demands, limited time, and scarce resources severely restricts the viability of using external gauges for performance assessment, unless those gauges are readily available. Few local government officials or interested citizens can afford to spend the hundreds of hours necessary to track down applicable standards or suitable interjurisdictional performance indicators. Desire to make external comparisons—or lack of that desire—is not necessarily the issue; time, resources, and other practical constraints may simply constitute too much of a hurdle, if a reservoir of relevant comparison information is not at hand.

That is where this volume comes into play. Between the covers of this book, local government managers, elected officials, and citizens will find standards and comparison statistics intended for ready use. Countless contacts with local government officials, professional associations, and trade organizations, coupled with careful scouring of budgets, financial statements, and other performance-reporting documents from local governments across the United States, have produced a collection of municipal benchmarks that will enhance a community's ability to answer the question, "How are we doing?"

Included in the pages that follow are "standards," "norms," and "rules of thumb" offered by professional associations, trade organizations, and other groups with a stake in local government. Readers are cautioned here—and will be reminded elsewhere—that the motives of such groups in prescribing standards may range from civic-minded service to self-serving protection of the status and working conditions of association members. It is also important to realize that standards may in some cases be intended to imply minimum acceptable levels of performance, in other cases to designate norms, and in still other cases to identify targets toward which local governments should aspire.

Collections of actual performance indicators have been gleaned from the documents of more than 100 local governments of various sizes scattered across the nation. The set of municipalities chosen for this project does not constitute a random sample; each municipality was included because it measures performance and because its reporting documents were made available. Most of the municipalities included were selected from a list of recent recipients of the Governmental

Finance Officers Association's (GFOA; 1993, p. 4) Distinguished Budget Presentation Award, an award that includes as a nonmandatory criterion the reporting of performance results. The input of GFOA award recipients was supplemented by that of several other cities that were also found to document performance in a manner conducive to cross-jurisdictional comparison.

The performance indicators reported in this volume are neither workload measures nor unit costs. Workload measures were excluded because of their limited usefulness in cross-jurisdictional comparisons and their inability to answer the "how well" and "how efficiently" questions. Unit costs have been omitted because of their extreme vulnerability to inflation (e.g., during periods of high inflation unit costs may become quickly outdated unless "constant dollar" calculations are made), economic differentials (e.g., regional variations in the cost of labor could produce erroneous judgments on the relative efficiency of labor-intensive operations), and accounting vagaries (e.g., inconsistent accounting practices across jurisdictions for overhead, employee benefits, capital acquisition, and depreciation, to name a few, could distort comparisons). Retained is a collection of indicators that easily traverse jurisdictional contexts or that otherwise have been standardized in a manner that removes the most obvious effects of differences in population. Measures such as "average response times to public safety emergencies," "average return on municipal investments," "inspections per electrical inspector," "park acreage per 1,000 persons," and "percentage of recreation program expenses recovered through fees" are examples.

Where performance indicators from actual cities conform in format to the standards promulgated by professional associations, they offer readers a "reasonableness review" for those standards. Actual performance of respected cities may influence the interpretation attached to a given standard—that is, whether it should be regarded as a minimum acceptable level, a norm, or a target of excellence. The performance records of recognizable and respected cities will help answer the practitioner's pragmatic questions:

"Is everybody else meeting this standard?"
"Is *anybody* meeting it?"

Contents, Format, and Uses

Following this introduction is a second chapter that encapsulates the major messages of some of the most prominent books, articles, and how-to manuals on performance measurement for local government. Although these publications instruct local officials on the development of ideal performance measures, they typically offer few, if any, relevant comparisons for use once those measures are crafted and operating statistics compiled. This book takes a different approach.

Following brief instructions on the development of good performance indicators, the focus of this volume turns toward the interpretation of performance information once it has been collected. Officials who have devoted a year or more to the development of a set of performance measures and the collection of relevant data often are disappointed to find no other jurisdictions reporting the same measure in the same fashion. They may be understandably frustrated to learn that they must collect the measure a second year before any relevant comparisons can be made— and then only with their own city's performance in the earlier year. With the information in this volume, officials will be able to make more immediate comparisons that will place local performance in an external context.

Following the chapter on performance measurement is a series of 22 chapters devoted to performance standards and cross-jurisdictional performance indicators pertinent to major functions common to municipal government. The concluding chapter, "Performance Milestones," will offer a concise set of suggested performance targets drawn from the standards developed by professional associations and performance indicators reported by cities included in this project.

A Few Comments Before Proceeding

The collection of benchmarks found in this volume can be a useful tool for gauging and improving municipal performance. Like other tools, the benchmarks are most effective in the hands of a craftsperson who knows how to use them and who understands their limitations.

ON BENCHMARKS AND BENCHMARKING

Comparing local performance statistics with selected benchmarks is a valuable step in evaluating municipal operations, but simple comparisons are not the same as full-fledged, corporate-style benchmarking. Benchmarking entails the analysis of performance gaps between one's own organization and best-in-class performers, the identification of process differences that account for the gap, and the adaptation of key processes for implementation in one's own organization in an effort to close the gap. A simple comparison of local performance with selected municipal benchmarks is more limited but also less expensive and more practical for a general assessment of a broad range of functions. Such a comparison places local performance in context and, where major performance gaps are detected, may suggest the need for additional analysis, perhaps leading to a full-scale benchmarking project.

REACTIVE VERSUS PROACTIVE MANAGEMENT

Many of the performance statistics included in this book reflect a reactive approach to service delivery—reporting, for example, how quickly potholes are filled or how well firefighters limit the spread of fire. In such cases, the emphasis is on speed or skill in responding to problems. As more local governments set their sights on proactive management that anticipates problems before they occur, we can expect in the future more performance statistics that measure effectiveness in preventing undesirable occurrences. Even then, however, high-performance municipalities will still be adept at responding to problems as they happen.

AGGREGATE STATISTICS AS CAMOUFLAGE

A municipality that performs well overall may nevertheless be deficient in delivering services to some neighborhoods or to a particular category of customers (for example, senior citizens, youth, minorities, or businesses). Simply comparing citywide performance statistics with municipal benchmarks would hide those deficiencies. Disaggregating statistics to reflect performance as it relates to various groups of customers may be more revealing.

UNAUDITED DATA

Much of the performance information in this book comes directly from municipal reporting documents. As such, it is largely unaudited, but it is information that each municipality has confidently reported of its own volition rather than in response to some questionnaire that could evoke as many estimates as real numbers. Of potentially greater concern to consumers of these municipal benchmarks is the possibility of inconsistent definition of terms from one city to another. Statistics are reproduced as reported. In some cases, analysts may wish to dig deeper—for instance, if a benchmark seems too good to be true.

TIME FRAME DIFFERENCES

Performance targets and statistics have been collected from different fiscal years and reporting periods.[2] Because inflation-sensitive expenditure statistics (e.g., total costs and unit costs) have been excluded, the measures that remain are, for the most part, little affected by reporting period variations or the passing of time. A library boasting a circulation of six items per capita in 1989, for example, was performing admirably then and could still be proud of that mark today. On the other hand, performance worthy of the label "commendable" in a few services may experience more rapid evolution. Especially in services such as data processing, in which changing technology is a major factor, appropriate caution should be exercised in the application of benchmarks more than 5 or 6 years old.

Audiences for This Book

This volume is intended to serve three audiences. It is designed for elected officials who desire a context within which to judge municipal operations, for local government managers and department heads who wish to improve their performance measurement systems and who want to supplement internal with external performance gauges, and for citizens who wish to know whether their community's public services really measure up.

Notes

1. Keillor, author of *Lake Wobegon Days* (1985), *Leaving Home* (1987), and other works, fondly describes Lake Wobegon as a place "where all the women are strong, the men are good-looking, and all the children are above average" (1987, p. xvii).

2. Hyphenated fiscal years (e.g., FY 1991-1992) are reported in this book simply as the year in which the fiscal period ends (e.g., 1992).

2

Performance Measurement and Benchmarking

Aggressive businesses and industries measure performance meticulously. In their popular book on excellence in corporate America, Tom Peters and Bob Waterman (1982) characterized top companies as "measurement-happy and performance-oriented" (p. 240). They found that the *very best of the best* complement their measurement systems with an action orientation that separates them from the pack. Outstanding companies carefully monitor and record the relevant dimensions of performance—they know the important facts and figures— and they act on that knowledge.

Municipalities and other public sector agencies confront a far different circumstance. Without the pressure of competition or the unforgiving bottom line of profit or loss, governmental units are apt to neglect performance measurement as they focus on more pressing matters. Carefully conceived systems of performance measurement are more than a little complex, often they are more than a little threatening to the status quo, and they do impose expenses on the organization from conception through administration. Consequently, performance measurement often is allowed to slip in priority, except among municipalities that have an extraordinary commitment to management information.

Recently, federal researchers examined successful private corporations in hopes of gleaning lessons for possible application in government. They found that most industry leaders were distinguished not only by management continuity and consistency, long-range vision, and customer orientation but also by their "systematic strateg[ies] for measuring performance" (Thompson, 1991, p. 4). Similarly, performance measurement and related matters are prominent in management expert Peter Drucker's (1975) list of characteristics common to successful service institutions:

- They clearly define the nature and scope of their function, mission, and activities.
- They set clear objectives.
- They set priorities, concentrate on the most important objectives and set standards of performance for themselves.
- They measure performance, analyze the results and work to correct deviations from established standards.
- They regularly conduct audits of performance to ensure that the management system continues to function properly. (pp. 158-159)

Why Measure Performance?

Properly developed and administered, a performance measurement and monitoring system can offer important support to a host of management functions (Decker & Manion, 1987, pp. 128-129; Glover, 1993; Hatry, 1980a; Osborne & Gaebler, 1992, p. 156; Thompson, 1991, p. 14; U.S. General Accounting Office, 1993, p. 1). Each function has important ramifications for cities that aspire to be excellent.

Accountability. Managers in top-performing organizations insist on accountability from their subordinates and, in turn, expect to be held accountable by their organizational superiors. Performance measures document what was done by various departments or units and, ideally, how well it was done and what difference it made. Through such documentation, outstanding departments and entire municipalities earn the trust of their clients and citizens as they demonstrate a good return in service provided for tax dollars received.

Planning/Budgeting. Cities with an objective inventory of the condition of public services and facilities, a clear sense of service preferences among their citizens, and knowledge of the cost of providing a unit of service at a given level are better equipped to plan their community's future and to budget for that future. Again, performance measurement—incorporating unit costs and indicators of citizen demand or preference—is key.

Operational Improvement. Municipalities that measure performance are more likely to detect operational deficiencies at an early stage. Furthermore, performance records enhance their ability to confirm the effectiveness of corrective actions.

Program Evaluation/MBO/Performance Appraisal. Carefully developed performance measures often provide valuable information for the systematic evaluation of program effectiveness. Management-by-objectives (MBO) programs and pay-for-performance systems for managerial employees, where they exist, typically are tied to performance measures. In some cases, employee performance appraisals or other forms of systematic performance feedback have been based, at least in part, on individual performance relative to established measures.

Reallocation of Resources. A clear indication of program effectiveness and unit costs—in essence, a scorecard on tax dollar investments and returns in various service functions—can aid decision makers in reallocation deliberations, especially in times of financial duress.

Directing Operations/Contract Monitoring. Managers equipped with a good set of performance measures are better able to detect operational strengths and weaknesses, to provide relevant feedback to employees and work units, and to deploy close supervision where it is needed most. Performance measures also provide evidence useful in determining whether the service quality specified in contractual arrangements is, in fact, being achieved.

Stated simply, performance measurement provides local governments with a means of keeping score on how their various operations are doing. As noted by Harry Hatry (1978), that scorekeeping function is vital:

Unless you are keeping score, it is difficult to know whether you are winning or losing. This applies to ball games, card games, and no less to government productivity. . . . Productivity measurements permit governments to identify problem areas and, as corrective actions are taken, to detect the extent to which improvements have occurred. (p. 28; reprinted by permission)

Types of Performance Measures

Efforts to identify different types of performance measures sometimes have yielded lengthy lists. Often, however, such lists are inflated by including measures of resource input and other pseudomeasures that reflect performance only indirectly, if at all (Hatry, 1980b).[1] True performance measures in local government generally may be categorized as one of four types:

1. Workload measures
2. Efficiency measures
3. Effectiveness measures
4. Productivity measures

Workload measures indicate the amount of work performed or the amount of services received. By comparing workload measures reporting, say, the number of applications processed by the personnel department, the number of sets of city council minutes prepared by the city clerk, the number of arrests by the patrol division of the police department, and the number of trees planted by parks crews with the corresponding records from a previous year, a city official or citizen can see whether workload volume is up or down. Although that information can be of some value, it reveals only how much work was done—not how well it was done or how efficiently.

Efficiency measures reflect the relationship between work performed and the resources required to perform it. Typically, efficiency measures are presented as unit costs, but they can take other forms as well.

Unit costs are calculated by dividing total costs of a service or function by the number of units provided. For example, if 2,000 feet of 8-inch sewer line are installed by municipal crews at a total cost of $80,000, then the unit cost of sewer line installation is $40 per foot. A reversal of the ratio—dividing the number of units by the resources consumed—reveals the number of units produced per dollar and is also an efficiency measure (e.g., $3/10$ of an inch of sewer line per dollar). Other forms of efficiency measures reflect alternative types of resource input (for example, units produced per labor-hour)[2] or production relative to an efficiency standard. If meter readers complete only one half of their assigned rounds, if repairs by municipal auto mechanics take twice as long as private garage manuals say they should, or if expensive maintenance equipment is operated only 10% of the time it is available, questions of efficiency relative to a prescribed or assumed standard may be raised.

Effectiveness measures depict the degree to which performance objectives are achieved or otherwise reflect the quality of local government performance. Meter reading error rates of less than 0.5%, a consistent record of fire suppression with only minimal spread, and low return rates on auto repairs reflect effective operations. Response times and other measures of service quality are only indirectly related to effectiveness but typically are included among effectiveness measures for their presumed linkage.

Productivity measures combine the dimensions of efficiency and effectiveness in a single indicator. For example, whereas "meters repaired per labor-hour" reflects efficiency and "percentage of meters repaired properly" (e.g., not returned for further repair within 6 months) reflects effectiveness, "unit costs (or labor-hours) per *effective* meter repair" reflects productivity. The costs (or labor-hours) of faulty meter repairs as well as the costs of effective repairs are included in the numerator of such a calculation, but only good repairs are counted in the denominator—thereby encouraging efficiency *and* effectiveness by meter repair personnel.

Examples of each of the four types of performance measures for several common municipal functions are provided in Table 2.1. The value of workload measures is limited. Much greater insight into the performance of municipal operations may be gained from efficiency, effectiveness, and productivity measures.

TABLE 2.1 Examples of the Four Principal Types of Performance Measures

Municipal Function	Workload Measure	Efficiency Measure	Effectiveness Measure	Productivity Measure
City Clerk	Number of sets of city council meeting minutes prepared	Employee-hours per set of city council minutes prepared	Percentage of city council minutes approved without amendment	Percentage of city council minutes prepared within 7 days of the meeting and approved without amendment
Library	Total circulation	Circulation per library employee	Circulation per capita	Circulation per $1,000
Meter Repair	Number of meters repaired	Cost per meter repair	Percentage of repaired meters still functioning properly 6 months later	Cost per properly repaired meter (i.e., total cost of all meter repairs divided by number of meters needing no further repairs within 6 months)
Personnel	Job applications received	Cost per job application processed; cost per vacancy filled	Percentage of new hires/promotions successfully completing probation and performing satisfactorily 6 months later	Cost per vacancy filled successfully (i.e., employee performing satisfactorily 6 months later)

Criteria for a Good Set of Performance Measures

Properly developed sets of performance measures possess several distinctive characteristics (Bens, 1986; Hatry, 1980b; Hatry et al., 1992, pp. 2-3). Good sets include measures that are

- *Valid.* They measure what they purport to measure—that is, a high score on a given measure does, in fact, reflect possession of the underlying dimension or quality.
- *Reliable.* The measure is accurate and exhibits little variation due to subjectivity or use by different raters (for example, a measuring tape is a reliable instrument in that it is highly objective, and two different persons using the same instrument are likely to get similar measurements).
- *Understandable.* Each measure has an unmistakably clear meaning.
- *Timely.* The measures can be compiled and distributed promptly enough to be of value to operating managers or policymakers.
- *Resistant to perverse behavior.* The development of a performance measure raises the profile of the performance dimension being examined. A higher profile sometimes brings unin-

tended consequences or even strategies designed to "beat the system"—for instance, overzealous ticket writing if the police department is measured by that activity alone or the watering down of garbage to increase its weight if collection crews are rated solely by tons collected. The best sets of performance measures have little vulnerability to such actions because they have been devised carefully and also because they typically include multiple measures that address performance from several dimensions and thereby hold potentially perverse behavior in check.

- *Comprehensive.* The most important performance dimensions are captured by a good set of measures. Some minor facets of performance may be overlooked, but the major elements are addressed.
- *Nonredundant.* By favoring unique measures over duplicative measures, the best sets of performance measures limit information overload for managers, other decision makers, and consumers of municipal reports. Each measure contributes something distinctive.
- *Sensitive to data collection cost.* Most dimensions of municipal performance can be measured either directly or through proxies. In some cases, however, measurement costs may exceed their value. Good sets of performance measures include the best choices among *practical* measurement options.
- *Focused on controllable facets of performance.* Without necessarily excluding important, overarching, and perhaps relatively uncontrollable characteristics relevant to a particular function, good sets of performance measures emphasize outcomes or facets of performance that are controllable by policy initiatives or management action. For example, although a police department's set of performance measures might include the rate of domestic homicides in the jurisdiction, a good set of measures would also include indicators of public safety more widely considered controllable by police efforts.

Sources of Performance Data

Typically, performance data are secured from various combinations of the following sources:

- Existing records
- Time logs
- Citizen/client surveys
- Trained observer ratings
- Specially designed data collection processes

The simplest and most desirable source of performance measurement data is the set of records already maintained by a municipality. Workload counts, complaint records, and response times for various services are common, even among cities with modest performance-reporting practices. Such information can serve as the foundation of a performance measurement system and sometimes can be converted to higher-level measures. Depending on its level of precision, for example, workload data might be combined with expenditure information to yield efficiency measures.

Time logs completed on either a comprehensive or random basis provide resource-input information for labor-related efficiency measures (e.g., work units per employee-hour) or for measures comparing actual work completed in a given amount of time with the amount of work expected on the basis of engineered work standards. Apart from vehicle maintenance and perhaps a handful of other services, however, engineered work standards are rarely available for common municipal functions.

The surveying of users of particular services or citizens in general is a feedback mechanism used by many city governments. Survey use may get an additional boost from the customer focus of the currently popular total quality management (TQM) movement.

Surveys typically tap respondent perceptions regarding the adequacy of selected services, the nature of any perceived deficiencies, and the extent to which respondents avail themselves of various services and facilities. For a survey to contribute meaningfully to a city's performance measurement system, it must be conducted with sufficient scientific rigor to produce reliable results. Such rigor typically introduces costs that exceed those of tempting, low-cost alternatives, but surveys conducted "on the cheap" frequently are misleading and can be easily discredited. Carefully designed and properly randomized telephone surveys have been found to be widely accepted, moderately priced alternatives to more expensive face-to-face interviews and to low-cost, low-response-rate mail surveys.

Trained observer ratings have been used successfully in several cities for evaluating the condition of facilities and infrastructure. Typically, persons employed in positions outside the department responsible for facility maintenance are trained to serve on an intermittent basis as raters of street cleanliness or parks maintenance, for example. Instructed on the finer points of distinguishing between various grades of facility condition and customarily armed with photographic depictions of those grades, trained observers can provide a more systematic evaluation of condition than can usually be drawn from records of citizen compliments or complaints, or even from citizen surveys. The effective administration of trained observer programs requires careful attention to training and issues of reliability. Supervisors often spot-check facilities to corroborate ratings and, where two or more raters assess the same facility within a short period, supervisors usually reinspect if the grades of raters are more than one unit apart.[3]

Unfortunately, information already on hand or readily available through customary means does not always meet current performance measurement needs. Analysts may discover, for example, that existing workload counts and random time logs have been based on estimates, rather than on precise figures. The citizen survey everyone in town likes to quote may turn out to be suspect because it was conducted using questionable methods—perhaps relying on a tear-out questionnaire in the local newspaper. Even federal figures, presumed to be rock solid, may be less comprehensive and conclusive than originally thought; FBI crime statistics, for example, include only *reported* crimes—not all crimes actually committed. Typically, at least a few new data collection procedures must be introduced to support the development of a good set of performance measures.

Status of Performance
Measurement in City Government

A little more than a half century ago, a pair of esteemed observers noted the availability of a number of "rough and ready measurement devices" but nevertheless conceded the limited development of performance measurement among city governments.

> It is probably true that in the present state of our knowledge the citizen can often better judge the *efficiency* of his local government by the political odor, be it sweet or foul, which emanates from the city hall than by any attempt at measurement of services. (Ridley & Simon, 1943, p. ix; reprinted by permission)

Through the years, various movements to improve management practices, rationalize decision making, and enhance accountability have called for improved performance measurement. Respected professional associations, including the American Society for Public Administration, the Governmental Accounting Standards Board (GASB), the Government Finance Officers Association (GFOA), and the International City/County Management Association (ICMA), have promoted the practice. Some cities have responded to that encouragement. Many of the writings on performance measurement in local government repeat the names of familiar municipalities as exemplars: Sunnyvale and Palo Alto, California; Phoenix, Arizona; Charlotte, North Carolina; Dayton, Ohio; Savannah, Georgia; Randolph Township, New Jersey; Dallas, Texas; Aurora, Colorado; New York City; Alexandria and Charlottesville, Virginia; and Portland, Oregon (see, for example, Bens, 1986, p. 1; Hatry et al., 1992, p. 1; Osborne & Gaebler, 1992, p. 349; Smith, 1992, p. 8). These cities, however, are far from typical in the advanced status of their performance measurement systems. Despite survey responses suggesting widespread and fairly sophisticated performance measurement in local government, more exacting research involving the examination of actual performance reporting documents reveals far more limited development (Ammons, 1995; Grizzle, 1987; LeGrotte, 1987; MacManus, 1984; Usher & Cornia, 1981).[4] Not all cities even engage in performance measurement. Among those that do, most rely heavily on workload measures. Some report a few efficiency or effectiveness measures for various functions. Rarely are productivity measures reported.

Although performance measurement is well developed and still improving in some city governments, most cities have a long way to go. Speaking of his own city in the not-too-distant past, one Philadelphia official remarked, "If you were to ask how much it cost to pick up a ton of trash, I hope you weren't waiting for an answer the same year" (Barrett & Greene, 1994, p. 44). The situation, he noted, had improved greatly in recent years. Perhaps the remarks of an Indianapolis official put the development of municipal performance measurement and program evaluation systems in proper perspective: "We're coming from an era of cleaning our clothes on rocks, and now we have come up with a hand-cranked washing machine. It's a much better life, but of course we'd rather have an automated washing machine, eventually" (p. 44).

Overcoming Resistance to Performance Measurement

The likelihood is great that some form of resistance will be encountered in efforts to develop or enhance performance measurement systems. Although the precise source often is not predictable, *some* resistance almost inevitably will come from *somewhere.* The proper starting point for developing a coping strategy is to try to understand the reasons for that resistance.

Typically, performance measurement is seen by various groups or individuals in an organization as a threat to their status. Some employees may fear that it is the first step in a process that will lead to tougher work standards and a forced speedup of work processes, perhaps to be followed by the layoff of workers who are no longer needed or who cannot keep up.

Supervisors and managers may feel threatened by what they perceive to be the insinuations of a performance measurement drive. Believing that executive and legislative satisfaction with their performance would generate no such movement, they may sense an accusation of poor performance. They may fear loss of status or even loss of employment if performance measures confirm those accusations, and they feel certain of a loss of discretion as the upper echelons gain yet another means of looking closely over their shoulders.

Even some top city officials, who seem to have so much to gain from improved performance measures, may resist their development. Especially prone to do so are those who believe that by virtue of their own political adeptness or membership in the dominant political coalition, their preferences are more likely to prevail in political negotiations without the influence of performance facts and figures.

Usually, such fears are overblown. Rarely are performance measurement efforts inspired by sinister or mean-spirited motives. Although detractors often fear worker exploitation or loss of supervisory discretion, the most commonly prescribed blueprints for measurement system development call for the involvement of those parties, rather than for their exclusion. Measurement initiatives typically are most successful when they secure the input and support of frontline employees and supervisors. When that happens, a new measurement and reporting system— and the clarification of work unit objectives and priorities that accompanies the system's development—is likely to earn accolades from those parties, rather than charges of exploitation.

Persons who fear that politics will be displaced in local decision making through the rationalizing influence of objective performance measures need not be overly concerned. Advances in management practices may, at most, supplement political considerations in decision making. In a democratic setting, politics can never be displaced—nor should it be.

Resistance to advances in performance measurement may emanate from a variety of concerns and may be manifested in different forms. A few things in this regard, however, are fairly predictable. Proponents of improved performance measures should expect three common declarations from opponents.

"You Can't Measure What I Do!" If the performance of a department or office has not been measured in the past, it should not be surprising that incumbents might reason that their activities must be unmeasurable. If their performance could have been measured, it already would have been. Offices or departments making this declaration often are characterized by nonroutine work and the absence of an existing data collection system. Both factors can make measurement difficult. Rarely is it impossible.

Sometimes a bit of creativity is necessary to devise either a suitable direct measure or a proxy that measures performance indirectly. Often, interviewing skills are even more important. "If your office closed shop for a few weeks, I know you would be missed," the interviewer might suggest. "But who would suffer the greatest impact, and what aspect of your work would they miss the most?" Gradually, with cooperation from employees and departmental officials, a widely accepted set of performance measures can emerge.

"You're Measuring the Wrong Thing!" Once again, the involvement of service providers is a key to resolving their complaints or calming their fears. It could be that relatively insignificant performance dimensions *are* being measured and that more important dimensions are being ignored. If so, the remedy is to replace the former with the latter. The involvement of service providers can ensure corrective action.

On the other hand, it is possible that the debate over appropriate performance measures may have uncovered a fundamental problem. A difference of opinion may exist between upper management and the service delivery unit over desired elements of service, or the unit may simply have labored under a long-standing misconception about what management wants. Such disagreements or misunderstandings should be resolved.

Yet another group of stakeholders in measuring the right thing is the citizen-customers. Their involvement, typically through strategic planning processes, focus groups, and citizen commissions, can enhance the likelihood that the right things are being measured.

"It Costs Too Much and We Don't Have the Resources!" Understandably, departmental officials who already feel that their resources are stretched too thin may be reluctant to tackle new, time-consuming measurement and reporting tasks that may siphon program resources. Those officials are a little like the overbur-

dened logger who, facing stacks of uncut logs, felt he could not spare the time to sharpen his dull saw. Well-meaning but reluctant officials must be persuaded of the value of performance measurement as a tool for improving services and making better use of scarce resources.

There is no reason to embark on performance measurement improvements unless better measures are expected to lead to improved services, to make services more efficient, or to make them more equitable. Service providers must be assured of management's commitment to those ends and be convinced of management's resolve to use performance measurement to improve, rather than to drain resources from, services to the public. It may be helpful to point out to skeptics that most local government officials in jurisdictions that have performance measurement systems report that their systems have been worth the expense (McGowan & Poister, 1985).

Developing a Performance Measurement and Monitoring System

Performance measures will not lead inevitably to improved performance. Although the act of measuring an operation for its results draws attention to that function and may thereby inspire greater efforts and improved performance, such results cannot be guaranteed. Performance measurement is merely a tool. If wielded properly, it can identify areas of performance adequacy and areas of performance deficiency; however, it can neither explain the former nor prescribe remedies for the latter. Reliable explanations and appropriate prescriptions require subsequent analysis of those targeted operations.

Performance measurement and monitoring systems may be developed in a variety of ways. The steps outlined in Table 2.2 have been gleaned from the experience of several municipalities and distilled from writings on the topic, but many variations on this pattern are possible and, indeed, perhaps desirable in a given setting.

Securing management's commitment—at the highest level possible—is an important first step. Not only will that commitment help to overcome resistance at lower levels, but also management interest will increase the likelihood that resulting measures will be used in subsequent decisions, thereby reinforcing and reinvigorating the process. Although the commitment of the city's chief executive is highly desirable, the bare essential for performance measurement at the service unit level is the commitment of that service unit's administrator or supervisor.

Decisions regarding functional responsibility for coordinating the development of performance measures and identifying functions to be measured should be made at an early stage. Depending on the measurement foundation already in existence, some cities may be capable of establishing a solid performance measurement

TABLE 2.2 Steps in the Development and Administration of a Performance Measurement and Monitoring System

1. Secure managerial commitment.
2. Assign responsibility (individual or team) for spearheading/coordinating departmental efforts to develop sets of performance measures.
3. Select departments/activities/functions for the development of performance measures.
4. Identify goals and objectives.
5. Design measures that reflect performance relevant to objectives:
    - Emphasize service quality and outcomes rather than input or workload.
    - Include neither too few nor too many measures.
    - Solicit rank-and-file as well as management input/endorsement.
    - Identify the work unit's customers and emphasize delivery of services to them.
    - Consider periodic surveys of citizens, service recipients, or users of selected facilities.
    - Include effectiveness and efficiency measures.
6. Determine desired frequency of performance reporting.
7. Assign departmental responsibility for data collection and reporting.
8. Assign centralized responsibility for data receipt, monitoring, and feedback.
9. Audit performance data periodically.
10. Ensure that analysis of performance measures incorporates a suitable basis of comparison.
11. Ensure a meaningful connection between the performance measurement system and important decision processes (e.g., goal setting, policy development, resource allocation, employee development and compensation, and program evaluation).
12. Continually refine performance measures, balancing the need for refinement with the need for constancy in examining trends.
13. Incorporate selected measures into public information reporting.

system covering all departments simultaneously; however, some experts recommend a more gradual approach, beginning with one or two departments or activities as demonstration or pilot projects (Bens, 1986). Whichever strategy is adopted, it is reasonable to anticipate that some departments will be more receptive to performance measurement than others. It is also important to recognize that some departments perform functions more conducive to easy measurement than others.

Performance measures should be consistent with and pertinent to the city's goals and objectives. If such goals and objectives have not been articulated, their development or clarification should become an early step in the process. Mission statements and goals are broad expressions of general purpose. Objectives are more precise, declaring specifically what is going to be accomplished and when. Performance measures track the city's progress in achieving those objectives, ideally documenting the efficiency with which they are pursued and the extent of their achievement.

Performance measures should emphasize the quality of services and the outcomes that those services produce rather than merely reporting the resources consumed or providing workload or activity figures. Indicators that simply record compliance with prescribed processes rather than outputs or results should be deemed acceptable substitutes only when more direct indicators of effectiveness are impractical.

Measurement systems should focus on neither too few nor too many performance indicators. Having too few could cause important service dimensions to be overlooked or could inadvertently encourage perverse behavior in the overzealous pursuit of high marks on the featured dimension. Having too many measures can contribute to information overload and disuse of the system.

The proper approach may be the development of a system with a moderate number of performance indicators to be tracked by departmental officials, with only a subset of key measures published in performance reporting documents designed for higher-level officials or the public (Hatry et al., 1992, p. 3). To reduce the tendency to swamp readers by reporting relatively unimportant as well as vital information, for instance, the city of Milwaukee limits departments to reporting only their five most important outcome measures in the budget (Barrett & Greene, 1994, p. 42).

Performance measures should be customer sensitive, emphasizing effectiveness in meeting customer expectations as well as efficiency in service delivery. The early involvement of rank-and-file employees and service recipients along with management in the design of appropriate performance indicators will minimize the likelihood of overlooking important performance dimensions.

Establishing the logistics of performance measurement and monitoring is important. Assigning responsibility within the service delivery units for data collection and reporting, determining the frequency of reporting (e.g., monthly, quarterly, or semiannually), and assigning centralized responsibility for data receipt, monitoring, and feedback are essential for ensuring the viability of the system and sustaining interest in its continuation. Maintenance of the system should also include provisions for periodic audit of performance data to ensure their accuracy.

For performance data to be useful, periodic analysis should examine the performance of service delivery units in comparison with performance in previous periods, standards, targets, comparable jurisdictions, or other suitable bases of comparison. Although continuity of performance measures is desirable for identifying trends, performance measures should be refined or replaced, as appropriate.

A meaningful connection should be established between the performance measurement system and the organization's decision processes. For example, the design of performance measures should be influenced by a municipality's previous goal-setting efforts. Performance measurement results may, in turn, influence future goal-setting processes, policy development, resource allocation, and employee development and compensation decisions. Furthermore, a performance measurement system may be instrumental in a municipality's subsequent program evaluation efforts, as well as its accountability and public information initiatives.

Benchmarking

To determine whether a municipality's performance is favorable or unfavorable, it is necessary to compare that jurisdiction's performance marks against some relevant peg. Among city governments that monitor their own performance, many compare current performance with figures for the same measures in previous reporting periods. Some compare the performance measures of different units in the same jurisdiction providing similar services or compare performance records with predetermined targets. Until recently, relatively few have used national or state standards, private sector performance, or the performance records of other jurisdictions as benchmarks for gauging their own jurisdiction's performance (Hatry, 1980b).

In the private sector, benchmarking has been widely acclaimed as a technique that contributes to performance improvement. By identifying best-in-class performers and the practices that make them so, industries may refine their own processes in a quest to meet or exceed the benchmarks set by outstanding performers.

Public sector application of the term *benchmarking* has been a bit broader. In some cases, objectives or targets set arbitrarily by a state or local government itself have been labeled *benchmarks*. In other cases, the term has been used in a manner more consistent with private sector application. In this volume, the term *benchmark* will be reserved for anticipated or desired performance results anchored either in professional standards or in the experience of respected municipalities.

Municipalities desiring to identify suitable benchmarks for their operations confront two major issues. One is data availability. This volume will overcome that problem by providing information on standards and the experience of several other municipalities for a host of local government functions. The second issue is comparability. In assembling suitable benchmarks, each jurisdiction must be vigilant in identifying factors that make some jurisdictions suitable for comparison and others unsuitable. Cost comparisons are especially vulnerable; differences in reporting periods, accounting practices, and cost of living may confound simple comparison. Social and economic factors that may influence the difficulty of a given jurisdiction's incoming workload may similarly set it apart from its counterparts and limit the value of comparisons. Care is therefore in order as communities select their benchmarks.

Cautionary notes are appropriate for jurisdictions embarking on benchmarking, but so are words of encouragement. The principle of accountability demands that department heads and managers be able to tell city councils how their operations are doing and that city councils, in turn, be able similarly to inform the public. How a city government is doing in comparison with last year is interesting, but not as interesting as how it is doing in comparison with national standards or with others in the same field. Such comparisons are sometimes difficult and often require explanations. The willingness of some public officials to grapple with that difficulty and to provide explanations when they are needed may well distinguish those truly committed to the principle of accountability from others who are not.

Notes

1. Hatry (1980b) identifies 10 types of performance measures: cost measures, workload-accomplished measures, effectiveness/quality measures, efficiency/productivity measures, actual unit-cost to workload standard ratios, efficiency measures and effectiveness quality, resource utilization measures, productivity indices, pseudomeasures, and cost-benefit ratios. Although this is an excellent list of the types of indicators that are presented as performance measures in various reports, some may be dismissed as impostors (e.g., cost measures and pseudomeasures), and others may be incorporated into the four major types of performance measures identified in this chapter.

2. *Labor-hour* is one of several terms used in this volume to depict the equivalent of an hour of work performed by one employee or a combination of employees (e.g., two employees working $\frac{1}{2}$ hour apiece on a given project is also 1 labor-hour). *Man-hour,* the term historically used in local government, will be used interchangeably in this volume with *labor-hour, staff-hour,* and other similar terms.

3. For more details on the use of trained observer ratings, see Hatry et al. (1992) and O'Connell (1989).

4. When simply *asked* in surveys if they measure performance, local government officials tend to respond affirmatively, with high percentages claiming to use not only workload measures but also efficiency and effectiveness measures (see, for example, O'Toole & Marshall, 1987; O'Toole & Stipak, 1988). More exacting research, requiring actual evidence of performance measurement, often reveals less extensive measurement activity and only limited use of efficiency, effectiveness, and productivity measures.

3

Animal Control

nimal control problems may not be the number one aggravation of every mayor and city manager in America, but they easily make most "top ten" lists. In many cases, problems erupt over matters that seem trivial in relation to much bigger issues confronting the community. A neighbor's dog barks too much; someone's roaming cat assaults songbirds at the backyard feeder; wandering dogs—or even dogs being walked by their owners—foul the lawn; a dog "protecting" its owner's front yard frightens bicyclists and pedestrians; someone's flower bed was destroyed last night; the garbage was tipped over, *again*. Monumental crises? In most instances, no; but on some occasions, animal control problems rise beyond the nuisance and aggravation level to pose serious threats to health and safety. Dog attacks are now the greatest reportable childhood public health problem in the nation—greater than measles, mumps, and whooping cough combined (Handy, 1993, p. 2).

Whether threatening or merely annoying, disputes over pets can be serious and often quite emotional for the persons involved. Prompt and effective action is important.

Evaluating Animal Control Operations

Communities secure animal control services in a variety of ways. Some contract for those services from other local government units or from a local humane society. Others provide animal control using their own employees, sometimes organized as a separate animal control department but often as a component of the public health or police department.

Whether secured by contract or through the efforts of local government workers hired for that function, a city that *pays* for animal control services needs some means of judging the adequacy of those services. Typically, a city will simply compare workload statistics for 2 or 3 years and assume that the operation is doing all right if the numbers are up for licenses, apprehensions, and adoptions and if complaints to the mayor and city council members are down. As informative as those numbers may be, they compare the jurisdiction only with itself in earlier periods. If it was performing well in the past, it is doing even better today. But what if previous performance was woefully inadequate? Even if performance is better today, is it where it should be? The answer to that question requires comparison with relevant external pegs.

Officials wishing to assess the adequacy of an animal control operation should consider criteria focusing on two important performance dimensions:

1. compliance with standards for the proper operation of an animal control service, especially as they pertain to humane animal treatment, and
2. operational efficiency and effectiveness.

Standards of Proper Operation and Humane Treatment

The Humane Society of the United States (HSUS) has been among the national leaders in the development of standards for animal control operations. HSUS standards address topics ranging from animal shelter operating hours, fees, and policies (Table 3.1) to operating procedures (Table 3.2), estimated kenneling needs (Table 3.3), and shelter size and design. Local policies, procedures, and facilities may be assessed by comparing them with HSUS guidelines, if deemed applicable and appropriate by community officials.[1] Local compliance with such standards may be a more revealing indicator of program adequacy than year-to-year comparisons of workload statistics.

HSUS contends that most successful animal control programs emphasize legislation, education, and sterilization. Common elements include

- An ordinance establishing differential licensing with lower fees for spayed or neutered pets
- A low-cost spay/neuter clinic or equivalent program
- Mandatory sterilization of animals adopted from the public shelter
- A public education program (Handy, 1993, p. 2)

Sterilization is an important component of animal control programs not only because it limits the animal population but also because it tends to curb animal

TABLE 3.1 Policy Standards for Animal Control Operations

Animal Shelter Operating Hours

- Shelter should receive sick or injured animals 24 hours per day.
- Shelter should be open for adoption/redemption Monday through Friday with extended hours to 7 p.m. at least one weekday and open at least 4 hours on Saturday and Sunday.

Adoption Standards

- Adoption programs should strive to find responsible lifelong homes for animals.
- Animals should be placed only with adults who intend to keep them as household pets; never as gift to another person; never to a transient individual; dogs should never be placed with an adopter without a properly fenced yard or suitable means of exercise.
- Puppies/kittens should be at least 4 months old before being placed in a home with children less than 6 years of age.
- Only domestic animals should be placed as pets.

Sterilization of Animals Placed by Shelter

- All animals placed by shelter should be sterilized prior to release or within 30 days of adoption. Any purebred animal should be sterilized before leaving the shelter.

Nonadoption Policy for Fighting Dogs

- Pit-bull terriers and other dogs bred and conditioned for fighting purposes should be euthanatized as soon as legally possible rather than placed for adoption.

Policies on Fees

- There should be no fee for picking up or receiving animals.
- Adoption fee for purebred or especially desirable animals should be no higher than for others; adoption fees for dogs should be no higher than for cats.

SOURCES: Adapted by the Humane Society of the United States (HSUS) from the following publications to reflect current HSUS policy: *HSUS Guidelines for Animal Shelter Policies* (Washington, DC: HSUS, April 1982); *HSUS Shelter Guidelines Pertaining to Potentially Dangerous Dogs* (Washington, DC: HSUS, November 1988); *HSUS Guidelines for Responsible Pet Adoptions* (Washington, DC: HSUS, n.d.). Used by permission.

aggression. According to HSUS, unsterilized animals are twice as likely as spayed or neutered animals to bite humans and, furthermore, account for 95% of all fatal maulings (Handy, 1993, p. 11). Some communities require that animals impounded more than once be spayed or neutered, some require that cats allowed to roam outdoors be sterilized, and many follow the HSUS recommendation of offering lower licensing fees for such pets (Handy, 1993).

Operational Efficiency and Effectiveness

In addition to standards for facilities and recommended policies, national performance statistics and rules of thumb developed by professional associations and municipal leagues provide pegs for comparison with local performance marks. One such rule of thumb, useful for estimating shelter requirements or for documenting a greater-than-normal animal control problem, suggests that the number of animals entering the shelter during a year will equal approximately 7% to 10% of the human population of the community (HSUS, 1990, p. 1). Another rule of thumb—although

TABLE 3.2 Summary of Guidelines for Animal Shelter Operations

Facility

- Facility size and design should permit necessary animal separation and minimize stress (e.g., dogs from cats, sick or injured from healthy, puppies from adults, aggressive from others, etc.).
- Sealed concrete (or other nonporous) floors should slope toward drains.
- Nonporous walls at least 4 feet high and topped with at least 2 feet of chain link or wire mesh should separate kennels.
- Runs should be covered with fence fabric or wire mesh.
- Heavy-duty drainage and plumbing system should include drainage for each run to prevent cross contamination.
- Facility design should provide adequate heating, cooling, humidity control, and air circulation (guidelines: floor temperature of at least 65°F for adult animals, at least 75°F for infant animals; circulation sufficient to permit exchange with outside air 8 to 12 times per hour).
- Cages for indoor holding of dogs should have floor space of at least 12 square feet for small breeds, 20 square feet for medium breeds (36-50 pounds), and 24 square feet for large breeds.
- Dog kennels with runs should include at least 56 square feet combined (24 sq. ft. for kennel and 32 sq. ft. for run). For shared kennels, allow a similar amount of space per dog.
- Cages for cats should provide floor space of at least 9 square feet—with assignment of one cat per cage, except for mothers with kittens.
- Euthanasia room and dead-animal storage should be accessible from the kennel but separated from public view.

Procedures

- Feeding three times per day for puppies/kittens less than 12 weeks of age; twice per day if 12 weeks to 1 year old.
- Potable water should be available at all times.
- Self-feeders, if used, should be mounted to prevent contamination.
- Each cat cage should contain a litter pan and a perch (or resting shelf).
- Animals should be inspected for disease on arrival and daily thereafter.
- All kennels, cages, and runs should be scrubbed daily with hot water and disinfectant—with animals placed in separate holding area and remaining there until enclosure is completely dry.
- Dogs confined in cages should be exercised twice a day.
- Euthanasia should be performed only by trained personnel using humane methods (e.g., sodium pentobarbital, pentobarbital plus local anesthetic, and bottled carbon monoxide are acceptable under specified conditions and application; hot, unfiltered carbon monoxide from an engine is not).
- Disposal of dead animals should be by incineration, burial in a landfill, or another method approved by the community.

Records

- Thorough and easily accessible records should be maintained for all animals entering the shelter and for all receipts (e.g., donations and fees).

Vehicles

- Vehicles should be well marked and fully equipped with animal control equipment and should include a protected, well-ventilated, and separate enclosure for each animal carried.

SOURCE: Adapted by the Humane Society of the United States from HSUS, *HSUS Guidelines for the Operation of an Animal Shelter* (Washington, DC: HSUS, September 1990), to reflect current HSUS policy. Used by permission.

of dubious value because it is an input standard—is HSUS's declaration that adequate funding of an effective animal control program requires $3 to $5 per capita (human) annually (Handy, 1993, p. 5).

TABLE 3.3 Projected Kenneling Needs for Dogs and Cats

| Human Population | Minimum Kenneling Needed for Number of Animals to Be Handled: | |
	Dogs and Puppies	Cats and Kittens[a]
Under 50,000	0-50	0-20
50,000-100,000	50-75	20-30
100,000-200,000	75-100	30-35

SOURCE: Phyllis Wright, Barbara A. Cassidy, and Martha Finney, "Local Animal Control Management," *Management Information Service Report,* Vol. 18, No. 7 (Washington, DC: International City Management Association, July 1986), p. 12. Reprinted by permission.

a. The calculations given for cat kenneling do not take into account the increasing number of laws that include cat licensing and rabies vaccinations. If there is a local ordinance that requires cats be held for 5 days or more, more kenneling for the species would be needed.

Useful benchmarks for assessing operational effectiveness may be drawn from national studies and from the performance targets and experience of respected municipalities. A municipality forced to euthanatize 80% of the cats entering its animal shelter, for example, may be surprised to discover its rate to be about the national average (Table 3.4). A shelter that avoids high euthanization rates by placing in new homes more than one out of every five dogs and cats it receives is performing quite well by national standards.

Shelter statistics should be interpreted cautiously. Increasing workload at the animal shelter in the form of greater numbers of apprehensions and other intakes indicates greater animal control activity and may suggest increased effectiveness. At some point, however, increased effectiveness in animal control—including reduction in the number of the animals at large and reduction of unwanted pets as a result of effective education and sterilization programs—should *reduce* the number of animals entering the animal shelter.

Responsiveness is an important dimension of public service. How prompt should responses to animal control calls be? The response times and response time targets

TABLE 3.4 Estimated Shelter Entry and Exit Statistics: 1988 National Averages

	Dogs	Cats
Animals Entering Shelter		
Relinquished	41%	46%
Strays	59%	54%
	100%	100%
Exit		
Placed	19%	19%
Redeemed by owner	15%	2%
Euthanatized	66%	79%
	100%	100%

SOURCE: Carol Moulton, Phyllis Wright, and Kathryn Rindy, "The Role of Animal Shelters in Controlling Pet Overpopulation," *Journal of the American Veterinary Medical Association, 198* (April 1, 1991), p. 1173, citing the American Humane Association's *Animal Shelter Reporting Study, 1985-1988.* Reprinted by permission.

TABLE 3.5 Prompt Response to Animal Control Calls: Performance Targets and Experience of Selected Cities

Irving, TX
  Target: Average of less than 20 minutes to animal bite incidents
  Actual: 16-minute 5-second average (1991)

Overland Park, KS
  Target: Average response time of 20 minutes to calls for service concerning animal control problems

Fayetteville, AR
  Target: Respond to 85% of calls within 2 hours
  Actual: 24-minute average response time (1992)

Lubbock, TX
  Actual: 25-minute average for emergency responses; 70-minute average for routine responses (1991)

Charlotte, NC
  Targets: Respond to (a) 90% of all emergency calls within 45 minutes; (b) 100% of all emergency calls within 1 hour; (c) respond, investigate, and resolve 98% of nonemergency complaints within 1 workday; (d) 100% of all nonemergency complaints within 2 workdays
  Actual: (a) 94%; (b) 100%; (c) 97%; (d) 100% (1987)

Boston, MA
  Actual: 74% of responses within 24 hours of emergency complaint (1992)

of several cities compiled for this chapter suggest that an average emergency response time of 20 minutes or less is achievable by many agencies (Table 3.5). A set of standards developed by the League of California Cities (LCC) distinguishes between different degrees of response urgency (Table 3.6). Calls about threats to human life or safety should receive highest priority and immediate response; slower response times for routine calls are generally acceptable.

The LCC's standards (1994, p. 20) also offer rules of thumb for judging animal control officer workload. Field officers in high-service departments may be expected to handle fewer than 1,000 calls per year, those in medium-service departments are likely to deal with 1,000 to 2,500 calls each, and officers in low-service

TABLE 3.6 Response Time Standards: League of California Cities

	Agency Service Level		
Incident	*High*	*Medium*	*Low*
Nuisance animal	Less than 4 hours	4 to 72 hours	More than 3 days
Aggressive animal	Less than 20 minutes	20 to 120 minutes	More than 2 hours
Neglect or cruelty	Less than 1 hour	1 to 24 hours	More than 24 hours
Confined or trapped animal	Less than 90 minutes	1.5 to 24 hours	More than 24 hours
Dead animal pickup	Less than 4 hours	4 to 36 hours	More than 36 hours
Sick or injured animal	Less than 15 minutes	15 minutes to 4 hours	More than 4 hours
Animal endangering human life or safety	Immediate action only acceptable performance		

SOURCE: League of California Cities, *A "How To" Guide for Assessing Effective Service Levels in California Cities* (Sacramento: League of California Cities, 1994), pp. 18-19. Reprinted by permission.

departments may be required to respond to more than 2,500 calls for service annually. According to the LCC's judgment, field officers in high-service agencies are likely to spend more than half their time on patrol, compared with 30% to 50% in medium-service departments and less than 30% in low-service departments.

User fees for public services—primarily license, shelter, and adoption fees in the case of animal control operations—are important policy issues in many communities. How much of the expense of animal control operations should, for instance, be recovered through license fees? In some cities, a substantial portion of expenses are recouped, but those cities stand as exceptions. A 1990 survey conducted by the Colorado Municipal League revealed that license fees covered at least three quarters of the animal control costs in only 11% of the responding cities in that state and less than one fourth of the cost in most others (Handy, 1993, p. 6).

Note

1. A model ordinance for the control and regulation of animals has been developed by the Humane Society of the United States (HSUS, *Responsible Animal Regulation,* 1986). For more information, contact the HSUS, 2100 L St., NW, Washington, DC 20037.

4

City Attorney

Standards addressing the responsible and ethical conduct of attorneys have been articulated by the American Bar Association. Standards of *performance,* however—benchmarks that would aid a mayor, city manager, or citizen wishing to assess the performance of the city attorney's office—are more elusive.

Benchmark Possibilities

Performance standards are more easily developed for routine functions than for nonroutine work. Because so much of a city attorney's role falls into the latter category, the prospects for precise standards of output that would be both credible and broadly applicable are limited. Nevertheless, careful scrutiny of performance-reporting documents of selected municipalities reveals performance targets and actual experience that offer an external perspective on several key dimensions of attorney performance. That perspective may be helpful to cities wishing to evaluate their own operations.

PREPARATION OF LEGISLATION AND LEGAL OPINIONS

As the municipality's legal expert, the city attorney is expected to prepare—or at least to approve—proposed ordinances placed before the city council for

TABLE 4.1 Prompt Preparation of Legislation and Legal Opinions: Performance Targets and Experience of Selected Cities

Chandler, AZ
 Target: 4 workdays for routine legal opinions
 Actual: 4-day average (1991)

Fort Collins, CO
 4-day average turnaround on legal service requests (1991)

Fayetteville, AR
 Response to 100% of requests for legal services within 5 days (1991)

Bakersfield, CA
 Target: Respond orally or in writing to all requests for legal advice immediately or within 10 days if research is required

Orlando, FL
 100% of ordinances/resolutions prepared within 10 workdays of request when accompanied by backup material (1991)

Dayton, OH
 Target: Complete 95% of legislation within 2 weeks or within the time designated

Oakland, CA
 Target: Respond to at least 85% of requests for legal assistance within 21 days
 Actual: 85% within 21 days (1992)

Decatur, IL
 Target: Respond to at least 85% of written legal inquiries within 30 days
 Actual: 79% within 30 days (1991)

Overland Park, KS
 Target: Oral opinions on routine matters within 24 hours, on nonroutine matters within 30 days; written opinions on routine matters within 15 days, on nonroutine matters within 30 days

Wilmington, DE
 Target: Respond within 30 days to at least 90% of requests for legal opinions

Milwaukee, WI
 Target: Complete at least 25% of opinions and assignments within 30 days; 50% within 90 days
 Actual: 15% completed within 30 days; 40% within 90 days (1992)

adoption. The attorney also is expected to render legal opinions on all manner of issues coming before the city government. Cities tend not to report on the quality of the ordinances crafted by the city attorney or on the quality of opinions rendered, but several do report on the speed with which those services are provided.

Among the examined municipal documents that report on city attorney responsiveness, most indicate that requested legislation or legal opinions are expected within 2 or 3 weeks (Table 4.1). A few expect more rapid response—notably, Chandler, Arizona, and Fort Collins, Colorado, with 4-day average turnaround for legal opinions. On the other hand, some cities set their response target at 30 days.

Fairly precise performance targets have been set for city attorney responsiveness in Overland Park, Kansas. The attorneys there are expected to render oral opinions on routine matters within 24 hours and on nonroutine matters within 30 days. Written opinions on routine matters are expected within 15 days and on nonroutine matters within 30 days.

RESPONSIVENESS TO
REQUESTS FOR LEGAL ACTION

How quickly does the city attorney file legal documents, review pending contracts, and perform other important tasks? Municipal attorneys in Orlando, Florida, file arrest documents within 15 days (Table 4.2). The city attorney in Decatur, Illinois, files most demolition cases for unsafe structures within 30 days and most

TABLE 4.2 Responsiveness to Requests for Legal Action: Selected Cities

PROMPT FILING OF ARREST DOCUMENTS

Orlando, FL
100% of charging documents for city code violations filed within 15 days (1991)

PROMPT FILING OF CASES

Decatur, IL
95% of demolition cases (unsafe structures) filed within 30 days of council approval; 84% of environmental and nuisance cases filed within 21 days of request for prosecution (1991)

PROMPT REVIEW/PREPARATION OF CONTRACTS

Fayetteville, AR
100% reviewed within 24 hours (1991)

Boston, MA
9 days (average) for review of noncompetitively bid contracts; 5 days (average) for competitive contracts (1992)

Orlando, FL
100% prepared within 15 workdays of request when accompanied by backup material (1991)

PROMPT PREPARATION OF CITY CODE MATERIAL

Orlando, FL
100% prepared within 60 workdays of ordinance adoption (1991)

---

environmental and nuisance cases within 21 days. Reported review times for municipal contracts ranged from 24 hours to an average of 9 days. The time required to prepare a contract is longer—for example, 15 workdays in Orlando.

SUCCESS RATIOS

Several aspects of the city attorney's performance lend themselves to the calculation of success rates: convictions as a percentage of all cases, collection rates on claims and hot checks, ratio of cases won to cases lost, percentage of lawsuit settlements, percentage of litigation avoided, and amount of judgment or settlement as a percentage of the amount demanded of the municipality by claimants, to name several (Table 4.3). If the number of cases used in the calculation of a given ratio is large, that ratio may be adopted as a benchmark more confidently than if the number of cases is small. If, for example, a conviction rate is based on a large number of cases—as the 90% marks of Greenville, South Carolina, and Dayton, Ohio, undoubtedly are—that rate is less likely to be uncharacteristically high or low simply as the result of a streak of good or bad courtroom luck. Such distortions can more easily occur when the percentage is based on only five or six cases.

Should every city attorney be expected to meet or exceed the ratios reported in Table 4.3? Probably not. Several factors beyond the control of the city attorney could intervene to influence the success ratio in one direction or the other—for example, the difficulty of the cases, the practice of contracting out the most difficult ones, the city council's willingness to accept the city attorney's recommendations regarding litigation, and the municipality's insurance coverage and its ramifications for litigation or settlement. Furthermore, without a larger set of reporting municipalities,

TABLE 4.3 Success Ratios for City Attorneys: Performance Targets and Experience of Selected Cities

CONVICTION RATE	**RATIO OF LAWSUIT WINS TO LOSSES**
Shreveport, LA	**San Antonio, TX**
Estimates: 95% traffic conviction rate; 95% DUI[a] conviction rate (1992)	Actual: 40 to 3 (1991)
	PROMPT RESOLUTION OF SMALL CLAIMS CASES
Overland Park, KS	
Actual: 94.7% DUI[a] conviction rate (1991)	**Decatur, IL**
Greenville, SC	Actual: 51% resolved within 120 days of request for prosecution (1991)
Actual: 90% (1990)	**LITIGATION AVOIDANCE**
Dayton, OH	**Charlotte, NC**
Target: 90% in housing cases; 85% in prostitution-related cases	Actual: 67% of cases settled by negotiation (1987)
Alexandria, VA	**Milwaukee, WI**
Actual: 85% conviction rate for drug cases (1992)	Actual: 54% of cases resolved by pretrial evaluations (1994)
CLAIMS/HOT CHECK COLLECTION RATE	**Oakland, CA**
Denton, TX	Target: Settle at least 30% of prelitigation claims Actual: 30% (1992)
Actual: 73% of claims collected; 78% of hot checks collected (1991)	**SUCCESSFUL SUBROGATION**
Lubbock, TX	**Alexandria, VA**
Actual: 60.2% of claims collected (1991)	Percentage of city property damage losses collected from third parties: 68% (1990); 72% (1991); 79% (1992); 81% (1993)
PROMPT SETTLEMENT OF LAWSUITS	**JUDGMENT/SETTLEMENT AS PERCENTAGE OF DEMAND**
Oakland, CA	
Target: Settle at least 39% within 1 year of filing Actual: 35% (1992)	**Shreveport, LA**
Fayetteville, AR	Estimate: 15% (1992)
Actual: 19% of current year's litigation cases closed; 72% of prior year's litigation cases closed (1991)	

a. Driving under the influence (DUI); alternatively, driving while intoxicated (DWI).

it is impossible to know whether these numbers are at the high or low end of a range of reasonable performance expectations. Even so, they offer a useful external context for performance assessments. When the local city attorney is boasting about having won 20 out of 30 cases, it is useful to know that the city attorney in San Antonio, Texas, won 40 out of 43 cases one year—whether that deflating information is shared with the local city attorney or not.

5

City Clerk

Duties of city clerks—often titled "city secretary"—are varied. In some communities, the city clerk is the chief appointed official and serves as coordinator or manager of most municipal functions. In other communities, the chief executive or administrative director role is performed by another official—typically the mayor, city manager, city administrator, or executive assistant to the mayor—and the primary duties of the city clerk are more narrowly defined. In most cases, basic responsibilities of the office include providing secretarial services to the mayor and city council, preparing minutes of city council meetings, serving as custodian of official records and ensuring access to those records, and serving as a principal contact for citizen and business inquiries. In many municipalities, the city clerk has a major role in coordinating the assembly and delivery of city council agenda packets that provide background information on items being considered at upcoming city council meetings.

Performance Benchmarks From Other Cities

Perhaps because the role of city clerk varies from one community to another, incumbents in that office sometimes assume that meaningful performance comparisons across cities are impossible. Variations in population and differences in the scope and scale of duties assigned to the city clerk reduce the likelihood that the work done in one community closely matches the work done in another.

Although it is true that the performance of most municipal clerks is reported only in the form of workload statistics (e.g., city council meetings attended, minutes prepared, and number of requests for official records) and that raw figures from one community are of little use or interest elsewhere, several cities do report information that *is* of use as potential benchmarks. That information focuses on the fundamental aspects of the job of city clerk—that is, the responsibilities that tend to fall to the city clerk whether the remainder of assigned duties are narrow or broad—and the information is presented in a standardized way that negates much of the relevance of community size and other variations. In such cases, the performance targets and experience of other municipalities can help a city frame its own expectations as it ponders answers to these and other questions regarding the performance of the city clerk:

- How far in advance of a meeting should the city council reasonably expect to receive the meeting agenda and background information?
- How soon after a meeting should the city council expect to receive the minutes?
- If one assumes that custodial duties for official records imply ensuring general accessibility, does that mean that official action and documents should be indexed? If so, how up-to-date should the index to be?
- When the city council takes action that must then be processed by the city clerk, what level of promptness is reasonable to expect?
- When inquiries are made that involve official city records, how quickly should the city clerk be expected to respond?

In some cases, the answers to these questions are established by state law or local ordinance. In other communities, affected parties debate these and other questions of responsiveness and proceed to establish or reaffirm local expectations—often without the benefit of relevant information from other jurisdictions. "It's always taken about a month to prepare the minutes; it's unreasonable to expect them sooner," for example, is an argument based entirely on local experience, deriving no benefit from external information.

ADVANCE MATERIAL FOR UPCOMING MEETING

City clerks responsible for assembling and distributing city council agenda packets often take the heat when a last-minute agenda announcement reveals that a controversial or otherwise unwelcome topic will be taken up at the next meeting or when insufficient lead time is provided for reading and considering critical background information and staff analysis. How much lead time is reasonable? Preparing background information for delivery too far in advance of a meeting may deny council members important late-breaking information or may encourage the practice of tacking on one agenda addendum after another as the meeting date approaches. On the other hand, failure to provide adequate lead time restricts the

TABLE 5.1 Timely Notice of Council Agenda and/or Issuance of Agenda Packet: Performance Targets of Selected Cities

Anaheim, CA
Provide notice of council agenda by noon of the Thursday preceding the meeting
Boston, MA
Distribute council meeting agendas 24 to 48 hours before all meetings
Sunnyvale, CA
Deliver agenda packets to council members at least 4 days prior to meeting

ability of the city council to perform its role. Information reported by some cities suggests a preference for providing advance notice of at least 2 or 3 days, ideally extending over a weekend (Table 5.1).

PROMPTNESS OF MINUTES PREPARATION

Considerable variation exists in the turnaround time for city council meeting minutes. Cities such as Duncanville, Texas, and Sunnyvale, California, expect minutes to be prepared within 2 days following a meeting, whereas some other cities apparently are comfortable projecting much longer turnaround periods—even a full month (Table 5.2). If prompt preparation of city council minutes is a high priority only to the city clerk and only as a matter of professional pride, the wisdom of that priority may be debatable in light of other demands. If, on the other hand, city council members and other consumers of those minutes desire more rapid turnaround, preparation time of, say, 2 weeks is well within the band of reasonableness. Turnaround that is even quicker may be difficult but is not unprecedented.

INDEXING OF COUNCIL DOCUMENTS AND ACTIONS

Many progressive city clerks consider the indexing—sometimes the electronic indexing—of city council documents and actions to be part of their responsibilities as custodian of official records. Indexing increases the likelihood that persons seeking information on a given topic will find that information and will find it quickly. Oklahoma City set the speed record among cities examined, reportedly indexing council actions within 2 or 3 hours of their occurrence (Table 5.3). Other cities that reported prompt indexing as a priority established targets in the 1-week to 6-week range.

PROMPT PROCESSING OF OFFICIAL DOCUMENTS

Once action has been taken at a city council meeting, the responsibility for seeing that official documents are executed, perhaps published, and copies distributed often rests with the city clerk. How quickly should those duties be performed?

TABLE 5.2 Prompt Preparation of City Council Minutes: Performance Targets and
Experience of Selected Cities

Duncanville, TX
Target: Within 48 hours of meeting
Actual: 99% within 48 hours (1990)

Sunnyvale, CA
Target: Within 2 days of meeting
Actual: 89% within 2 days (projected, 1994)

Irving, TX
Actual: 100% within 9 days of meeting (1991)

St. Petersburg, FL
Target: By Friday of the following week
Actual: 100% by Friday of the following week
(1990)

Reno, NV
Target: Prepared *and approved* within 2 weeks
of meeting
Actual: 95% prepared *and approved* within 2
weeks (1990)

Oakland, CA
Target: Within 14 days of meeting
Actual: 90% within 14 days (1991); 100% within
14 days (1992)

Ocala, FL
Target: Within 2 weeks of meeting

Overland Park, KS
Target: Within 15 working days

Anaheim, CA
Target: Available by second subsequent meeting

Alexandria, VA
Target: Within 1 month of meeting
Actual: 95% within 1 month (1991); 100% within
1 month (1992); 100% within 1 month (1993)

Tallahassee, FL
Target: Within 1 month of meeting
Actual: 35% within 1 month of meeting (1991)

Processing times among reporting cities ranged from a prompt 1 or 2 days in
Alexandria, Virginia, and Lubbock, Texas, to as many as 10 working days as the
targeted maximum (Table 5.4).

RETRIEVAL OF RECORDS AND INFORMATION

Speed of information retrieval depends in part on the diligence of clerical
employees but also on the adequacy of filing, storage, indexing, and information

TABLE 5.3 Prompt Indexing of Council Documents and Actions: Selected Cities

Oklahoma City, OK
Target: Within 2 hours of adjournment
Actual: 2.5-hour average (1992)

Anaheim, CA
Target: Within 5 workdays

Oakland, CA
Target: 80% within 7 working days

Hurst, TX
Target: Within 2 weeks
Actual: 75% within 2 weeks (1990)

Ocala, FL
Target: Within 2 weeks

Lubbock, TX
Actual: 80% within 1 month (1991)

Tallahassee, FL
Actual: 35% within 6 weeks (1991)

TABLE 5.4 Prompt Processing of Official Documents: Selected Cities

Alexandria, VA
  Target: Complete resolutions within 2 days following council meeting; submit ordinances to city
  attorney for completion within 1 day following meeting; complete action docket within 1 day following
  meeting

Lubbock, TX
  Actual: 100% of documents processed within 1.5 days of council meeting (1991)

Oakland, CA
  Actual: 85% of legislation numbered and distributed within 3 days (1992)

Ocala, FL
  Target: Execution of legal documents within 3 days of official action

Reno, NV
  Actual: 95% of documents executed within 5 days (1990)

Decatur, IL
  Actual: 100% of documents processed within 7 days (1991)

Anaheim, CA
  Target: Transmit documents within 8 days

Oak Ridge, TN
  Target: Execute/publish/file 90% of official documents within 10 days of adoption/receipt/authorization
  Actual: 85% within 10 days (1991); 70% within 10 days (1992)

Irving, TX
  Actual: 99% of documents processed and distributed within 10 workdays (1991)

retrieval systems. Rapid retrieval often is a testimony to those systems. Among cities reporting targets for actual retrieval times, a few reported average retrieval in a matter of minutes—typically 5, 15, or 30 minutes—but several targeted their efforts toward guaranteeing the availability of requested records within 1 or 2 working days (Table 5.5). In some instances, allowances were made for longer response time in the event that requested information was of a sensitive nature (e.g., falling under Freedom of Information Act provisions) or involving information stored in a remote or otherwise less accessible location.

OTHER POSSIBLE BENCHMARKS

The availability of comparable statistics from several cities on a given performance dimension lends a degree of confidence to the reasonableness of a reported performance range. Although less compelling, the reported performance of even a singular jurisdiction may sometimes be of use. Corpus Christi, Texas, for example, reports that 88% of its minutes were approved without correction; Sunnyvale attempts to prepare its council action digest within 24 hours of council meetings; and Oakland, California, reports that 80% to 90% of all necessary referrals are made within 2 business days following council action (Table 5.6). Other singular benchmarks address the issuance of solicitation permits and selected output statistics.

TABLE 5.5 Prompt Retrieval of Municipal Records and Information: Selected Cities

Irving, TX
Actual: 98% within 5 minutes (1991)

Duncanville, TX
Actual: 12-minute average

Reno, NV
Actual: 95% within 15 minutes (1990)

Nashville-Davidson County, TN
Target: 95% within 30 minutes

St. Petersburg, FL
Target: Available same day requested (records requested after 3 p.m. available by 8 a.m. next workday)

Oakland, CA
Target: 90% within 1 working day

Peoria, AZ
Target: 80% within 24 hours

Orlando, FL
Target: Within 48 hours

Tallahassee, FL
Actual: 100% within 2 days (1991)

Sunnyvale, CA
Target: 95% within 2 days; microfilmed or inactive records within 10 working days

Alexandria, VA
Target: Process requests under Freedom of Information Act within 5 days
Actual: Average of 10 days (1990); 5 days (1991); 5 days (1992); 5 days (1993)

TABLE 5.6 Odds and Ends for City Clerks: Miscellaneous Benchmarks From Selected Cities

QUALITY OF CITY COUNCIL MINUTES
Corpus Christi, TX
88% of minutes required no corrections (1991)

PROMPT PREPARATION OF COUNCIL ACTION
DIGEST FOR NEWSPAPER PUBLICATION
Sunnyvale, CA
Target: Prepare at least 95% within 24 hours following meeting (1994)

PROMPT ISSUANCE OF COUNCIL REFERRALS
Oakland, CA
90% within 2 working days of meeting (1991); 80% (1992)
Oak Ridge, TN
90% of council actions followed up within 5 days (1992)

PROMPT ISSUANCE OF NONPROFIT SOLICITATION PERMITS
Reno, NV
100% within 1 week (1990)

OUTPUT STATISTICS: PROCESSING TIME FOR VITAL STATS
Boston, MA
17 minutes per record (birth/marriage/death certificate) (1992)

OUTPUT STATISTICS: MICROFILMING
Reno, NV
333 pages microfilmed per hour (1990)

6

Courts

β e careful talking about performance standards and performance measurements around municipal court judges. Courts, they are likely to declare, are
about due process and justice; they are *not* about revenues and efficiency. And who
wants to argue with a judge?

Courts *are* about justice and due process, but measures of effectiveness can be
devised that are consistent with that mission. Furthermore, measures of efficiency need
not be considered an affront to the court's higher mission. Prompt—that is, efficient—
judicial action backed by efficient support services is a worthy objective.

Possible Benchmarks

The nature of the municipal court function renders quite slim the prospect of
extremely precise and widely accepted standards of performance, at least insofar
as the nonroutine elements of the court's work are concerned. The more routine
support services attached to court operations could be another matter. Many of
those services could be the focus of engineered standards or performance benchmarks. Unfortunately, few cities report municipal court performance statistics in
a fashion that makes them viable candidates for performance benchmarks.

The relative scarcity of performance indicators for individual municipalities,
other than workload measures, admittedly reduces confidence that the small set of

TABLE 6.1 Prompt Scheduling and Disposition of Court Cases

Scheduling	*Disposition*
Fort Worth, TX Target: Each case scheduled on a docket within 7 days of receipt	**Milwaukee, WI** 45% of nonpriority cases tried within 90 days after intake date (1993)
Savannah, GA Target: Schedule misdemeanor cases for preliminary hearing on the court business day after date of arrest; schedule preliminary hearing for felony cases within 1 week of arraignment	**Oklahoma City, OK** Percentage of criminal cases disposed within 90 days: 86% (1991); 88% (1992) Percentage of jury cases disposed within 90 days: 84% (1991); 85% (1992) Percentage of traffic cases disposed within 90 days: 91% (1991); 100% (1992)
Reno, NV Target: Schedule all trials within 60 days of arrest Actual: 100% (1990)	**Overland Park, KS** Targets: Dispose of minor traffic cases within 45 days; major traffic cases within 60 days; misdemeanor and driving under the influence (DUI) cases within 100 days
Fayetteville, AR Target: Set criminal cases for trial within 90 days	**Savannah, GA** Targets: Average time of 5 days between arrest and disposition for misdemeanor defendants; 10 days for felony defendants Actual average for misdemeanor defendants: 5.0 days (1988); 5.0 days (1989); 4.8 days (1990); 5.0 days (1991) Actual average for felony defendants: 10.0 days (1988); 10.0 days (1989); 8.0 days (1990); 10.0 days (1991) Target: Average of 30 days for processing traffic citations from filing to disposition

reported measures fully and accurately reflects reasonable performance levels, but the small number does not render the reported measures valueless. They simply must be applied more cautiously.

PROMPT SCHEDULING AND DISPOSITION

The right to a speedy trial is a fundamental principle of the U.S. judicial system. Some trials, however, are much speedier than others. Fort Worth's municipal court, for example, acts quickly—within 7 days—to set each case on a future docket; Reno, Nevada, attempts to set all trials within 60 days of arrest (Table 6.1). Most cases in Oklahoma City and almost half the nonpriority cases in Milwaukee are disposed within 90 days; action in Overland Park, Kansas, and Savannah, Georgia, is typically even quicker.

PROCESSING PAPERWORK

Administrative and clerical tasks associated with municipal court operations—from inputting case information into the court's record system to the processing of

TABLE 6.2 Prompt Preparation and Processing of Court Paperwork

INPUTTING OF CASE INFORMATION	NOTIFICATION OF COURT DATE	PROCESSING OF APPEALS
Fort Worth, TX Target: Within 7 days (criminal cases)	**Oklahoma City, OK** Percentage of notifications mailed within 12 hours: 100% (1991); 100% (1992)	**Oklahoma City, OK** Percentage of appeals processed within 24 hours of receipt: 98% (1991); 98% (1992)
Oklahoma City, OK Percentage of criminal case information input within 24 hours of receipt: 100% (1991); 100% (1992) Percentage of jury case information input within 24 hours of receipt: 100% (1991); 100% (1992) Percentage of traffic case information input within 24 hours of receipt: 90% (1991); 95% (1992)	**PROVIDE TRIAL DATA TO CITY ATTORNEY** **Reno, NV** Target: Provide scheduled trial data to city attorney within 5 days of arraignment Actual: 95% (1990)	**Overland Park, KS** Target: Forward notice of appeal within 10 days of disposition
	UPDATING OF CASE RECORDS	**Reno, NV** Target: File appeals in district court within 10 days of municipal court conviction date Actual: 95% (1990)
Houston, TX Target: Enter 95% of tickets within 2 days of receipt	**Oklahoma City, OK** Percentage updated within 12 hours: 95% (1991); 97% (1992)	**RECORDING OF CASE DISPOSITION, COMPLETION OF PAPERWORK**
UPDATING OF PARKING CASE RECORDS	**PROCESSING OF BOND FORFEITURE NOTICES**	**Overland Park, KS** Target: Within 3 workdays
Oklahoma City, OK Percentage updated within 24 hours of receipt: 100% (1991); 100% (1992)	**Oklahoma City, OK** Percentage processed within 8 hours: 97% (1991); 98% (1992)	**Reno, NV** Target: Enter sentencing data within 5 days of trial
Orlando, FL Percentage of newly issued tickets input within 2 workdays: 100% (1991)	**ISSUANCE OF WARRANTS**	**DEPOSIT COURT REVENUES**
Reno, NV Target: Enter traffic citations into court's computer database within 48 hours of receipt Actual: 100% (1990)	**Oklahoma City, OK** Average number of days between request for arrest warrant and issuance: 3 days (1992) Average number of days between failure to appear and issuance of bench warrant: 15 days (1992)	**Reno, NV** 100% deposited daily (1990)
		Milwaukee, WI 100% posted within 24 hours of receipt (1993)
Boston, MA (DOT) Percentage of parking tickets processed within 72 hours of receipt from ticket-issuing agencies: 100% (1992) Percentage of payments entered into system within 48 hours of receipt: 75% (1992)	**Fayetteville, AR** Target: Within 10 workdays	**Houston, TX** Target: Process 95% of mailed payments within 1 day of receipt
	Overland Park, KS Target: Within 20 days	**Orlando, FL** Target: Process mailed parking fines within 2 workdays of receipt Actual: 81% (1991)
Fort Worth, TX Target: Within 7 days	**ISSUANCE OF OVERDUE NOTICE ON UNPAID PARKING CITATIONS** **Oklahoma City, OK** Within average of 15 days following due date (1992)	

appeals—constitute important components of court workload. The performance targets and experience of other jurisdictions in completing those tasks offer an external context in which to assess local performance (Table 6.2).

TABLE 6.3 Court Employee Production Ratios

CASE WORKLOAD

Shreveport, LA
  1,500 criminal cases per month per judge; 427 civil cases ($15,000 or less) per month per judge; 6,888 traffic cases per month per judge (1991)

Alexandria, VA
  Number of cases heard and processed per employee-year: 444 (1989); 403 (1990); 434 (1991); 468 (1992)

CLERICAL WORKLOAD

Alexandria, VA
  Number of documents recorded and processed per year per staff member: 3,528 (1989); 3,841 (1990); 3,412 (1991); 4,645 (1992)

WARRANT WORKLOAD

Oklahoma City, OK
  Average number of warrant cases served and cleared per day per employee: 18 (1991); 17 (1992)

COURT EMPLOYEE PRODUCTION RATES

Selected production benchmarks for court employees are provided in Table 6.3. Because of the limited number of cities reporting such information, these benchmarks should be applied cautiously.

COLLECTION RATE FOR FINES

What is a reasonable collection rate for fines imposed by the municipal court? Figures provided by selected cities suggest that a reasonable rate lies between 50% and 80% (Table 6.4).

TABLE 6.4 Collection Rates for Fines Imposed by Court: Selected Cities

Savannah, GA
  Actual: 80% (1992); 82% (1993)

Auburn, AL
  Target: 80% of all parking fines (1993)

Fayetteville, AR
  Actual: 69% (1991); 70% (1992)

Indianapolis, IN
  Actual: 65.3% for traffic tickets (1991)

Wilmington, DE
  Target: At least 55%
  Actual: 50% (1992)

7

Data Processing

In most cases, municipal data processing is an internal service operation. Its customers are other city departments rather than citizens. As those customers become increasingly reliant on a variety of automated systems for conducting their operations, they grow to expect quick and reliable service—and to become increasingly impatient when speed and reliability are absent.

Benchmarks

Many cities report precise performance statistics for their data processing operations. Among those examined, several reported system response times, response times for dealing with computing problems, and statistics on system downtime.

SYSTEM RESPONSE TIME

Several cities tabulated system response times or declared their response targets, sometimes subdivided between local and remote transactions (Table 7.1). Most expected their system to respond to users within 2 seconds.

TABLE 7.1 Computer System Response Time: Selected Cities

Knoxville, TN
Maximum response time: Less than 0.001 second (1991)

Reno, NV
Actual: Average response time of 0.9 second (1990)

Savannah, GA
Target: Mainframe response time of less than 2 seconds 90% of time; less than 5 seconds 98% of time
Actual: Less than 2 seconds 90% of time; less than 5 seconds 98% of time (1991)

Anaheim, CA
Target: Average on-line terminal response time of less than 2 seconds

Fort Worth, TX
Target: Average terminal response time of less than 1 second
Actual: 1 to 3 seconds (1991)

Oakland, CA
Percentage of responses within 2 seconds: 90% (1990); 90% (1991); 98% (1992)

Corpus Christi, TX
Percentage of responses within 2 seconds: 75% (1991)

Nashville-Davidson County, TN
Target: 2.5 seconds or less for 90% of local transactions; 3.5 seconds or less for 90% of remote transactions

Jacksonville, FL
On-line transaction response time: Less than 10 seconds (estimate, 1992)

Cincinnati, OH
Target: Maximum acceptable response time of 20 seconds

TABLE 7.2 Prompt Response to Computer System Problems: Selected Cities

Savannah, GA
Percentage of service requests completed within 2 hours: 97% (1988); 95% (1989); 97% (1990); 96% (1991)

Sunnyvale, CA
Target: 90% of in-house repairs made within 4 hours of notification

Charlotte, NC
Target: Initial response within 1 hour; 90% resolved within 24 hours
Actual: 99.5% response within 1 hour; 97.5% resolved within 24 hours (1987)

Dayton, OH
Target: Respond to 90% of requests for host network and PC equipment repair within 4 working hours and resolve the reported problem within 24 working hours

Anaheim, CA
Target: Respond to calls for assistance within 4 hours

Orlando, FL
Target: Respond to help calls/requests for assistance within 4 hours

Winston-Salem, NC
Percentage of reported minor system problems resolved within 24 hours: operating, 98%; development, 96% (1992)

Fort Worth, TX
Target: 95% of PC repairs within 24 hours

Portland, OR
Target: Resolve 90% of computer program problems within 1 day

Alexandria, VA
Percentage of service requests completed within 24 hours: 70% (1989); 90% (1990); 90% (1991)

Oakland, CA
Percentage of help desk calls responded to within 24 hours: 80% (1992)
Percentage of help desk calls resolved immediately: 57% (1992)

Corpus Christi, TX
Percentage of reported problems resolved within 48 hours: 93.75% (1991)

St. Petersburg, FL
Targets: (a) Respond to 95% of production problems within 24 hours; clear 80% within 48 hours; (b) respond to 95% of network problems within 24 hours; clear 80% within 48 hours
Actual: (a) 91.7% response to production problems within 24 hours; 91.7% cleared within 48 hours; (b) 91.9% response to network problems within 24 hours; 84.2% cleared within 48 hours (1990)

Columbus, OH
2-day average turnaround for equipment repair

PROMPT RESPONSE TO PROBLEMS

Greater variation could be found in the performance targets and performance statistics for the promptness of data processing department response to computer system problems (Table 7.2). Some emphasized the speed with which they answered calls for assistance; some tracked their speed in resolving those problems; others reported both. Most expected all but the most difficult problems to be resolved within 1 or 2 days.

DOWNTIME

Most of the examined cities reported high computing system reliability—several claiming system downtime of less than 1% (Table 7.3). Most expected, and achieved, system availability in the vicinity of 98% or greater.

TABLE 7.3 Minimizing Downtime: Selected Cities

Chandler, AZ
Unscheduled downtime: 0.14% (1991)

Fort Worth, TX
Target: Central computer uptime of 99.8% or greater
Actual: 99.8% uptime (1991)

Oakland, CA
Percentage of scheduled time computer available: 99.7% (1992)

Reno, NV
Actual: 99.6% availability (1990)

St. Petersburg, FL
Target: 95% availability of on-line systems during user work shifts
Actual: 99.5% availability (1990)

Winston-Salem, NC
Percentage of scheduled system availability achieved: 99.5% (1992)

Nashville-Davidson County, TN
Target: At least 98% system availability
Actual: 99.3% availability (1991)

Boston, MA
Mainframe downtime: 0.96% (1992)

Alexandria, VA
Percentage of computer time available for staff use: 98% (1991); 99% (1992); 99% (1993)

Columbus, OH
Target: 99% composite on-line system uptime; 97% for single-user applications

Little Rock, AR
Computer uptime: 99% (1991)

Raleigh, NC
System availability, excluding planned downtime: 98% (1991)

Savannah, GA
Average monthly incidence of downtime: 3.4 (1990); 10 (1991)
Average hours of downtime per incident: 3.66 hours (1990); 1.5 hours (1991)
Total percentage of downtime (mainframe): 2% (1990); 2% (1991)

Hurst, TX
Actual: 98% operational (1990)

Irving, TX
Actual: 98% mainframe availability (1991)

Oak Ridge, TN
Target: 98% system availability during normal work hours
Actual: 96% availability (1991); 98% availability (1992)

Sunnyvale, CA
Target: 98% availability

Wilmington, DE
Target: 98% uptime

Tallahassee, FL
Actual: 97% mainframe uptime (1991)

Charlotte, NC
Target: On-line systems available at least 95% of users' normal workdays
Actual: 96% (1987)

Cincinnati, OH
Target: At least 94% system availability
Actual: 96% availability (1991)

Denton, TX
Actual: Mainframe uptime of 95% (1991)

Dayton, OH
Target: Maintain 98.5% uptime on the on-line system on 90% of workdays

TABLE 7.4 Data Processing Odds and Ends: Miscellaneous Benchmarks From Selected
Cities

DATA ENTRY ACCURACY

Knoxville, TN
Data entry error rate: 0% (1991)

Raleigh, NC
Accuracy in keying payroll: 99.7% at first edit (1991)

ON-TIME PRODUCTION RUNS

Anaheim, CA
Target: At least 98% on schedule

ACCURATE PROJECTIONS OF PROJECT COSTS/TIMETABLES

Anaheim, CA
Target: Within 15% of cost projections

Alexandria, VA
Percentage of operating system and utility upgrades completed on schedule: 70% (1989); 75% (1990);
85% (1991)
Percentage of application programming completed within promised time frame: 44% (1989); 38%
(1990); 55% (1991)

ODDS AND ENDS

A few cities reported statistics on dimensions of performance that were ignored
by most others. Although less conclusive than statistics derived from multiple
cases, singular performance targets or the experience of even one respected city
may nevertheless be of interest to other municipalities (Table 7.4). Reported
statistics suggest that customers of data entry services should expect error rates
well below 1%. Furthermore, data processing customers should expect high adher-
ence to production run schedules. On the other hand, it is unreasonable to expect
the same high level of precision when it comes to projecting costs and timetables
for the development of new computer applications.

8

Development Administration

lthough most municipal departments have a role to play in their community's development, a handful are crucial members of the cast for local development activities—especially the community's physical development. Department labels differ from place to place, but the functions clustered in this chapter under the heading "Development Administration" commonly are called planning, inspection, code enforcement, and zoning.

Development Climate

Some communities enjoy reputations for favorable development climates. Others sometimes find themselves under attack from real estate persons, builders, land developers, bankers, economic development specialists, or the local chamber of commerce for their *unfavorable* climate for development. In part, a community's development climate is determined by official policies that encourage or restrict industrial, commercial, or residential growth. Policies adopted to preserve the heritage of the community, to protect the integrity of neighborhoods, to rebuff smokestack industries, to encourage upscale development, or to prevent spot zoning or strip shopping centers often are criticized as antidevelopment—even if proponents of those policies prefer that they be cast more favorably as controlled growth or quality development policies.

In addition, administrative and operational factors shape the climate for local development. The responsiveness of departmental employees to the needs of developers and builders for prompt action on plans, inspections, and a host of other regulated activities influence development costs and profitability. Even within a community with restrictive development policies, a highly responsive bureaucracy can mitigate many developer complaints. Conversely, the development climate in a community with strong progrowth policies may be dampened by inspection delays, slow approvals, and general lack of responsiveness among operating personnel.

Benchmarks

Some cities set performance targets for various aspects of development administration. Those targets, plus the records established by others, offer useful guidance to communities attempting to establish their own performance expectations.

COMMUNITY PLANNING

Many work elements of community planning departments are nonroutine in nature and therefore difficult—although not necessarily impossible—to measure. Workload counts for one community—the number of inquiries received, the number of planning commission meetings, the number of zoning map updates, and so forth—are of little relevance as benchmarks for another community. More useful as benchmarks are indicators that measure the quality of various planning actions. For example, Portland, Oregon, reported that 95% of its planning staff decisions in one recent year were upheld on appeal by various boards and hearing officers. Chandler, Arizona, reported that only 13.6% of its staff recommendations required alteration or otherwise failed to receive approval by the planning commission. As an indication of the thoroughness of its planning staff work, Raleigh, North Carolina, reported that during a recent year its city council took action on 42% of all zoning cases and 73% of its site plan or subdivision cases as consent agenda items, meaning that little or no elaboration beyond that provided in advance by the staff was needed to reach a decision.

Processing speed is a much more commonly reported aspect of performance quality. Promptness is documented by several cities for a host of functions:

- Reno, Nevada, reviewed 98% of one recent year's sign permit applications within 48 hours; Orlando, Florida, reviewed 90% within 4 days.
- Oakland, California, tries to prepare at least 90% of initial environmental review studies within 21 days and 100% within 30 days.
- The planning department of Cincinnati, Ohio, has a target of providing advice on matters referred from the city council or city administration within 2 weeks. The planning staff of Fayetteville, Arkansas, responded to 90% of all requests for special information from boards

TABLE 8.1 Time Frame for Planning Department Review and Recommendation on Zoning, Planned Unit Development, and Subdivision Requests: Selected Cities

Oak Ridge, TN
Target: Process subdivision plats within 14 days (50% of statutory maximum)
Actual: 14-day average (1991); 14-day average (1992)

Sacramento, CA
Target: Within 20 days

Nashville-Davidson County, TN
Target: Within 28 days

Charlotte, NC
Target: Review subdivision plans, final plats, multifamily projects, rezoning applications, and special use permits within 30 days
Actual: 100% within 30 days (1987); 100% of subdivision plans and final plats within 30 days, rezonings within 45 days (5 months, 1994)

Phoenix, AZ
Target: Review at least 85% of all preliminary subdivision plats within 30 days

Wilmington, DE
Target: Completion within 30 days

Hurst, TX
Target: Completion time of 6 weeks or less for subdivision plats and zoning cases
Actual: 6-week average (1990)

Iowa City, IA
Target: Review rezoning applications, subdivision plats, and site development plans within 45 days

Oakland, CA
Target: Secure 90% of subdivision application decisions within 45 days; 100% within 50 days

Portland, OR
Target: Land use cases (e.g., subdivisions, conditional uses, design, landmark, environmental, adjustments, and master plans) reviewed within 120 days
Actual: 100% (1991)

and committees within 2 weeks and 75% of its citizen requests within 1 week during a recent year. The planning department of Reno, Nevada, addressed 96% of all inquiries from the public within 48 hours.

- The Overland Park, Kansas, planning department strives to update zoning maps and records within 1 week of each zoning change.

Although performance information reported by one or two municipalities on a given dimension of performance is likely to be of interest to other communities, compilations involving greater numbers of jurisdictions offer a higher degree of confidence for setting reasonable parameters for performance expectations. In that vein, several cities have established performance targets and have reported actual time frames for planning department review of zoning, planned unit development, and subdivision requests (Table 8.1). Their experience generally validates the standards recommended by the League of California Cities (LCC; 1994, p. 32) for the processing of proposed zone changes, variances, and development permits. High-service-level departments, the LCC contends, process these applications and make recommendations within 30 days; medium-service-level departments, within 45 days; and low-service-level departments require more than 60 days.

TABLE 8.2 Review Time for Building Permits: Selected Cities

Commercial Buildings	Residential Buildings
Denton, TX Target: 4 workdays Actual: 45% within 4 workdays (1991)	**Anaheim, CA** Target: Same-day turnaround on minor residential and tenant improvement building permits
Ocala, FL Target: Within 1 week	**Fayetteville, AR** Actual: 100% within 8 working hours (1991)
Peoria, AZ Target: Within 1 week Actual: 100%	**Orlando, FL** Actual: 100% within 1 workday
Savannah, GA Target: 7 days (if 3 stories or less); 10 days (4-7 stories); 14 days (if high-rise) Actual (averages): 7/10/14 days (1990); 7/10/14 days (1991)	**Savannah, GA** Target: Within 1 day for single-family and duplex; 7 days for triplex and quad; 10 days for apartments Actual: 99%/99%/100% (1991); 99%/85%/85% (1992)
Tallahassee, FL Target: Within 8 days Actual: 71% (1991)	**Winston-Salem, NC** Target: Within 1 day Actual: 100% (1991); 100% (1992)
Winston-Salem, NC Target: Within 9 days Actual: 90% (1991); 86% (1992)	**Denton, TX** Target: Within 2 workdays Actual: 97% (1991)
Greenville, SC Target: Within 10 days	**Ocala, FL** Target: Within 2 days
Irving, TX Target: Within 10 working days Actual: 95% (1991)	**Raleigh, NC** Target: Within 2 days Actual: 2 days (1991)
Lubbock, TX Actual: 100% within 10 working days (1991)	**Grants Pass, OR** Target: 90% within 2 days
Orlando, FL Actual: 74% within 10 working days (1991)	**Lubbock, TX** 100% within 3 workdays
Fayetteville, AR Target: Within 14 working days Actual: 100% (1991)	**Irving, TX** Target: Within 3 workdays Actual: 98% (1991)
Fort Collins, CO Target: Within average of 3 weeks Actual: 3-week average (1991)	**Hurst, TX** Target: Review plans and issue permit within an average of 3 days Actual: 3-day average (1990)

Because of variations in the nature and complexity of planning cases, staff time requirements may fluctuate from one instance to another. Nevertheless, a pair of municipalities tabulated and reported professional staff-hour averages required for a variety of planning cases in their jurisdictions. On the basis of experience in 1990 and 1991, Wichita, Kansas, found that approximately 17 professional staff-hours were required for zoning cases, approximately 20 hours for plats, approximately 8 hours for lot splits, and approximately 12 hours for planning cases in general. On the basis of its experience in 1989-1990, Reno, Nevada, reported that approxi-

TABLE 8.2 *Continued*

Commercial Buildings	Residential Buildings
Phoenix, AZ Target: Initial construction plan review on at least 95% of major projects within 20 workdays; on at least 95% of small projects within 10 workdays	**Greenville, SC** Target: Within 4 days
	Reno, NV Target: Complete review of plans for single-family residences within 4 days, for residential remodeling within 2 days
Anaheim, CA Target: Turnaround of 20 workdays or less on all plans; 2-week turnaround on small commercial/industrial permits	**Portland, OR** Target: Within 5 days Estimate: 90% (1993)
Reno, NV Target: Complete review of plans for commercial buildings within 5 weeks; for commercial remodeling within 3 days	**Peoria, AZ** Target: Within 1 week Actual: 100% (1991)
	Fort Collins, CO Target: Within average of 1.5 weeks Actual: 1-week average (1991)
Tacoma, WA Target: Complete 75% of commercial building plan reviews within 6 weeks; 95% within 10 weeks	**Cincinnati, OH** Target: Within 10 workdays, if 20 dwelling units or less Actual: 100% (1991)
Gresham, OR Target: Within 8 weeks	**Tallahassee, FL** Target: Within 10 days Actual: 23% (1991)
	Phoenix, AZ Target: Review custom home plans within 10 workdays
	Gresham, OR Target: Within 2 weeks
	Tacoma, WA Target: Issue 90% of permits for single-family structures within 4 weeks

mately 36 staff-hours were required for each planning commission application, 25 hours for each board of adjustment application, 6 hours per final map, 3 hours per parcel map, 25 hours per site plan review, 6 hours per multiresidential, commercial, or industrial building permit, 45 minutes per single-family residential building permit, and 30 minutes per sign permit.

PLAN REVIEW AND PERMITS

Careful review of construction plans prior to the issuance of a building permit is an important protection of the public interest. From the perspective of builders and their clients, however, a slow review can also mean expensive delays in the construction process. Some communities are more adept than others at minimizing such delays and their associated expenses to builders.

New York City's building code requires examination of plans within 40 days of receipt, but plans for new buildings were examined within an average of 19 working days in 1992. Despite comfortably beating the legal deadline, however, New York City's delay for plan checks was considerably longer than that of many other communities. Houston, Texas, for example, reported reviewing 60% of all plans in 1 day or less, 80% within 7 days, and 90% within 14 days. Oakland, California, reviewed 90% of all new construction plans within 20 days during a recent year, 85% of the plans for major additions or alterations within 15 days, and 95% of the plans for minor additions or alterations within 10 days. Corpus Christi, Texas, reported an average cycle time for its permit and plan review process of 3 to 10 days, with 92% completed within 7 days.

Cincinnati, Ohio, attempts to maintain an average plan review time of 7 work-days or less and reported an average of 4.2 days in 1991. An average plan check turnaround time of 5 days was reported by Iowa City, Iowa, and Decatur, Illinois. The average plan check backlog in Long Beach, California, dropped from 7.5 days in 1991 to 3.95 days the following year.

Most of the plan review times reported in the preceding paragraphs represent an average for all plan reviews. None of them makes any distinction between plan reviews for commercial structures and the review of generally less complicated residential buildings. Many cities, however, make that distinction in the perfor-mance targets they set and the statistics they report (Table 8.2). On the basis of performance figures from several cities, it appears reasonable to expect building permit review within 4 weeks for most commercial projects and within 2 weeks for most residential buildings.

Review of construction plans for fire safety purposes may be performed by building officials but often is performed by fire department personnel. All reporting municipalities indicated completion of most such reviews within 2 weeks (Table 8.3).

Other, less commonly reported dimensions of performance include promptness of attention to persons arriving at the building permit service counter and the percentage of expenditures recouped through permit fees and other revenues. Cincinnati, Ohio, strives to attend to persons at its service counter within 2 minutes of arrival, and Oakland, California, reported serving 85% to 90% of its customers within 15 minutes. Recent annual figures placed revenues at 104% of expenditures in Winston-Salem, North Carolina; 116% of expenditures in Fort Collins, Colo-rado; and 147% of expenditures in Iowa City, Iowa.

INSPECTIONS

Ideal benchmarks for the building inspection function would document the effectiveness of inspections. Were building flaws or noncomplying construction practices spotted by inspectors and thus prevented? A few cities do report the number of inspections rejected by their inspectors, but how many others should have been rejected that slipped by? Inspectors in Raleigh, North Carolina, for example, rejected an average of 448 building inspections per inspector in 1991,

TABLE 8.3 Prompt Fire Safety Review of Building Plans: Selected Cities

---

Vancouver, WA
  Target: Review within 3 workdays

Reno, NV
  Target: At least 75% reviewed within 72 hours

Raleigh, NC
  Target: Within 2 days
  Actual: 7-day average; 50% reviewed within 5 days (1991)

Charlotte, NC
  Target: Review at least 75% of all multifamily and commercial plans within 5 days
  Actual: 89% (1987)

Oakland, CA
  Actual: 50% reviewed within 1 week (1991); 90% (1992)

Cincinnati, OH
  Target: Review by fire department within 7 days
  Actual: 94% (1991)

Corvallis, OR
  Target: Review all fire protection plans within 7 days
  Actual: 100% (1991); 100% (1992)

Oak Ridge, TN
  Target: Review by fire department within 7 days
  Actual: 96% (1991); 95% (1992)

Long Beach, CA
  Target: At least 95% reviewed within 2 weeks

---

366 electrical inspections per inspector, 297 fire inspections per inspector, 249 inspections per mechanical inspector, and 255 inspections per plumbing inspector. Knowing Raleigh's rejection rates may help officials in other cities judge whether charges of overzealous enforcement against one of their own inspectors are credible or not, but the applicability of one city's rates to other communities is questionable, and the extent to which those rates represent the detection of all construction flaws in Raleigh remains unknown. In the absence of more desirable measures of effectiveness, other dimensions of inspection service quality and quantity may be gauged. The performance targets and experience of several cities, for example, indicate that it is reasonable to expect building inspections to be performed within 2 workdays from the time requested (Table 8.4). More aggressive inspection operations often can perform building inspections on the day requested.

Electrical, plumbing, and other specialty inspections also are typically performed within 2 workdays of the request (Table 8.5). Lubbock, Texas, reported that most of its electrical, mechanical, and plumbing inspections during a recent year were performed within 4 work hours of the request.

Average inspector workload varies by inspection specialty among the municipalities reporting that statistic (Table 8.6). On the basis of reported figures, a workload of 10 to 16 general building inspections per day seems reasonable, as do slightly lower numbers of electrical, mechanical, or plumbing inspections.

TABLE 8.4 Prompt Response to Requests for Building/Construction Inspections:
           Selected Cities

Fort Collins, CO
Actual: 3.5-hour average response time; 100%
completed on day requested (1991)

Ocala, FL
Target: Same-day service

Corpus Christi, TX
Actual: 99% performed on day requested (1991)

Corvallis, OR
Actual: 98% within requested time frame (1992)

Hurst, TX
Target: Within 4 hours of notice
Actual: 95% within 4 hours (1990)

Lubbock, TX
Actual: 95% within 4 work hours (1991)

Oak Ridge, TN
Target: 95% within 4 hours of request
Actual: 95% within 4 hours; 75% within 2 hours
of request

Raleigh, NC
Actual: 95% on date requested (1991)

Vancouver, WA
Target: 100% within 24 hours of request; at least
95% on the day of request

Grants Pass, OR
Target: At least 90% on day of request

Anaheim, CA
Target: Within 24 hours of request

Chandler, AZ
Target: Within 24 hours of request
Actual: 100% within 24 hours (1991)

Orlando, FL
Actual: 100% within 24 hours (1991)

Overland Park, KS
Target: Within 1 day of request

Peoria, AZ
Actual: 100% within 24 hours of request (1991)

Fayetteville, AR
Actual: 98% within 8 work hours of request
(1991)

Irving, TX
Actual: 98% within 1 workday of request (1991)

Long Beach, CA
Target: At least 98% within 24 hours of request
Actual: 98.6% (1991); 98.4% (1992)

Winston-Salem, NC
Actual: 100% within 1 day of request (1991);
97% (1992)

Cincinnati, OH
Target: At least 95% within 1 day of request

Portland, OR
Target: At least 95% within 24 hours of request
for commercial, combination, and emergency
housing inspections; 95% of nonemergency
housing inspections within 5 days

Reno, NV
Actual: 95% within 1 day of request (1990)

Oakland, CA
Actual availability: Next-day inspections
available: 87% (1990); 80% (1991); 85% (1992)

Phoenix, AZ
Target: At least 80% within 1 day of request

Burbank, CA
Target: Within 48 hours of request
Actual: Most within 24 hours of request

Gresham, OR
Target: 95% within 48 hours of request

Savannah, GA
Target: 95% within 2 workdays of request; 100%
within 3 workdays
Actual: 95% within 2 workdays; 100% within 3
workdays (1988-1993)

The amount of time an inspector spends on each inspection may, of course, influence daily workload. By holding down the inspector-time per inspection, inspector workload may be increased. Quantity gains, however, may be negated by quality losses if inspector speed results in faulty inspections. It may be reassuring, therefore, to fall within the band of average inspection times among other cities (Table 8.7). Lying outside those bounds at either extreme may be a cause for concern—too slow and perhaps wasteful at one end and too fast and perhaps error-prone at the other.

TABLE 8.5 Prompt Response to Requests for Electrical, Plumbing, and Other Specialty Inspections: Selected Cities

Electrical Inspection	Fire Inspection	Mechanical/ HVAC[a] Inspection	Plumbing Inspection
Lubbock, TX 95% within 4 work hours of request (1991)	**Raleigh, NC** 93% on date requested (1991)	**Lubbock, TX** 80% within 4 work hours of request (1991)	**Lubbock, TX** 80% within 4 work hours of request (1991)
Raleigh, NC 90% on date requested (1991)	**Chandler, AZ** 90% within 48 hours of request (1991)	**Raleigh, NC** 100% on date requested (1991)	**Cincinnati, OH** Target: 100% on date requested
Anaheim, CA Target: Within 24 hours of request		**Anaheim, CA** Target: Within 24 hours of request	**Raleigh, NC** 92% on date requested (1991)
Winston-Salem, NC Within 1 day of request: 94% (1991); 94% (1992)		**Winston-Salem, NC** Within 1 day of request: 96% (1991); 95% (1992)	**Anaheim, CA** Target: Within 24 hours of request
Gresham, OR Target: At least 95% within 48 hours of request		**Gresham, OR** Target: At least 95% within 48 hours of request	**Winston-Salem, NC** Within 1 day of request: 92% (1991); 98% (1992)
			Gresham, OR Target: At least 95% within 48 hours of request

a. HVAC is heating, ventilation, and air conditioning.

The LCC (1994) contends that an inspector of single-family residential units working for a high-service-level department should be able to complete 12 framing or foundation inspections per day (pp. 21-22). Fewer than 12—or more than 20—may signal problems and a low service level.

CODE ENFORCEMENT ACTIVITIES

General code enforcement activities typically focus on existing neighborhoods, commercial areas, and other already-developed properties rather than on new construction projects. Greenville, South Carolina, for example, directs its code enforcement efforts in part to such offenses as "illegal materials on private property, overgrown vacant lots, abandoned autos, and illegal materials at curbsides," defined as *environmental violations,* and records the number of violations per block face (City of Greenville, *1991 Year-End Performance Report,* p. 10). Greenville reported 0.45 environmental violations per block face in 1991, down from 0.76 the preceding year.

Similarly, Savannah, Georgia, gauges code enforcement progress by tabulating the percentage of occupied housing that complies with that city's housing code. In 1991, 91% of all occupied housing was in compliance, up from 75.7% in 1988.

TABLE 8.6 Inspector Workload: Inspections per Inspector in Selected Cities

Building/Construction Inspector	Electrical Inspector	Fire Inspector	Mechanical/HVAC Inspector	Plumbing Inspector	General Code Enforcement Inspector
INSPECTIONS PER DAY					
Irving, TX 16.3 (1991)	**Irving, TX** 12.3 (1991)	**New York City** 6.1 (1992)	**Irving, TX** 33 (1991)	**Irving, TX** 17.5 (1991)	**Greenville, SC** General inspector: 8.3 (1989); 7.7 (1990) Sign inspector: 7.5 (1989); 6.7 (1990) Zoning inspector: 10.4 (1989); 17 (1990)
Shreveport, LA 16 (1991)	**Shreveport, LA** 10 (1991)	**Portland, OR** 2.12 (1991); 3.16 (estimate, 1992)	**Shreveport, LA** 11 (1991)	**New York City** 12.3 (1991); 11.2 (1992)	**Savannah, GA** Target: 15 code inspections/reinspections per day Actual: 17 (1989); 18 (1990); 19 (1991)
Greenville, SC 8.2 (1989); 12.4 (1990)	**New York City** 9.6 (1992)		**Greenville, SC** 6.7 (1989); 6.3 (1990)	**Shreveport, LA** 8 (1991)	**Shreveport, LA** 12 (1991)
	Greenville, SC 9.3 (1989); 7.0 (1990)			**Greenville, SC** 4.3 (1989); 5.9 (1990)	**Charlotte, NC** Target: 10 per inspector-day Actual: 10.4 (1987)
INSPECTIONS PER YEAR					
Alexandria, VA 4,566 (1991); 3,487 (1992); 3,512 (1993)	**Raleigh, NC** 1,858 (1991)	**Shreveport, LA** 4,037 (1991)	**Raleigh, NC** 2,072 (1991)	**Raleigh, NC** 2,363 (1991)	**Corpus Christi, TX** 3,354 (1991, average per inspector includes building trade inspectors and general code enforcement)
Corpus Christi, TX 3,354 (1991)		**Phoenix, AZ** 1,376 (estimate, 1990)			
Raleigh, NC 2,378 (1991)		**Raleigh, NC** 987 (1991)			

TABLE 8.7 Inspector Speed: Average Inspector Time per Inspection in Selected Cities

Building/Construction Inspector	Electrical Inspector	Fire Inspector	Mechanical/HVAC Inspector	Plumbing Inspector	General Code Enforcement Inspector
Fort Collins, CO 24 minutes (1991)	**Wichita, KS** 14 minutes (1990); 13 minutes (1991)	**Winston-Salem, NC** 0.63 inspector-hours per nonresidential inspection	**Houston, TX** 37 minutes (1992)	**Houston, TX** 21 minutes (1992)	**Wichita, KS** Zoning inspection: 20 minutes (1990); 29 minutes (1991) Housing inspection: 51 minutes (1990); 60 minutes (1991)
Wichita, KS 40 minutes (1990); 24 minutes (1991)	**Houston, TX** 22 minutes (1992)	**Charlotte, NC** 1 inspection per inspection-hour		**Wichita, KS** 22 minutes (1990); 21 minutes (1991)	**Houston, TX** Occupancy inspection: 33 minutes (1992)
Houston, TX 35 minutes (1992)					
Largo, FL 44 minutes (1990); 67 minutes (1991)					

As with inspections related to new construction, general code enforcement activities are judged in several cities partially by the promptness of response by inspectors (Table 8.8). In general, responses are a bit slower than in the case of new construction inspections, but that is understandable given the generally less pressing financial ramifications of modest delay.

An important dimension of code enforcement performance is the rate of compliance by offending parties following notice of code violation. What is a reasonable compliance rate? A city with compliance rates of less than 50% for various categories of code violation can find company among some of the jurisdictions reporting such figures, but many report compliance rates in the 70%, 80%, and even 90% range (Table 8.9).

TABLE 8.8 Prompt Response to Requests for Zoning and Miscellaneous Code Enforcement Inspections: Selected Cities

Zoning	Weed Abatement	Existing Housing	Noise	Abandoned/Junk Vehicles	Miscellaneous/General
Overland Park, KS Target: Respond to alleged violations within 24 hours	**Corvallis, OR, Fire Department** Hazard and weed complaint responses within 3 days: 100% (1991); 100% (1992)	**Charlotte, NC** Target: Respond to 95% of all housing code complaints within 8 work hours. Actual: 97.4% within 8 work hours (8 months, 1987)	**Alexandria, VA** Responses to noise complaints within 1 day: 77% 1989); 79% (1990); 66% (1991)	**Winston-Salem, NC** Responses to *routine* abandoned vehicle complaints within 3 days: 91% (1991); 84% (1992) Responses within 1 day to notification of abandoned vehicles along major thoroughfares: 95% (1991); 99% (1992)	**Overland Park, KS** Target: Respond to complaints within 24 hours
Oklahoma City, OK Responses within 24 working hours of complaint: 94% (1991); 98% (1992)		**Cincinnati, OH** Target: Respond to at least 80% of housing code compliance complaints within 10 workdays		**Wilmington, DE** Target: Remove unregistered abandoned vehicles within 4 days following identification; registered abandoned vehicles within 10 days	**Peoria, AZ** Complaints investigated within 48 hours: 100% (1991)
Winston-Salem, NC Responses within 3 days of zoning enforcement request: 90% (1991); 92% (1992)				**Charlotte, NC** Target: Respond to abandoned auto complaints within 7 days Actual: 100% (1987)	**Hurst, TX** Target: Respond promptly and supply notice of disposition within 4 workdays Actual: 95% within 4 workdays
Orlando, FL 100% of site investigations of public complaints completed within 4 workdays of receipt (1991)				**Oakland, CA, Police Department** Target: Process and remove abandoned vehicles within 10 days Actual: 8-day average (1990); 10-day average (1991); 10-day average (1992)	**Phoenix, AZ** Target: Investigate at least 95% of alleged sign code violations within 5 days

continued

TABLE 8.8 *Continued*

Zoning	*Weed Abatement*	*Existing Housing*	*Noise*	*Abandoned/Junk Vehicles*	*Miscellaneous/General*
Fayetteville, AR 90% of complaints answered within 1 week (1991)				**Lubbock, TX** 90% of abandoned vehicles towed within 10 days of notification (1991)	**Anaheim, CA** Target: Investigate all alleged code violations within 1 week
Tacoma, WA Target: Investigate at least 70% of complaints within 2 weeks of receipt				**Boston, MA** 92% of abandoned vehicles removed within 14 days of notification sticker being applied to vehicle (1992)	**Fayetteville, AR** Complaints answered within 1 week: 90% (1991)
					Charlotte, NC Target: Investigate and resolve 90% of all service requests by residents and violations noted by inspectors within 15 days Actual: 90% (1987)

62

TABLE 8.9 Compliance Rates for Code Enforcement Actions: Selected Cities

			Compliance Following Notice of Violation				
General	Fire Code Violations	Grass and Weeds	Housing Violations	Junk Vehicles	Miscellaneous	Trash and Debris	Zoning Violations
Denton, TX 90% (1991)	**Winston-Salem, NC** 81.3% (1992)	**Denton, TX** 98% (1991)	**Charlotte, NC** Target: 100% of serious violations within 120 days	**Denton, TX** 90% (1991)	**Denton, TX** Compliance following notice of sign ord. violation: 80% (1991)	**Charlotte, NC** 92% within 30 days (5 months, 1994)	**Fort Collins, CO** Resolved voluntarily: 98% (1991)
Anaheim, CA Target: Resolve 90% of all documented violations within 30 days	**Reno, NV** Target: At least 75% within 30 days	**Lubbock, TX** 98% (estimate, 1992)	**Lubbock, TX** 87.4% (1991)	**Lubbock, TX** 83.7% (1991)	**Decatur, IL** Percentage of nuisance cases brought to compliance: 90% (1991)	**Denton, TX** 90% (1991)	**Lubbock, TX** 93.9% (1991)
Lubbock, TX 88% (estimate, 1992)	**Oak Ridge, TN** Compliance within 180 days: 81% (1991); 70% (1992)	**Hurst, TX** Corrected within 10 days: 90% (1990)	**St. Petersburg, FL** Target: 85% within 180 days Actual: 67.7% (1990)		**Alexandria, VA** Percentage of alleged noise ordinance violations closed within 10 days: 35% (1991)	**Tallahassee, FL** Target: 70% compliance for care-of-premises violations	
Greenville, SC Compliance rate within 60 days: 88% (1990)		**Winston-Salem, NC** 90% (1992)	**Decatur, IL** 61% (1991)				
Peoria, AZ Violations resolved within 45 days: 85% (1991)		**Shreveport, LA** 50% (1991)	**Tallahassee, FL** Target: 50% compliance for substandard structures				

continued

63

TABLE 8.9 Continued

General	Fire Code Violations	Grass and Weeds	Compliance Following Notice of Violation				
			Housing Violations	Junk Vehicles	Miscellaneous	Trash and Debris	Zoning Violations
Charlotte, NC Target: At least 80% within 90 days Actual: 84% (1987)			**Shreveport, LA** Percentage of substandard structure violations corrected by owner (rehab/demolition): 17% (1991)				
Fayetteville, AR 79% (1991)							
Houston, TX Target: At least 70%							
Tallahassee, FL Average time for compliance: 87 days (1991)							
Savannah, GA Target: Code compliance within 6 months Actual: Average 24 months for compliance (1991); 6 months (estimate, 1992)							

9

Emergency Communications

The primary function of emergency communications is to receive messages from persons in need of emergency services and to summon assistance on their behalf. In some cases, emergency communications focus on, and are assigned within, a single department—typically, a fire or police department. In other cases, the emergency communications operation covers a combination of police, fire, ambulance, and perhaps other emergency services and may be performed by one of the participating departments on behalf of all or, alternatively, may be established as a separate operational entity.

Standards

Communities seeking standards applicable to emergency communications will be interested in a pair of guidelines addressing police communications. The first pertains to equipment capability and is a standard promoted in Pennsylvania that calls for police radio systems to be "engineered to produce a 12-decibel or greater SINAD ratio to the radio receivers in 95 percent of the service area" (Southwestern Pennsylvania Regional Planning Commission [SPRPC], 1990, p. V-7).

The other is a standard developed by the Commission on Accreditation for Law Enforcement Agencies (CALEA; 1993) that specifies the information to be recorded when a request for service is received: date and time of request, name and

TABLE 9.1 Dispatcher Workload: An Example From a Single Municipality

Task	Average Hourly Workload per Dispatcher
Radio communications	24.2
Telephone communications	9.5
Alarms dispatched	1.0

SOURCE: Statistics for the city of Santa Ana, California, 1986-1987, from City of Santa Ana, *Annual Budget 1988-89*, p. 2.82.

address of complainant (if possible), type of incident reported, location of incident reported, time of dispatch, time of officer arrival, time of officer return to service, and disposition or status of reported incident (p. 81.3). Managers of emergency communications units may wish to focus more directly on dispatcher workload and key dimensions of dispatcher performance.

Dispatcher Workload

One municipality's analysis of its emergency communication dispatchers' workload yielded the set of average hourly activities identified in Table 9.1. Workload statistics are apt to vary from one community to another, depending on population, dispatcher responsibility for a single function or for multiple emergency functions, and variation among communities in the rate of emergency incidents. Even within a single community, workload will vary by hour of the day and by day of the week. Nevertheless, Santa Ana's figures indicate substantial capacity for a single operator.

Benchmarks

Three performance dimensions are especially critical to emergency communications:

1. speed,
2. accuracy, and
3. good judgment.

Response to callers should be prompt and emergency units should be dispatched quickly and to the proper location—but ideally, only if emergency personnel are really needed.

SPEED

To a frantic caller reporting an emergency, precious seconds are critical, and each unanswered ring seems like an eternity. Time *is* critical in every phase of

TABLE 9.2 Prompt Answer of 911 Calls: Selected Cities

Houston, TX
Target: Within 5 seconds
Actual: 99% within 5 seconds (1992)

Fayetteville, AR
Actual: 4-second average (1991)

Shreveport, LA
Actual: 5-second average (1991)

Reno, NV
Target: Within 6 seconds
Actual: 6.2-second average (1990)

Tucson, AZ
Target: Within 8 seconds

Jacksonville, FL
Target: Within 10 seconds

Oakland, CA
Target: Within 10 seconds
Actual: 96% within 10 seconds (1992)

Long Beach, CA
Target: At least 93% within 10 seconds
Actual: 95.2% within 10 seconds (1992)

Orlando, FL
Actual: 93% within 10 seconds (1991)

Phoenix, AZ
Target: At least 89% within 10 seconds
Actual: 88% within 10 seconds (1989); 86% (1990); 86% (1991)

Raleigh, NC
Target: Within 2 rings (12 seconds)
Actual: 95.6% within 12 seconds (1991)

Oklahoma City, OK
Target: Within 15 seconds
Actual: 97% within 15 seconds (1991)

Wichita, KS
Actual: 10.5-second average (1990); 12.0-second average (1991)

Overland Park, KS
Target: 3 rings or less

Peoria, AZ
Target: Within 3 rings

Cincinnati, OH
Actual: 7-second average; 98.7% within 20 seconds (1991)

Portland, OR
Target: Within 20 seconds
Actual: 94.5% within 20 seconds (estimate, 1992)

New York City
Actual: 99.7% within 30 seconds (1992)

---

emergency response, and many cities measure the performance of communications personnel in quickly answering 911 telephone calls (Table 9.2). Most set their targets at answering emergency calls somewhere between a maximum of 10 seconds and a maximum of 20 seconds. The "quick draw" leader among cities examined was Houston, Texas, reporting response to 99% of its 911 calls within 5 seconds in 1992.

How quickly, then, should an operator proceed from answering the call to actually dispatching emergency units? Although a few of the cities reported average dispatch times of 3 minutes or longer, most set targets or reported actual experience at 2 minutes or less for emergency dispatches (Table 9.3). Duncanville, Texas, set the speed record here, reporting a 45.3-second average for high-priority calls.

Some cities have established priority tiers not only for urgency of response by field units to various types of calls but also for promptness of dispatch. For example, a five-tier call prioritization system was developed for handling calls to the Dallas (Texas) police department (City of Dallas, 1989). Priority 1 calls included shootings, knifings, officers in need of assistance, emergency blood transfers, and felonies in progress and had a performance target of dispatch within 1 minute. Priority 2 calls had a performance target of dispatch within 3 minutes;

TABLE 9.3 Rapid Dispatch: Selected Cities

Duncanville, TX
Actual: 45.3-second average for high-priority
calls (1990)

Raleigh, NC
Target: Within 60 seconds for "in-progress" calls

Portland, OR
Target: Within 60 seconds for priority 1 police
calls; within 80 seconds on code 3 EMS calls
Actual: 90% within target (estimate, 1992)

Wichita, KS
Actual: 80% of priority calls within 1 minute
(1991)

Liberty, MO
Target: Within 60 seconds
Actual: 87-second average (1992)

Corvallis, OR
Target: Dispatch to all life-threatening calls
within 1 minute
Actual: 94% of all calls dispatched within 3
minutes (1992)

Alexandria, VA
Actual: 1.1-minute average for police
emergency; 5.4-minute average for police
nonemergency; 137-second average for fire/EMS
(1993)

Lubbock, TX
Actual: 1.5-minute average for police emergency
calls; 30-second average for fire calls; 15-second
average for EMS transfer calls (1991)

Corpus Christi, TX
Actual: 1-minute 39-second average for police
emergencies; 13-minute 23-second average for
routine police calls (estimate, 1992)

Tucson, AZ
Target: Within 2 minutes for extreme emergency
calls

Savannah, GA
Target: Within 2 minutes for emergency calls; 4
minutes for "immediate" calls

Houston, TX
Target: 2-minute average for fire and EMS calls

Grand Prairie, TX
All calls (actual): 5.4-minute average, police;
1.4-minute average, fire
Priority 1 calls (actual): 2.4-minute average,
police; 1.0-minute average, fire (1992)

Little Rock, AR
Actual: 2.5-minute average (1.2-minute
telephone processing time plus 1.3-minute
dispatch) (1991)

New York City
Actual: 3.0-minute average for
"crime-in-progress" calls; 42-second (Brooklyn)
to 60-second (Manhattan) averages for fire calls
(1992)

Sunnyvale, CA
Target: 85% of emergency/urgent calls within 3
minutes

Charlotte, NC
Target: 95% of emergency calls within 3 minutes
and 100% within 5 minutes; 85% of
"immediate" calls within 3 minutes and 95%
within 5 minutes
Actual: 91% of emergency calls within 3
minutes and 98% within 5 minutes; 72% of
"immediate" calls within 3 minutes and 94%
within 5 minutes (1987)

College Station, TX
Actual: 6.4-minute average (all calls, police,
1992)

priority 3, within 8 minutes; priority 4, within 60 minutes. Priority 5 calls were to be handled entirely by telephone without dispatch, if the caller agreed to that approach.

The city of Fort Collins, Colorado, measures dispatch speed from a slightly different perspective. That city tabulates the percentage of calls delayed or "stacked" more than 5 minutes waiting availability of an appropriate unit. Fort Collins reported 28% of all calls delayed in that fashion in 1991.

Yet another aspect of emergency communication speed pertains not to call dispatching but instead to promptness in providing requested information to field

TABLE 9.4 Dispatch Accuracy: Error Rates for Three Cities

City	Dispatcher Error Rate
Duncanville, TX	0.4% (1990)
Charlotte, NC	0.5%[a] (1987)
Chandler, AZ	1.3%[b] (1991)

a. A reported transmission accuracy rate of 99.5% for Charlotte was based on a sample of 18,484 fire emergency dispatches (City of Charlotte, *FY88-FY89 Objectives*, p. 22).
b. The city of Chandler, Arizona, reported an accuracy rate of 98.7% for 1990-1991 in "prioritizing of calls for service, dispatching police and fire units, and communications paperwork" (City of Chandler, *Annual Budget 1992-1993*, p. 148).

officers regarding wants and warrants. Overland Park, Kansas, sets its performance target at response within 1 minute, if possible—and within 5 minutes in every case.

ACCURACY

Dispatch errors can be disastrous. Sending a fire or police unit to the wrong location can result in the loss of precious minutes at a time when there are none to spare.

Dispatcher error rates reveal the frequency of accuracy lapses. They may pertain strictly to the inaccurate radio transmission of critical information or, as in the case of Chandler, Arizona, they may incorporate other aspects of the job, such as improper prioritization of service calls or inaccuracies in communications paperwork (Table 9.4). In either case, the critical nature of dispatch accuracy demands the consistent achievement of extremely low error rates.

JUDGMENT

Communications personnel must decide quickly whether to dispatch field units. Some cities report impressive contributions to the effective deployment of police, fire, and other emergency resources through the ability of emergency communications officers to deflect calls that can be handled effectively by other means. Alexandria, Virginia, for example, reported an estimated 29% reduction in field officer workload in 1991 by directing appropriate calls to a telephone reporting unit rather than dispatching field personnel. Phoenix, Arizona, expects half its calls to be handled without dispatch of emergency units; Orlando, Florida, sets its target at two thirds of all calls. Properly trained and supervised dispatchers who consistently make the right decisions can contribute mightily to efficient and effective emergency operations. On the other hand, dispatchers who have inappropriately delayed calls for emergency units have landed their cities in embarrassing legal and public relations fiascoes. Clearly, good judgment is crucial.

10

Emergency Medical Services

Amunicipality's emergency medical service (EMS) operation performs the critical life-or-death function of stabilizing patients—often the victims of serious illness or injury—and transporting them quickly to a hospital. The adequacy of a community's EMS operations can be assessed in a variety of ways.

The Need for Speed

Among the many key aspects of EMS performance, a crucial dimension is speed of response. For seriously stricken patients or critically injured accident victims, delays of even a few minutes can be catastrophic.

Knowing what to do on arrival is crucial to effective EMS action. Because even the most proficient stabilization efforts provided too late may be of little value, however, response speed is the most frequently reported EMS performance indicator. Paraphrasing comedian Woody Allen only slightly, 90% of success in this business is just showing up . . . on time.

TABLE 10.1 Emergency Medical Service Response Times Among Major Cities, as Reported in 1989

City	Time From Receipt of Call Until Dispatch (minutes)	Time From Dispatch Until Arrival (minutes)	Total Response Time (minutes)
St. Louis	.50	6.00	6.50
Los Angeles	.60	6.00	6.60
Oklahoma City	.70	7.18	8.04
Kansas City	.80	4.00	4.80
Honolulu	1.00	7.00	8.00
Memphis	1.00	5.00	6.00
Phoenix	1.00	5.00	6.00
Philadelphia	1.00	5.00	6.00
Cleveland	1.00	5.50	6.50
New Orleans	1.00	6.00	7.00
Nashville	1.00	6.00	7.00
Baltimore	1.00	6.00	7.00
El Paso	1.00	6.30	7.30
Houston	1.00	6.50	7.50
San Antonio	1.00	6.70	7.70
San Francisco	1.14	5.26	6.40
San Jose	1.30	7.00	8.30
Chicago	1.50	4.70	6.20
Boston	1.50	7.00	8.50
Washington, D.C.	3.10	6.90	10.00

SOURCE: City of Washington, D.C., *Improving Emergency Medical Services Communications* (Office of the City Administrator, Productivity Management Services, June 1989), p. 6.

Benchmarks

RESPONSE TIME

A 1989 study of EMS communication services in Washington, D.C., reported EMS response times among major cities ranging from 4.8 minutes in Kansas City, Missouri, to 10 minutes in Washington, D.C. (Table 10.1). A broader array of cities examined for this chapter reported generally quicker EMS responses (Table 10.2). Several cities reported response times in the 3- to 5-minute range. It should be noted, however, that not only does this second set of cities include smaller communities with potentially shorter and less congested travel distances, but also self-reporting by these cities may exclude the time from receipt of a call until dispatch, a component of total response time explicitly included in Table 10.1. Even so, the response times reported by the second set of cities were quicker even than the time from dispatch until arrival reported for most cities of the first set.

TABLE 10.2 Emergency Medical Service Response Times Among a Broader Array of Cities: Performance Targets and Actual Experience

Duncanville, TX
3-minute average (1990)

Gresham, OR
3.5-minute average (1992)

Burbank, CA
3.78-minute average (1989)

Phoenix, AZ
Target: Respond to at least 64% of ALS calls within 5 minutes
Actual EMS response time (average): 3 minutes, 54 seconds (1988); 3 minutes, 57 seconds (1989); 3 minutes, 52 seconds (1990)

Richmond, VA
4-minute 4-second average (January-June 1991)
3-minute 58-second average (January-June 1992)

Vancouver, WA
Target: 4 minutes or less

Tacoma, WA
Target: Within 4 minutes for BLS and 8 minutes for ALS at least 80% of time

Fullerton, CA
80% within 4 minutes (1989)

Portland, OR
77.6% in 4 minutes or less (1991)

St. Petersburg, FL
4.37-minute average (1990)

Huntington Beach, CA
4-minute 20-second average first-in response to medical aid calls; 86% within 5 minutes; 54% of paramedic unit responses within 5 minutes; 97% of paramedic unit responses within 10 minutes (1988)

Ocala, FL
Target: Rescue unit response in less than 4 minutes on at least 80% of calls
Actual: 4.5-minute average (1992)

Tucson, AZ
Target: 4.6-minute average

Irving, TX
4.9-minute average (1991)

College Station, TX
Target: Less than 5 minutes

Corvallis, OR
Targets: (a) provide BLS prior to arrival of paramedic ambulance, within 5 minutes from dispatch; (b) provide ALS within 8 minutes of notification (in city)
Actual: 95% of ALS responses (in city) within 8 minutes of notification (1992)

Long Beach, CA
Target: ALS average response time of 5 minutes or less

Dallas, TX
5.17-minute average (1989)

San Antonio, TX
5.2-minute average within city; 6.3-minute average for suburbs; 11.6-minute average for county (1991)

Alexandria, VA
6-minute average (estimate, 1993)

League City, TX
6.06-minute average (1990); 6.42-minute average (1991); 6.4-minute average (1992)

Corpus Christi, TX
6.5-minute average; 88% of responses within 8 minutes (1991)

Blacksburg, VA
6.9-minutes, in town; 10.2 minutes, outlying county; 2.3 minutes, Virginia Tech campus (1994)

Norfolk, VA
6.9-minute average

Boston, MA
100% of EMS arrivals within 7 minutes of time call entered into CAD system (includes dispatch and response time); 100% of responses within 5 minutes of receipt of call from dispatcher (1992)

Nashville-Davidson County, TN
7-minute average (1991)

Shreveport, LA
7.51-minute average medic response time; 73% of ALS responses within 8 minutes; 58% BLS responses to ALS calls within 4 minutes; 4.2-minute average response by firefighters to EMS calls (1991)

According to the City of Washington, D.C. (1989) study, the national medical community and the EMS industry have defined a two-part standard for EMS responsiveness: "90 percent of [EMS] responses should be within four minutes,

TABLE 10.2 *Continued*

Chesapeake, VA 8-minute average (1991)	**Santa Ana, CA** 89.96% within 8 minutes (1987)
Grand Prairie, TX Target: At least 90% within 8 minutes Actual: 93.5% within 8 minutes (year-to-date, 12/31/92)	**New York City** 8-minute 21-second average (1992) **Charlotte, NC** Target: At least 98% within 9 minutes
Avon, CT Target: At least 90% within 8 minutes	Actual: 99.3% within 9 minutes (1987)
Portsmouth, VA Target: At least 85% within 8 minutes Actual: 92.4% within 8 minutes (1993)	**Houston, TX** Target: At least 90% within 10 minutes

NOTE: BLS is basic life support; ALS is advanced life support. CAD is computer-assisted dispatch.

and 90 percent of paramedic responses should be within eight minutes" (p. 101). A standard promoted in Pennsylvania calls for an 8-minute response time on at least 90% of all emergency calls (SPRPC, 1990, p. VII-1). Reported performance targets and experience of the cities examined here suggest that an 8-minute standard may be realistic but that a 4-minute standard lies well beyond the reach of many communities.

WORKLOAD

Much as in the case of police officers and firefighters, some communities attempt to gauge the adequacy of their EMS services by considering the number of paramedics per 1,000 population. Such indicators can be misleading because they fail to take into account a given community's distinctive needs for various emergency services. In reality, such measures are not really performance indicators at all because they do not measure performance. They measure resource inputs—not performance outputs—and can vary greatly from one community to another. For example, at 0.44 paramedics per 1,000 population, the reported paramedic rate in College Station, Texas, in the early 1990s was almost twice that of Chesapeake, Virginia's 0.226 per 1,000 population.

More useful indicators of workload may be secured through direct measurement, rather than by population ratio. In that regard, Chesapeake, Virginia, reported 967 responses per paramedic during a recent year and Sedgwick County, Kansas, reported an average of 810 calls per crew.

EFFECTIVENESS

College Station's EMS units strive to provide treatment and achieve stabilization within 15 minutes. Shreveport, Louisiana, reported that in 76% of its responses in 1991, prehospital time for major trauma cases was held to less than 36 minutes.

TABLE 10.3 Fee Collection Rates and Cost Recovery Targets for EMS Services

COLLECTION RATE	COST RECOVERY THROUGH FEES

Burbank, CA
  63% (1988); 54% (1989)
	College Station, TX
	  Target: 40% to 70%

Duncanville, TX 52%	
Shreveport, LA 50% (1991)	

Corpus Christi, Texas, reported that during one recent year, 22% of its cardiac arrest patients were resuscitated, whereas Sedgwick County, Kansas, reported a resuscitation success rate of 28%.

Yet another gauge of effectiveness is the rate of complaints. In 1992, complaints about EMS services received by the city of Boston represented only 0.03% of all ambulance responses.

COLLECTION RATES AND COST RECOVERY

Fees charged to recipients of EMS services often go unpaid. Reported collection rates among examined cities ranged from 50% to 63% (Table 10.3). Among a variety of proposed performance targets for EMS services in Washington, D.C., (Exhibit 10.1) was an ambulance fee collection rate of 60%.

Recognition of the nonpayment factor as well as the high cost of service leads most cities to anticipate a municipal subsidy of the EMS operation. College Station, for example, sets as its target the recovery of 40% to 70% of EMS costs through fees.

Item Number	Indicators	Last Month	This Month	YTD	Target
	EMS PERFORMANCE REPORT Month:_____ 19___				
1.1	Number of incidents handled				—
2.1	Average call to scene time for first firefighting unit (in minutes)				4.0
2.2	Average call to scene time for first EAB unit (in minutes)				N/A
2.3	Average call to scene time for first responding unit (in minutes)				4.0
2.4	Average time from call to EAB unit dispatch (in minutes)				0.5
3.1	Percent of firefighting units responding within 4 minutes				90%
3.2	Percent of calls receiving a response within 4 minutes				90%
3.3	Percent of urgent calls receiving an ALS response within 8 minutes				90%
3.4	Percent of urgent calls for which an ALS unit is *not* dispatched				0
4.1	Percent of incidents upgraded in priority on the scene				10%
4.2	Percent of incidents downgraded in priority on the scene				10%
5.1	Percent of EAB runs evaluated in field				25%
5.2	Percent of fire unit medical runs evaluated				90%
6.1	Ambulance for collection rate				60%
6.2	EAB absentism rate				5%

Exhibit 10.1. Performance Targets and EMS Performance Report Proposed for Washington, D.C.

SOURCE: City of Washington, D.C., *Improving Ambulance Operations in Washington, D.C.: A Blueprint for Change* (Office of the City Administrator, Productivity Management Services, March 1989), p. 106.

11

Finance and Budgeting

Financial management and budgeting in local government are complex topics. Attempting to assess the performance of those functions in an individual community is no easy task. No simpler is the task undertaken here of compiling a relatively comprehensive set of financial management and budgeting benchmarks and confining those benchmarks to a single chapter.

Assessing the Financial Health of a Municipality

A "quick-and-dirty" way of gauging the quality of financial management in a given municipality is by noting that city's general financial health. One popular method of doing so—perhaps because of its speed and simplicity—is by checking the local bond rating to see how industry analysts judge the municipality's credit-worthiness. Other methods, sometimes more revealing because they address specific aspects of financial management more fully controllable by local officials, are available as well.

BOND RATINGS

Capital improvements typically are financed by the issuance of bonds—general obligation (GO) bonds, if the funds are used for traditional, non-revenue-generating functions such as street construction and park development and the city is willing to place its full faith and credit (i.e., taxing authority) behind the issuance to ensure repayment of investors; revenue bonds, when revenue from public projects such as airports or water treatment facilities can be used to pay off their own debt; or industrial development bonds. Prior to the issuance of such bonds, municipal credit rating agencies—the largest being Moody's Investors Service and Standard & Poor's—perform a credit risk evaluation of the municipality to advise potential investors of the city's creditworthiness. That rating is a shorthand declaration of the city's financial health.

In rating municipalities for the issuance of general obligation debt, Standard & Poor's (1995) reviews four factors that are considered especially relevant to a government's "capacity and willingness" to repay its debt:

1. the local economic base, including local employment, taxes, and demographics (for example, age, education, income level, and skills of the local population);
2. financial performance and flexibility, including accounting and reporting methods, revenue and expenditure structure and patterns, annual operating and budgetary performance, financial leverage and equity position, budget and financial planning, and contingency financial obligations, such as pension liability funding;
3. debt burden; and
4. administration, including local autonomy and discretion regarding financial affairs, background and experience of key administrative officials, and frequency of elections. (pp. 20-23; used with permission)

Rating labels for Moody's differ from those of Standard & Poor's. A chart produced by the city of Fort Collins, Colorado, however, suggests general comparability across the rating services (Table 11.1). Higher ratings (for example, BBB

TABLE 11.1 Moody's and Standard & Poor's Credit Ratings for Municipal Bonds

Rating		Description
Moody's	*Standard & Poor's*	
Aaa	AAA	Best quality; extremely strong capacity to pay principal and interest
Aa	AA	High quality; very strong capacity to pay principal and interest
A	A	Upper-medium quality; strong capacity to pay principal and interest
Baa	BBB	Medium-grade quality; adequate capacity to pay principal and interest
Ba and lower	BB and lower	Speculative quality; low capacity to pay principal and interest

SOURCE: City of Fort Collins (Colorado), *1993 Annual Budget: Volume I. Budget Overview,* p. 95.
NOTE: Within groups, Moody's designates those bonds with the strongest attributes with a "1." For instance, A1 or Aa1 would be of slightly higher quality than A or Aa. Standard & Poor's attaches a "+" or a "-" to show slight variation within the rating groups. Examples would be AA- or A+ to indicate a credit better than an A but less than an AA.

TABLE 11.2 Early Warning Guidelines: Bond Rating Checklist

Warning signs:

- Current-year operating deficit
- Two consecutive years of operating fund deficit
- Current general fund deficit (two or more years in last five)
- Short-term debt (other than BAN) at end of fiscal year greater than 5% of main operating fund revenues
- Short-term interest and current-year debt service greater than 20% of total revenues
- Total property tax collections less than 92% of total levy
- Declining market valuations—two consecutive years on a three-year trend
- Overall net debt ratio 50% higher than four years ago
- Current-year operating deficit larger than previous year's deficit
- General fund deficit in the current year—balance sheet—current position
- Two-year trend of increasing short-term debt outstanding at fiscal year end
- Property taxes greater than 90% of tax limit
- Net debt outstanding greater than 90% of debt limit
- A trend of decreasing tax collections—two consecutive years on a three-year trend
- Overall net debt ratio 20% higher than previous year

SOURCE: Standard & Poor's, *Credit Overview: Municipal Ratings* (New York: Standard & Poor's, 1983), p. 119. Reprinted by permission.
NOTE: BAN = bond anticipation notes.

and above for Standard & Poor's) are considered "investment grade"; lower ratings (BB and below) are considered "speculative grade." Lower ratings for Standard & Poor's include B, CCC, CC, C, and D.

A checklist of "early warning guidelines" developed by Standard & Poor's is provided in Table 11.2. Recent indications suggest that in addition to basic financial criteria, such as those included in the checklist, socioeconomic factors may become increasingly important in rating a local government's creditworthiness (Crowell & Sokol, 1993).

ICMA'S FINANCIAL TREND MONITORING SYSTEM

In 1980, the International City Management Association (ICMA) published a set of handbooks that described a Financial Trend Monitoring System (FTMS) for tracking local financial conditions (Groves & Godsey, 1980). The FTMS relies on 36 indicators shown in Table 11.3. Minimum standards are not declared for most indicators. Instead, potential "warning trends" are identified and suggestions for analysis are offered. In a few cases, however, relevant "credit industry benchmarks" are noted by the FTMS authors:

> Credit rating firms assume that a local government normally will be unable to collect from 2 to 3 percent of its property taxes within the year that the taxes are due. If uncollected property taxes rise to more than from 5 to 8 percent, rating firms consider

TABLE 11.3 Financial Trend Monitoring Indicators

	Indicator	*Formula*	*Number*	*Indicator*	*Formula*
1.	Revenues Per Capita	$\dfrac{\text{Net Operating Revenues in Constant Dollars}}{\text{Population}}$	20.	Debt Service	$\dfrac{\text{Net Direct Debt Service}}{\text{Net Operating Revenues}}$
2.	Restricted Revenues	$\dfrac{\text{Restricted Operating Revenues}}{\text{Net Operating Revenues}}$	21.	Overlapping Debt	$\dfrac{\text{Overlapping Long–Term Debt}}{\text{Assessed Valuation}}$
3.	Intergovernmental Revenues	$\dfrac{\text{Intergovernmental Operating Revenues}}{\text{Gross Operating Revenues}}$	22.	Unfunded Pension Liability	$\dfrac{\text{Unfunded Pension Plan Vested Benefits}}{\text{Assessed Valuation}}$
4.	Elastic Tax Revenues	$\dfrac{\text{Elastic Operating Revenues}}{\text{Net Operating Revenues}}$	23.	Pension Assets	$\dfrac{\text{Pension Plan Assets}}{\text{Pension Benefits Paid}}$
5.	One-Time Revenues	$\dfrac{\text{One–Time Operating Revenues}}{\text{Net Operating Revenues}}$	24.	Accumulated Employee Leave Liability	$\dfrac{\text{Total Days of Unused Vacation and Sick Leave}}{\text{Number of Municipal Employees}}$
6.	Property Tax Revenues	Property Tax Revenues in Constant Dollars	25.	Maintenance Effort	$\dfrac{\text{Expenditures for Repair and Maintenance of General Fixed Assets}}{\text{Amount of Assets}}$
7.	Uncollected Property Taxes	$\dfrac{\text{Uncollected Property Taxes}}{\text{Net Property Tax Levy}}$	26.	Level of Capital Outlay	$\dfrac{\text{Capital Outlays from Operating Funds}}{\text{Net Operating Expenditures}}$
8.	User Charge Coverage	$\dfrac{\text{Revenues from Fees and User Charges}}{\text{Expenditures for Related Services}}$	27.	Depreciation	$\dfrac{\text{Depreciation Expense}}{\text{Cost of Depreciable Fixed Assets}}$
9.	Revenue Shortfalls	$\dfrac{\text{Revenue Shortfalls}^{a}}{\text{Net Operating Revenues}}$	28.	Population	Population
10.	Expenditures Per Capita	$\dfrac{\text{Net Operating Expenditures in Constant Dollars}}{\text{Population}}$	29.	Median Age	Median Age of Population
11.	Employees Per Capita	$\dfrac{\text{Number of Municipal Employees}}{\text{Population}}$	30.	Personal Income	$\dfrac{\text{Personal Income in Constant Dollars}}{\text{Population}}$
12.	Fixed Costs	$\dfrac{\text{Fixed Costs}}{\text{Net Operating Expenditures}}$	31.	Poverty Households or Public Assistance Recipients	$\dfrac{\text{Poverty or Public Assistance Households}}{\text{Households in Thousands}}$
13.	Fringe Benefits	$\dfrac{\text{Fringe Benefit Expenditures}}{\text{Salaries and Wages}}$	32.	Property Value	$\dfrac{\text{Constant Dollar Change in Property Value}}{\text{Constant Dollar Property Value Prior Year}}$
14.	Operating Deficits	$\dfrac{\text{General Fund Operating Deficit}}{\text{Net Operating Revenues}}$	33.	Residential Development	$\dfrac{\text{Market Value of Residential Property}}{\text{Market Value of Total Property}}$
15.	Enterprise Losses	Enterprise Profits or Losses in Constant Dollars	34.	Vacancy Rates	Vacancy Rates
16.	General Fund Balances	$\dfrac{\text{Unrestricted Fund Balance of General Fund}}{\text{Net Operating Revenues}}$	35.	Employment Base	• Rate of Unemployment • Number of Community Jobs
17.	Liquidity	$\dfrac{\text{Cash and Short–Term Investments}}{\text{Current Liabilities}}$	36.	Business Activity	• Retail Sales • Number of Community Businesses • Gross Business Receipts • Valuation of Business Property • Business Acres Developed
18.	Current Liabilities	$\dfrac{\text{Current Liabilities}}{\text{Net Operating Revenues}}$			
19.	Long-Term Debt	$\dfrac{\text{Net Direct Long–Term Debt}}{\text{Assessed Valuation}}$			

SOURCE: Sanford M. Groves, W. Maureen Godsey, and Martha A. Shulman, "Financial Indicators for Local Government," *Public Budgeting & Finance, 1* (Summer 1981), p. 12. Reprinted by permission.
NOTE: This chart is intended as a summary of the indicators, the data each requires, and their formulas. For a more detailed look at how each piece of data is to be derived, see Groves (1980).
a. Net operating revenues budgeted less net operating revenues (actual).

this a negative factor because it signals potential problems in the stability of the property tax base. An increase in the rate of delinquency for two consecutive years is also considered a negative factor.

A credit rating firm would regard a current-year operating deficit as a minor warning signal; funding practices and the reasons for the deficit would be carefully assessed before it would be considered a negative factor. The following situations, however, would be given considerably more attention and would probably be considered negative factors:

- Two consecutive years of operating fund deficits;
- A current operating fund deficit greater than that of the previous year.

The credit industry considers the following situations negative factors: (1) short-term debt outstanding at the end of the year exceeding 5 percent of operating revenues (including tax anticipation notes but excluding bond anticipation notes), and (2) a two-year trend of increasing short-term debt outstanding at the end of the fiscal year (including tax anticipation notes). (Groves & Valente, 1986, pp. 38, 63, 77; reprinted by permission)

FINANCIAL PLANNING STANDARDS
IN OVERLAND PARK

Some municipalities establish their own standards for financial planning and management. Overland Park, Kansas, is one such city (Table 11.4). Although some of the standards adopted by that community for capital planning purposes may not be deemed immediately achievable elsewhere, the outstanding bond ratings awarded to Overland Park by both major rating agencies suggests that its standards could serve as useful benchmarks.

SELECTED INDICATORS OF FINANCIAL HEALTH

Three facets of financial health that deserve additional attention, even in a concise overview of financial management benchmarks, are the following:

1. fund balance,
2. debt, and
3. enterprise funds.

Fund balance has been defined in general terms as "excess, surplus or un-budgeted money" (Carter & Vogt, 1989, p. 33). More precisely, it is "the cumulative difference of all revenues and expenditures from the government's creation. It can also be considered to be the difference between fund assets and fund liabilities, and can be known as fund equity" (Allan, 1990, p. 1).

A suitable fund balance provides a source of working capital, offers a cushion against revenue shortfalls and unanticipated expenditures, provides an accumulation of funds that some cities use for capital projects to avoid bond sales, helps

TABLE 11.4 Financial Planning Standards Used for Capital Plan Development in
Overland Park

Several financial standards have been developed as a basis for evaluating the financial soundness of the
5-year capital plan. These standards serve as guidelines in scheduling and balancing the composite of
financing methods for projects included in the plan. The standards are summarized below.

Percentage of Ending Cash to Operating Expenditures	15%
The General Fund's unreserved undesignated fund balance divided by the General Fund's operating expenditures (excluding transfers). (Source of standard: staff)	
Percentage of Pay-as-You-Go Funding to Total Program	20% to 30%
The amount of pay-as-you-go funding and General Improvement funds used to finance capital improvement projects divided by the total amount of the Capital Improvement Program. (Source of standard: staff)	
Percentage of Pay-as-You-Go Funding to City Funds	40% to 50%
The amount of pay-as-you-go funding and General Improvement funds used to finance capital improvement projects divided by the total amount of the city funds financing the Capital Improvement Program.	
Percentage of Debt to Total Program	35% to 50%
The amount of the city's general obligation debt incurred to fund capital improvement projects divided by the total amount of the Capital Improvement Program.	
Percentage of Debt to City Funds	45% to 55%
The amount of the city's general obligation debt incurred to fund capital improvement projects divided by the total amount of the city funds used to finance capital improvement projects. (Source of standard: staff)	
Total Direct Debt per Capita	$436
The amount of per capita direct bonded debt (debt for which the city has pledged its full faith and credit) issued by the city and overlapping debt of other governmental entities within the city. (Source of standard: Moody's Investor Service)	
Percentage of Direct and Overlapping Debt to Market Value of Tangible Property	5% to 4%
The city's direct bonded debt and overlapping debt as a percentage of estimated market valuation of property within the city. (Source of standard: Moody's Investor Service)	
Percentage of Debt Service Cost to General Fund Operating Expenditures	Less than 25%
The city's annual payments to the Bond & Interest Fund for debt service cost divided by total operating expenditures (excluding transfers) of the General Fund. (Source of standard: ICMA)	
Mill Levy Equivalent of B & I Transfer	6 mills
The mill levy required to replace the General Fund's annual transfer to the Bond & Interest Fund. These monies are used to pay annual principal and interest payments on the city's maturing general obligation debt. (Source of standard: staff)	

SOURCE: City of Overland Park (Kansas), *1993 Budget*, pp. 6.6, 6.7.

others meet bond indenture requirements, and contributes to a favorable bond rating (Olson, Shuck, Vinyon, & Lilja, 1980). Appropriations from the fund balance of the general fund often are used to smooth peaks and valleys in property and sales tax revenues—with greater appropriations from the fund balance when current revenues are down and smaller appropriations when they are up—to reduce the need for frequent or erratic rate adjustment.

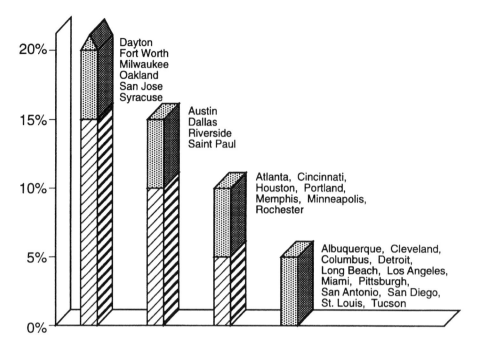

Figure 11.1. Major Cities' Cash Reserves at the End of 1990
SOURCE: City of Dayton (Ohio), *1991 Program Strategies,* 1991, p. 130.

A certain mystical quality surrounds the consideration of what constitutes an appropriate amount of fund balance. Some authors have suggested that the credit industry considers a general fund balance to be at a "healthy level" if it equals 10% of annual revenues or expenditures (Olson et al., 1980, p. 11). Others suggest that credit analysts view a fund balance of 5% as the minimum "prudent level," although lower or higher levels may be deemed appropriate in specific instances depending on long-term financial trends and other local factors (Allan, 1990, p. 6). The Local Government Commission of the North Carolina Department of State Treasurer recommends a general fund balance of at least 8% of expenditures (Carter & Vogt, 1989).

How much is enough? It is probably impossible to say with certainty for any given community, but a look at the amount of cash reserves held by other cities and a review of the fund balance policies actually in place among respected communities may be instructive to other municipalities that are attempting to map their own fund balance and reserve course. A graph produced by the city of Dayton, Ohio, depicts reported cash reserves in a host of major cities at the end of 1990 (Figure 11.1). Smaller cities, whose officials may feel especially vulnerable to the financial impact of a major emergency, have been found in recent studies to sometimes maintain fund balances equal to 50% of general fund expenditures or more (Carter & Vogt, 1989; Massey & Tyer, 1990). Even more informative, perhaps, are the policies of 49 cities among those examined for this book that prescribe what they believe to be appropriate levels of fund balance (Table 11.5). The cities may be divided fairly evenly between those prescribing fund balances of at least 10% and those prescribing lesser amounts as adequate minimums.

Debt is another useful indicator of a community's financial health. Although many aggressive and vibrant communities incur debt to secure capital for new infrastructure or public amenities, too much debt may signal impending problems. The cities of Portland, Oregon, and Largo, Florida, report the following, generally accepted warning signals for potentially burdensome debt:

- Overall net debt exceeding 10% of assessed valuation
- Overall net debt exceeding $1,200 per capita
- An increase of 20% over the previous year in overall net debt as a percentage of market valuation
- Overall net debt as a percentage of market valuation increasing 50% over the figure for 4 years earlier
- Overall net debt per capita exceeding 15% of per capita personal income
- Net direct debt exceeding 90% of the amount authorized by state law (City of Largo, 1993, p. 125; City of Portland, 1992, p. 33)

The first of the above-noted warning signals examines net debt in relation to assessed valuation. In applying that measure, it is important to note the distinction between the assessed value and the market value of property. Although in some places assessed value is intended to equal market value, in other places it is set at something less. Because a municipality's taxing authority is tied to and limited by assessed value, the previously declared warning signal—overall net debt exceeding 10% of assessed valuation—is a useful benchmark. A different and also useful reference is the full market value of taxable property. Some of the warning signals are tied to that value. Still another such measure is the ratio of a municipality's direct net debt, excluding short-term operating debt and bonds fully supported by enterprise revenues, to full market value of all taxable property within that community. Direct net debt in the 1% to 3% range is common (Table 11.6).

High levels of debt impose debt service burdens on municipal budgets, forcing undesirably high percentages of local resources to be devoted to the payment of principal and interest on that debt. According to the City of Portland, Oregon (1992), "the credit rating industry considers net direct debt service exceeding 20 percent of operating revenues a potential problem. Ten percent is considered acceptable" (p. 34).

Although Portland reported its debt service to be consistently below 3% of operating revenues, many other municipalities contend with higher amounts (Table 11.7). Many are reasonably content if no more than a dime or quarter out of every dollar goes toward debt service. Richmond, Virginia, and Fort Lauderdale, Florida, for example, try to keep debt service at 10% of operating expenditures or less. Iowa

TABLE 11.5 Policies of Selected Cities Regarding Fund Balances and Undesignated Reserves

City	Policy
College Station, TX	Maintain a fund balance of at least 15% of General Fund expenditures (55 days' operation). An additional contingency of 2.5% should be maintained. Furthermore, cash and investments should be the equivalent of 30 days' operating expenditures.
College Park, MD	Maintain a fund balance equal to at least 25% of General Fund expenditures.
Smyrna, GA	Maintain an unreserved, undesignated fund balance equal to at least 25% of the operating budget (3 months' operation).
Burbank, CA	Maintain a 15% unappropriated working reserve and a 5% unappropriated emergency reserve in the General Fund.
Fairfield, CA	Maintain ending cash and current receivable balances (including carryover and vehicle/leave reserves) of at least 20% of the approved budget.
San Luis Obispo, CA	Maintain a fund balance of at least 20% of operating expenditures in the General Fund, as well as water, sewer, and parking enterprise funds.
Southfield, MI	The "desirable" level of unreserved and undesignated fund balance to be maintained in each operating fund is to be 20% of net operating revenues for the year.
Hanover Park, IL	Contingency reserves for miscellaneous purposes and emergencies shall not exceed 20% of the General Fund.
Duncanville, TX	Maintain surplus and unencumbered funds in the Utility Fund equal to 2 months' expenditures.
Hurst, TX	Maintain a fund balance sufficient to cover at least 60 days' expenditures.
Grand Junction, CO	Long-range plan: Maintain a fund balance in the General Fund equal to 15% of expenditures and fund balances in enterprise funds equal to 5% of operating expenditures.
Greenville, SC	Maintain a fund balance of 15% of the general operating budget.
Dayton, OH	Maintain sufficient year-end reserves to cover 6 to 10 weeks of expenditures.
Chandler, AZ	Maintain an unrestricted contingency reserve of 11.5% of the General Fund budget.
Longview, TX	Maintain a cash balance of 10% for service delivery funds and a 15% balance in funds with major facilities and equipment.
Blacksburg, VA	Target the unrestricted fund balance at approximately 10% of net operating revenues.
Fayetteville, AR	Maintain an undesignated fund balance of at least 10% of current year operating expenditures for the General Fund and Street Fund.
Fort Smith, AR	Maintain a contingency reserve of at least 10% of general operating revenues.
Fort Worth, TX	Policy (proposed): Maintain an undesignated General Fund balance of 10% of current year budgeted expenditures.
St. Charles, IL	Maintain an available fund balance (working capital) of at least 10% of operating expenditures.
Sterling Heights, MI	Maintain an undesignated, unreserved General Fund balance equal to 10% of the general operating budget.
Portland, OR	Maintain a General Reserve fund at the level of 10%.
Winston-Salem, NC	Policies: (a) Maintain an unappropriated fund balance reserve equal to at least 10% of the General Fund budget; (b) allocation from reserve to balance the budget should be no more than 4% of revenues.
Sierra Vista, AZ	Maintain a reserve of $\frac{1}{12}$ of anticipated expenditures in each operating fund.

TABLE 11.5 *Continued*

City	Policy
Waukesha, WI	Maintain a working capital reserve of $\frac{1}{12}$ of the General Fund operating budget to provide funds for reasonable cash flow needs and unforeseen emergencies.
Denton, TX	Maintain an unreserved General Fund balance of 8% to 10% of annual expenditures.
Oklahoma City, OK	Maintain an unbudgeted General Fund reserve of 6% to 11% of 1 year's operating expenditures (in addition to a budget contingency of 2% or greater).
Chesapeake, VA	Maintain a cash flow and emergency fund reserve of at least 6% of General Fund revenue.
Gresham, OR	Maintain a General Fund unappropriated balance equal to 6% of the operating budget.
Cincinnati, OH	Maintain a working capital reserve of 5.2% to 8% of general operating revenues.
Avon, CT	Town manager's recommendation: Maintain a fund balance at 5% to 10% of budgeted expenditures.
Boise, ID	Maintain an undesignated, unreserved fund balance of at least 5% of expenditures.
Columbus, OH	Maintain a "rainy day" account of at least 5% of General Fund appropriations.
Irving, TX	Maintain an unreserved cash balance in the General Fund at not less than 5% of expenditures, not including debt service.
Lewiston, ID	Maintain a fund balance equal to at least 5% of budgeted expenditures.
Nashville-Davidson County, TN	Maintain an undesignated fund balance of at least 5% of expenditures.
Savannah, GA	Maintain a fund balance between 5% and 10% of General Fund revenues.
Shreveport, LA	Maintain an operating reserve of $5 million or 5% of revenue, whichever is greater.
Sioux Falls, SD	South Dakota law requires that cities maintain at least 5% of annual appropriations as an unencumbered reserve (labeled the "5 Percent Unexpendable" in the Sioux Falls CAFR).
Ogden, UT	State law requires a fund balance of at least 5%.
Greenbelt, MD	City manager's recommendation: Maintain an unreserved General Fund balance of 5% of current year expenditures.
Jacksonville, FL	Mayor's policy: Maintain a "rainy-day" reserve of 5% of operating budget.
Largo, FL	Maintain an undesignated General Fund balance, including contingency account, equal to 5% of General Fund revenues.
Little Rock, AR	Maintain an emergency or working capital reserve at approximately 5% of current operating budget appropriation.
Long Beach, CA	City manager's recommendation: Maintain reserves equal to approximately 5% of General Fund appropriations.
Richmond, VA	Establish/maintain an undesignated General Fund balance as a "rainy day reserve" at 5% of the budget.
St. Petersburg, FL	Maintain a cash-working fund balance of 5% of current year's General Fund operating appropriations.
Alexandria, VA	Maintain an unreserved General Fund balance equal to at least 4% (and preferably 5.5%) of General Fund revenue.
Loveland, CO	In addition to a designated reserve for emergencies, maintain a contingency of no less than 3% of operating expenses in the General Fund.

SOURCE: Adapted from David N. Ammons, "Local Policies on Municipal Fund Balances" (Unpublished manuscript, University of Georgia, Carl Vinson Institute of Government, 1996).
NOTE: CAFR = comprehensive annual financial report.

TABLE 11.6 Ratio of Municipal Debt to Market Value of All Taxable Property

Population	Direct Net Debt as a Percentage of Market Value of All Taxable Property (median)
500,000 or greater	2.2%
300,000 to 499,999	2.0%
200,000 to 299,999	2.7%
100,000 to 199,999	1.4%
50,000 to 99,999	1.5%
25,000 to 49,999	1.4%
10,000 to 24,999	1.3%
Less than 10,000	2.3%

SOURCE: Moody's Public Finance Department, *1988 Medians: Selected Indicators of Municipal Performance.* New York: Copyright © Moody's Investors Service, Inc., 1988, p. 9. Reproduced with permission granted by Moody's Investors Service, Inc. All rights reserved.
NOTE: Percentages shown exclude the debt of special districts and other overlapping jurisdictions, as well as short-term operating debt and bonds fully supported by enterprise revenues.

City, Iowa, tries to limit debt service to 25% of the general tax levy. Fayetteville, Arkansas, tries to hold debt service within 25% of operating revenues.

Enterprise funds offer yet another gauge of a municipality's financial health. An assessment of a municipality's financial well-being that focuses exclusively on general fund operations and ignores municipally owned utilities and other enterprise operations may lead to erroneous conclusions. A general sense of enterprise

TABLE 11.7 Debt Service Ratios: Selected Cities

Debt Service as a Percentage of Operating Expenditures	Debt Service as a Percentage of Operating Revenues
Sioux Falls, SD Actual: 0.57% to 1.99% (1982-1990); 0.35% (1991)	**Portland, OR** Actual: Less than 2% (1991)
Oak Ridge, TN Actual: 2% to 17% (1984-1993)	**Iowa City, IA** Target: Limit annual debt service to 25% of general tax levy
Iowa City, IA Actual: 10% (1991); 9% (1992)	**Fayetteville, AR** Target: Limit debt service to 25% of total operating revenues or less
Richmond, VA Target: 10% or less	
Fort Lauderdale, FL Target: Annual debt service requirements shall not exceed 10% of the annual budget	
Fort Collins, CO "Rules of thumb:" 10% to 15% is considered fair; more than 15% is generally considered poor Actual: 15.1% (1991)	
Carrollton, TX Actual: 15.8% to 24.4% (1984-1992); 19.7% (1993)	

TABLE 11.8 Financial Indicators for Municipal Enterprises

	Airport	Electric Distribution Systems	Electric Gen. and Trans.	Sewer	Water	Water and Sewer
			Municipal Enterprise Medians			
Operating ratio[a]	55.6%	88.5%	74.4%	60.6%	70.1%	63.7%
Net take-down[b]	51.3%	17%	27.8%	44.5%	34%	42%
Interest coverage[c]	2.25	4.93	4.05	3.45	4.08	3.58
Debt service coverage[d]	1.77	2.87	2.41	2.01	2.18	2.60
Debt service safety margin[e]	25.3%	8.25%	17.4%	30.1%	23.6%	24.3%
Debt ratio[f]	57%	39.8%	42.9%	22.1%	33.1%	20.5%

SOURCE: Moody's Public Finance Department, *1988 Medians: Selected Indicators of Municipal Performance.* New York: Copyright © Moody's Investors Service, Inc., 1988, pp. 10-12. Reproduced with permission granted by Moody's Investors Service, Inc. All rights reserved.
a. Operating and maintenance expenses divided by total operating revenues.
b. Net revenues divided by gross revenue and income.
c. Net revenues divided by interest requirements for year.
d. Net revenues divided by principal and interest requirements for year.
e. Net revenues less principal and interest requirements for year divided by gross revenue and income.
f. Net funded debt divided by the sum of net fixed assets plus net working capital.

health may be secured by comparing local status on several key indicators with national norms (Table 11.8).

Accuracy in Revenue Forecasting

The accurate projection of revenues is extremely important to the smooth operation of municipal government. The task of "crystal balling" future revenues often falls to the finance department.

Most cities take a conservative approach to the forecasting of revenues, hoping that any error is on the side of *underestimating* actual revenues. Ideally, the percentage variance between projected and actual revenues will be quite small. A review of performance documents suggests that many revenue forecasters do quite well. Revenue projections within 3% of accuracy appear fairly common (Table 11.9).

Financial Management

Finance departments perform an array of duties related to financial accountability and stewardship. They must account for revenues and expenditures, keeping local officials and operating departments informed about financial status on a timely basis; they must collect taxes; they must handle revenues efficiently, pay invoices on a timely basis, and manage cash and investments to the advantage of the municipality. But what constitutes good performance of these duties?

TABLE 11.9 Accuracy of Revenue Forecasts: Performance Targets and Actual Experience of
Selected Cities

City	Target	Actual Experience
Oakland, CA		0.7% variance between projected and actual General Fund revenue (1992)
Duncanville, TX		Actual revenues as percentage of projected revenue: 101% (1990)
St. Petersburg, FL	±2%	Actual revenues as percentage of projected: 100.2% (1990)
Savannah, GA		Actual General Fund revenues as percentage of projected: 102% (1988); 102% (1989); 103% (1990); 98% (1991)
Portland, OR	±3%	Actual General Fund revenues as percentage of projected: 104.7% (1989); 103.8% (1990); 101.8% (1991)
Knoxville, TN		3.4% variance between projected and actual revenues (1991)
Nashville-Davidson County, TN	±5%	

The Government Finance Officers Association (GFOA) has developed a set of
criteria for judging the adequacy of financial reporting in local government. Cities
and counties whose comprehensive annual financial reports meet those criteria
qualify for "certificates of conformance in financial reporting." Few standards have
been published for other finance department duties, but many cities have established their own performance targets.

Many cities target the release of monthly financial reports by the middle of the next
month; some even earlier (Table 11.10). Several municipalities target the completion
of their annual audits and release of their comprehensive annual financial reports
(CAFR) by the 5th or 6th month of the following fiscal year (Table 11.11).

Most of the municipal documents examined for this chapter reported current
property tax collection rates in the mid- to upper-90% range (Table 11.12). Those
that offered statistics on collection of delinquent taxes, such as Portland, Oregon,
typically provided evidence of diligent pursuit of delinquencies (Figure 11.2).

Aggressive cash management demands the prompt processing of newly received
revenues (Table 11.13) and the timely payment of invoices (Table 11.14). Some
municipalities consider timely payment to mean prompt payment—for instance,
within 2 or 3 days of receipt. Others regard timely payment to mean the timing of
payments to retain cash in the possession on the municipality as long as possible
for investment purposes but not so long as to result in penalties for late payment.
Cash management targets and performance indicators suggest two major dimensions of performance: (a) maximum prudent investment of idle funds and (b)
modest costs for performing the cash management function (Table 11.15).

TABLE 11.10 Prompt Issuance of Monthly Financial Reports: Selected Cities

City	Performance Target	Actual Experience
Raleigh, NC	Within 2 days of end of reporting period	
College Park, MD	By the 5th day of the following month	
Savannah, GA	Within 6 workdays of month's end	7.0 days (1988); 8.5 days (1989); 8.0 days (1990); 7.0 days (1991)
College Station, TX	Within 8 workdays of month's end	
Sunnyvale, CA	Within 8 workdays after end of accounting period	
Houston, TX	Within 10 days following month's end	50% within 10 days (1992, estimate)
Denton, TX	Within 9 workdays of month's end	
Greenville, SC	By the 10th workday of the following month	
Oakland, CA	Cash management reports issued by 15th of following month	100% by 15th (1992)
Oak Ridge, TN	By the 15th of the following month	84% by 15th (1991); 67% by 15th (1992)
Lubbock, TX	By the 18th of the following month	91% by 18th (1991)
St. Petersburg, FL	Within 15 workdays of end of reporting period	100% within 15 workdays (1990)
Fayetteville, AR	Monthly statements within 25 days; quarterly statements within 30 days	100% compliance (1991)
Grants Pass, OR	Within 4 weeks of month's end	
Shreveport, LA		Monthly closeouts within 10 workdays of month's end: 100% (1991); interim financial statements issued within 4 weeks of closeout: 67% (1991)

TABLE 11.11 Timely Audit and Financial Statement: Selected Cities

City	Performance Target
College Park, MD	Complete audit and release comprehensive annual financial report (CAFR) within 3 months after end of fiscal year
Anaheim, CA	Submit an unqualified financial statement within 5 months of end of fiscal year
Corvallis, OR	Publish CAFR within 5 months following end of fiscal year
Oakland, CA	Issue CAFR within 5 months of fiscal year close
Oak Ridge, TN	Complete audit process within 4 months after end of fiscal year; print and distribute CAFR by middle of 5th month
Chandler, AZ	Complete audit within 4 months after end of fiscal year; release CAFR by 20th day of 5th month
Portland, OR	Issue CAFR within 175 days after end of fiscal year
Raleigh, NC	Completion (with unqualified audit opinion) in 2nd quarter
Decatur, IL	Publish CAFR within 6 months following end of fiscal year; complete compliance audit within 8 months of end of fiscal year
Fayetteville, AR	Audit completed within 210 days of close of fiscal year

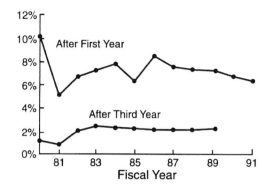

Figure 11.2. Delinquent Property Taxes: Portland, Oregon
SOURCE: City of Portland (Oregon), *An Evaluation of City Financial Trends: 1980-91,* May 1992, p. 10.

TABLE 11.12 Reported Collection Rates for Current Year Taxes

City	Collection Rate for Current Taxes
College Park, MD	99.62% of property tax levy (1991)
Iowa City, IA	Taxes collected, as percentage of property tax budget: 99.4% (1989); 99.1% (1990)
Grand Junction, CO	96.67% to 99.67% of property tax levy (1982-1990); 98.6% (1991)
Saginaw, TX	97.9% (1990); 98.2% (1991)
Irving, TX	96.9% to 98.46% (1983-1990); 98.01% (1991)
Portsmouth, VA	Percentage of real estate taxes collected in year of levy: 95.8% (1991); 97.0% (1992); 97.7% (1993) Personal property taxes: 84.3% (1991); 85.8% (1992); 84.9% (1993)
Duncanville, TX	97.5% (1990)
Richmond, VA	96.74% of real estate taxes (1989); 97.46% (1990); 96.99% (1991); 97.50% (1992)
Denton, TX	97.1% (1989); 96.6% (1990); 97.3% (1991)
Avon, CT	97.09% of property tax levy (1991)
Sioux Falls, SD	96.8% of property tax levy (1990); 97.1% (1991)
Fort Worth, TX	98% (1991); 97% (1992)
Shreveport, LA	97% of property taxes (1991)
College Station, TX	96.4% of property taxes (1990); 95.8% (1991); 96.7% (1992)
New York City, NY	96.2% of property tax levy
Boston, MA	96% of current year real estate and personal property taxes (1992)
Oak Ridge, TN	99% (1990); 93% (1991); 96% (1992)
League City, TX	Performance target: At least 96% of property taxes
Greenville, SC	93.8% of current year property taxes (1989); 94.7% (1990)
Newburgh, NY	94.48% (1991)
Portland, OR	Approximately 94%
Savannah, GA	Performance targets: At least 94% of property tax levy; 99% of other taxes (beer, liquor, wine, hotel-motel taxes) Actual: 91%, 99% (1988); 94%, 99% (1989); 91%, 99% (1990); 94%, 99% (1991); 94%, 99% (1992, estimate)
Rochester, NY	92.9% (1992)
Newnan, GA	85.1% to 91.9% of property taxes (1982-1991)

TABLE 11.13 Prompt Deposit of Revenues: Performance Targets and Actual Experience of Selected Cities

Raleigh, NC
Percentage of same-day deposits of utility bill payments: 79.8% (1991)

Tucson, AZ
Target: At least 50% same-day deposit; 100% by following morning

Orlando, FL
Percentage of revenue received by Revenue Collection that is processed within 1 workday of receipt: 100% (1991)

Reno, NV
Percentage of central cashier receipts deposited within 24 hours: 99% (1990)

Savannah, GA
Target: Process general fund and water and sewer receipts within 1 day at least 90% of the time
Actual: 90% within 1 day (1988-1991)

St. Petersburg, FL
Target: Promptly deposit all funds and forward revenue reports to accounting by 4 p.m. the next day
Actual: Deposit deadline missed on 2 days in 1990

The city of St. Petersburg, Florida, judges its municipal investments by comparing its rate of return with that of the Florida Investment Pool (Figure 11.3). Many other cities follow a similar strategy of comparing their investment returns with a fluctuating standard—typically, U.S. Treasury bills or, like St. Petersburg, their own state's investment pool (Table 11.16). By pegging their own performance to

TABLE 11.14 Timely Payment of Invoices: Selected Cities

Upper Arlington, OH
Turnaround time on payment vouchers: 2 days (1991)

Oakland, CA
Vendor payments processed within 3 days of receipt of payment request: 98.5% (1992)

Fort Worth, TX
Target: Process payment documents within 5 days of receipt
Actual: 8-day average (1991)

Shreveport, LA
Target: Process within 5 days
Actual: 75% within 5 days (1991)

Denton, TX
Target: Process vendor invoices within 5 workdays
Actual: 80% within 5 workdays (1991)

Fort Collins, CO
Days from voucher to check: 18.0 (1991); 11.5 (1992, estimate)

Raleigh, NC
Processed within 15 days: 99% (1991)

Portland, OR
Turnaround time from invoice to payment: 17-20 days (1991)

Boston, MA
Invoices processed within 21 days: 99.2% (1992)
Contracts processed within 4 days: 80% (1992)
Documents encumbered within 2 days: 80% (1992)

Corpus Christi, TX
Invoices processed within 4 weeks: 90% (1991)

College Station, TX
Target: Mail payment by 3 p.m. on the Friday prior to due date

Oklahoma City, OK
Target: Turnaround time of 30 days or less

Orlando, FL
Invoices paid within 30 days: 99.8% (1991)

Savannah, GA
Invoices paid within 30 days: 66% (1988); 79% (1989); 78% (1990); 90% (1991)

Houston, TX
Invoices paid within 45 days: 95% (1992)

TABLE 11.15 Cash Management: Selected Cities

INVESTMENT OF IDLE FUNDS

Rochester, NY
  Average daily invested cash as percentage of average daily cash ledger balance: 99.87% (1991)

Savannah, GA
  Idle cash as percentage of portfolio book value: 0.12% (1989); 0.18% (1990); 0.17% (1991); 0.15% (1992 estimate)

San Jose, CA
  Percentage of idle funds invested: 99.6% (1989)

Long Beach, CA
  Target: Maintain a fully invested position of at least 99.5% of all funds

Sunnyvale, CA
  Target: 99.5% of available funds placed in interest-bearing investments

Auburn, AL
  Target: Investment of 98.5% of all temporarily idle investable funds (1993)

Irving, TX
  Target: No more than 4.75% of available funds left idle
  Actual: 4.89% idle (1991)

New York City
  Total average daily resources invested: 93.4% (1991); 89.1% (1992)

CASH MANAGEMENT COSTS

Corpus Christi, TX
  Management cost per $1,000 of investment income: $0.024 (1991)

Fort Collins, CO
  Investment program cost as percentage of total portfolio: 0.06% (1991)

Gresham, OR
  Target: Cash management costs (e.g., personnel and bank service charges) should not exceed 0.25% of assets managed

---

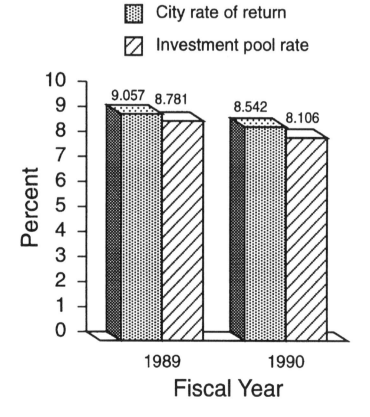

Figure 11.3. Performance of Investments: St. Petersburg, Florida
SOURCE: City of St. Petersburg (Florida), *Approved 1992 Program Budget*, p. 199.

TABLE 11.16 Favorable Return on Investments: Selected Cities

COMPARED WITH FEDERAL FUNDS

Alexandria, VA
Target: Average rate of return within 50 basis points of the federal funds rate[a]
Actual basis points above (+) or below (-) average federal funds rate: -4 (1989); +9 (1990); +44 (1991); +60 (1992); +66 (1993)

Oklahoma City, OK
Operating portfolio return: 26.7% over U.S. Treasury bill (T-bill) market (1991); 43.9% over T-bill market (1992)

Boise, ID
Portfolio yield to average federal funds rate: 133.4% (1991)

Phoenix, AZ
Target: Yield of at least 120% of the 90-day T-bill rate

Shreveport, LA
Average return on investments: 12% greater than 3-month T-bill rate (1991)

Rock Island, IL
Portfolio yield as percentage of T-bill rate: 112% (1990); 112% (1991)

Long Beach, CA
Target: Yield of at least 110% of 3-month T-bill rate
Actual: 128% of T-bill rate (1991); 145% of T-bill rate (1992)

New York City, NY
Ratio of rate of return on resources invested to T-bill discount rate: 1.09 (1991); 1.07 (1992)

Greenville, SC
Investment yield relative to 90-day T-bill rate: +1.04% (1991)

Corpus Christi, TX
Target: Meet or exceed rate on 6-month T-bill

Lubbock, TX
Target: Exceed the average return on 3-month T-bills or the average rate on federal funds, whichever is greater

Hurst, TX
Target: Rate of return exceeding the 3-month T-bill rate

Chandler, AZ
Target: Rate of return equal to or greater than 90-day T-bills

Boston, MA
City return compared with 90-day T-bill discount rate: 100% (1992)

Fort Worth, TX
Target: Meet or exceed the average return on 1-year T-bills

COMPARED WITH INVESTMENT POOLS

Milwaukee, WI
Rate of return on excess city funds as a percentage of rate of return on Wisconsin Local Government Investment Pool: 146.4% (6.94% vs. 4.74%, 1992); 100.3% (3.71% vs. 3.70%, 1993)

Corvallis, OR
Target: Yield at least 0.5% above state pool
Actual: 0.5% above state pool (1991)

Decatur, IL
Target: Average rate of return at least equal to that of the state investment pool

St. Petersburg, FL
Target: Meet or exceed the rate of the State of Florida Investment Pool

COMPARED WITH OTHER INVESTMENTS

Oakland, CA
Rate of return relative to market rates: +0.75% (1990); +0.87% (1991); +0.75% (1992)

Savannah, GA
Difference between investment yields and highest yields offered by large money center banks: -0.09% (1988); +0.16% (1989); -0.25% (1990); +0.30% (1991); +0.25% (1992, estimate)

Auburn, AL
Target: Average rate of return meeting or exceeding jumbo certificates of deposit, as quoted in *Wall Street Journal*

a. "A basis point is one-hundredth of one percent; the federal funds rate is the rate at which funds are traded between member banks of the Federal Reserve System." This comparison, according to the Alexandria budget, "is generally considered to be a demanding benchmark for measuring public fund portfolio performance" (City of Alexandria, Virginia, *Fiscal Year 1992-1993 Approved Budget*, p. 5.12).

a standard that floats upward or downward with market trends, their performance may be judged not simply by a raw interest rate but instead by how their interest rate compares with another avenue of investment available to the city.

TABLE 11.17 GFOA's Criteria for Exemplary Budget Documents

As a Policy Document

- The document should include a coherent statement of organization-wide financial and programmatic policies and goals that address long-term concerns and issues.
- The document should describe the organization's short-term financial and operational policies that guide the development of the budget for the upcoming year.
- The document should include a coherent statement of goals and objectives of organizational units (e.g., departments, division, offices, or programs).
- The document shall include a budget message that articulates priorities and issues for the budget for the new year. The message should describe significant changes in priorities from the current year and explain the factors that led to those changes. The message may take one of several forms (e.g., transmittal letter, budget summary section).

As a Financial Plan

- The document should include and describe all funds that are subject to appropriation.
- The document shall present a summary of major revenues and expenditures, as well as other financing sources and uses, to provide an overview of the total resources budgeted by the organization.
- The document shall include summaries of revenues, and other resources, and of expenditures for prior year actual, current year budget and/or estimated current year actual, and proposed budget year.
- The document shall describe major revenue sources, explain the underlying assumptions for the revenue estimates and discuss significant revenue trends.
- The document shall include projected changes in fund balances, as defined by the entity in the document, for governmental funds included in the budget presentation, including all balances potentially available for appropriation.
- The document should include budgeted capital expenditures and a list of major capital projects for the budget year, whether authorized in the operating budget or in a separate capital budget.
- The document should describe if and to what extent capital improvements or other major capital spending will impact the entity's current and future operating budget. The focus is on reasonably quantifiable additional costs and savings (direct or indirect) or other service impacts that result from capital spending.
- The document shall include financial data on current debt obligations, describe the relationship between current debt levels and legal debt limits, and explain the effects of existing debt levels on current and future operations.
- The document shall explain the basis of budgeting for all funds, whether GAAP, cash, modified accrual, or some other statutory basis.

Budgeting

Just as it has established criteria for excellence in financial reporting, the GFOA (1993) similarly has designed standards for the presentation of budgetary information. "Distinguished Budget Presentation Awards" are conferred on local governments whose budgets are deemed meritorious on four dimensions:

1. as a policy document,
2. as a financial plan,
3. as an operations guide, and
4. as a communication device. (pp. 1-6)

TABLE 11.17 *Continued*

As an Operations Guide

- The document shall describe activities, services, or functions carried out by organizational units.
- The document should provide objective methods (quantitative and/or qualitative) of measurement of results by unit or program. Information should be included for prior year actual, current year budget and/or estimate, and budget year.
- The document shall include an organization chart(s) for the entire organization.
- A schedule(s) or summary table(s) of personnel or position counts for prior, current, and budgeted years shall be provided, including descriptions of significant changes in levels of staffing or reorganizations planned for the budget year.

As a Communication Device

- The document should provide summary information, including an overview of significant budgetary issues, trends, and resource choices. Summary information should be presented within the budget document either in a separate section (e.g., executive summary) or integrated within the transmittal letter or other overview sections.
- The document should explain the effect, if any, of other planning processes (e.g., strategic plans, long-range financial plans, capital improvement plans) upon the budget and budget process.
- The document shall describe the process for preparing, reviewing and adopting the budget for the coming fiscal year. It should also describe the procedures for amending the budget after adoption. If a separate capital budget process is used, a description of the process and its relationship to the operating budget should be provided.
- Charts and graphs shall be used, where appropriate, to highlight financial and statistical information. Narrative interpretation should be provided when the messages conveyed by the graphs are not self-evident.
- The document should provide narrative, tables, schedules, cross-walks, or matrices to show the relationship between different revenue and expenditure classifications (e.g., funds, programs, organization units).
- The document shall include a table of contents to make it easy to locate information in the document.
- A glossary should be included for any terminology (including abbreviations and acronyms) that is not readily understood by a reasonably informed lay reader.
- The document should include statistical and supplemental data that describe the organization and the community or population it serves, and provide other pertinent background information related to the services provided.
- The document should be printed and formatted in such a way to enhance understanding and utility of the document to the lay reader. It should be attractive, consistent, and oriented to the reader's needs.

SOURCE: Excerpted from Government Finance Officers Association, *Distinguished Budget Presentation Awards Program: Awards Criteria* (Chicago: GFOA, August 1993), pp. 1-6. Reprinted by permission.
NOTE: GAAP = generally accepted accounting principles.

The GFOA does not attempt through its award program to prescribe a particular budget method or format, but it does regard the above functions to be key roles of the budget. Fairly specific criteria are articulated within each role, only some of those criteria being considered mandatory for recognition as a "distinguished budget" (Table 11.17).

12

Fire Service

More than most other municipal service units, fire departments revere tradition. And understandably so. Almost 300 years after the establishment of America's first public fire department, the roots of fire service tradition run deep (Roberts, 1991, p. 9.35).

Some fire department traditions are charming and contribute to an aura that fosters good will. A dalmatian living in the station house—or perhaps borrowed to ride on a gleaming fire truck in local parades—helps link the modern fire service with its colorful and heroic past. The Saturday morning ritual of firefighters washing their bright red fire engines—and most still are red despite scientific studies suggesting better visibility of other colors—and polishing their ample chrome is a scene the public enjoys. Where traditions collide with managerial pursuit of greater efficiency and effectiveness, however, the resulting confrontations often are less pleasurable.

Given their traditional bent, it is perhaps not surprising that many fire departments place considerable stock in long-standing rules of thumb that address everything from fireground tactics to prescribed levels of staffing. More than a few city executives with only a general background in fire services have met their match in the person of a fire chief able to cite with ease a battery of statistics, standards, and rules of thumb to bolster the department's position on a budgetary or operational issue. Understandably, nonspecialists feel at a disadvantage in exchanges of this type. So what's the best strategy? Mayors and city managers should enter such a fracas armed with a reasonable familiarity with relevant facts and figures themselves.

Overview

Various fire statistics offer potentially useful insights, but often they must be placed in proper context. Popular images of fire departments staffed with career firefighters square with reality in most cities of 25,000 population or greater but less accurately depict the reality of small-town America, where volunteer firefighters still play a major role (Table 12.1). Moreover, fighting fires is not the only thing—or even by some measures the *main* thing—that fire departments do. More than $\frac{1}{2}$ of all calls to the typical fire department are for medical aid, approximately $\frac{1}{10}$ are false alarms, and less than $\frac{1}{5}$ report actual fires (Figure 12.1). Time on the fireground fighting a real blaze typically consumes only a small portion of a firefighter's working day.

Among fire department personnel, the fire chief, a few other executive officers, and the department's clerical employees may have work schedules similar to other municipal personnel, but most firefighters do not. Workweeks for paid firefighters typically range from 40 to 56 hours, with the International City/County Management Association (ICMA) recently reporting a mean workweek of 51 hours ("Fast Facts," 1992). Workweeks of 56 hours usually indicate the use of staggered schedules calling for 24 hours on duty by a given firefighter followed by 48 hours off duty. Among departments with workweeks of 48 hours or less, a common schedule calls for a 10-hour day shift and a 14-hour night shift.

National statistics for fire departments often differentiate between departments with career firefighters and those that rely primarily on volunteers and, where staffing ratios are cited, between departments with 56-hour workweeks and those using other schedules. The prudent manager will look for those distinctions.

Typical Topics of Budgetary Debate

Three aspects of fire service typically draw considerable attention at budget time:

1. appropriate staffing level,
2. the hazardous nature of fire fighting and contentions that firefighter compensation should recognize that characteristic, and
3. the appropriate number and types of fire-fighting equipment.

Although each of these topics deals more directly with inputs than with performance, the frequency with which each becomes an issue in most communities prompts their brief discussion here before proceeding to actual benchmarks of performance.

TABLE 12.1 Department Type by Population Protected, 1991

Population Protected	Type of Department (Percentage)				
	All Career (%)	Mostly Career (%)	Mostly Volunteer (%)	All Volunteer (%)	Total (%)
1,000,000 or more	87.5	12.5	0.0	0.0	100.0
500,000 to 999,999	72.3	22.2	5.5*	0.0	100.0
250,000 to 499,999	79.5	17.6	2.9	0.0	100.0
100,000 to 249,999	77.6	14.1	5.9	2.4	100.0
50,000 to 99,999	68.5	19.8	9.6	2.1	100.0
25,000 to 49,999	46.8	23.6	21.2	8.4	100.0
10,000 to 24,999	18.5	17.2	40.2	24.1	100.0
5,000 to 9,999	2.5	2.8	33.1	61.6	100.0
2,500 to 4,999	0.7	1.1	15.4	82.8	100.0
Under 2,500	0.8	0.5	2.7	96.0	100.0
All Departments	6.6	4.1	13.6	75.7	100.0

SOURCE: NFPA Survey of Fire Departments for U.S. Fire Experience, 1991, as reported in Michael J. Karter, Jr., *U.S. Fire Department Profile Through 1991* (Quincy, MA: National Fire Protection Association, October 1992), p. 16. Reprinted with permission from National Fire Protection Association, Quincy, MA 02269.
NOTE: Type of department is broken into four categories. All-career departments are composed of 100% career firefighters. The mostly career category is composed of 51% to 99% career firefighters, whereas mostly volunteer departments are composed of 1% to 50% career firefighters. All-volunteer departments are composed of 100% volunteer firefighters.
*This statistic reflects the inclusion of several county fire departments that have a considerable number of volunteer firefighters on their staffs.

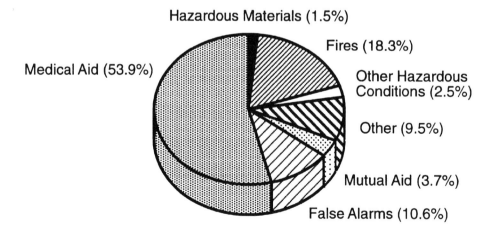

Figure 12.1. Nature of Fire Department Calls
SOURCE: John R. Hall, Jr., and Arthur E. Cote, "America's Fire Problem and Fire Protection," p. 1.22. In Arthur E. Cote and Jim L. Linville (Eds.), *Fire Protection Handbook,* 17th edition (Quincy, MA: National Fire Protection Association, 1991). Based on 1988 NFPA survey. Reprinted with permission from *Fire Protection Handbook,* 17th edition, Copyright © 1991, National Fire Protection Association, Quincy, MA 02269.

TABLE 12.2 Career Firefighters per 1,000 People by Workweek and Population Protected, 1989-1991

Population Protected	Career Firefighters per 1,000 People					
	Lowest Reported	Median	Highest Reported	40-45 hour workweek	46-51 hour workweek	52-60 hour workweek
1,000,000 or more	0.79	1.11	1.75	1.48	1.38	1.07
500,000 to 999,999	0.83	1.66	2.82	2.27[a]	1.73[a]	1.35[a]
250,000 to 499,999	0.61	1.32	2.63			
100,000 to 249,999	0.38	1.35	3.12	2.41	1.56	1.33
50,000 to 99,999	0.00	1.25	3.32	2.13	1.83	1.43
25,000 to 49,999	0.00	1.21	3.29	2.22	1.61	1.48

SOURCE: NFPA Fire Service Survey, 1989-1991, as reported in Michael J. Karter, Jr., *U.S. Fire Department Profile Through 1991* (Quincy, MA: National Fire Protection Association, October 1992), pp. 4 and 7. Reprinted with permission from National Fire Protection Association, Quincy, MA 02269.
NOTES: *The rates listed above are based on data reported to the NFPA and do not reflect recommended rates or some defined fire protection standard.*
The rates of a particular size of community may vary widely because departments face great variation in their specific circumstances and policies, including unusual structural conditions, type of service provided to the community, geographic dispersion of the community, and other factors.
Career rates are shown only for communities with more than 25,000 population in which departments are composed mostly of career firefighters.
a. This figure is for the combined population category 250,000 to 999,999.

STAFFING LEVELS

Staffing statistics compiled by the National Fire Protection Association (NFPA) reveal ratios of career firefighters per 1,000 population for various sizes of communities and different lengths of workweek (Table 12.2). Among cities of 100,000 to 249,999 persons, for example, the lowest reported ratio was 0.38 career firefighters per 1,000 persons, the highest ratio was 3.12 firefighters per 1,000 persons, and the median was 1.35. Among communities in that range, the average ratio for fire departments with a 52- to 60-hour workweek was 1.33 career firefighters per 1,000 people.

Staffing norms differ by region. Except among cities of 250,000 population or greater, northeastern municipalities tend to employ higher ratios of career firefighters than do other regions, with western municipalities employing the lowest ratios (Table 12.3).

From time to time, various ratios of firefighters to population are offered as prescriptions, as rules of thumb, or even as standards. Unfortunately, such ratios sometimes are applied without any thought given to important local factors such as the use of volunteers in a given community, the region in which the community is located, its population and density, and various other local characteristics that influence fire risk. Statistics such as those presented here that reveal low, median, and high ratios for various population categories, as well as variations by work-week and region, can be of greater value to managers and other public officials than can a "one size fits all" rule of thumb. By comparing the ratio in one's

TABLE 12.3 Median Rates of Career Firefighters per 1,000 People by Region and Population Protected, 1991

Population Protected	Northeast	North Central	South	West
250,000 or more	1.46	1.66	1.40	1.03
100,000 to 249,999	2.33	1.54	1.40	0.93
50,000 to 99,999	2.00	1.29	1.48	0.90
25,000 to 49,999	1.64	1.02	1.03	0.97

SOURCE: NFPA Survey of Fire Departments for U.S. Fire Experience, 1991, as reported in Michael J. Karter, Jr., *U.S. Fire Department Profile Through 1991* (Quincy, MA: National Fire Protection Association, October 1992), p. 8. Reprinted with permission from National Fire Protection Association, Quincy, MA 02269.
NOTES: *The rates listed above are based on data reported to the NFPA and do not reflect recommended rates or some defined fire protection standard.*
The rates of a particular size of community may vary widely because departments face great variation in their specific circumstances and policies, including unusual structural conditions, type of service provided to the community, geographic dispersion of the community, and other factors.
Career rates are shown only for communities with more than 25,000 population in which departments are composed mostly of career firefighters.

community with the relevant cluster, a more appropriate determination can be made as to whether local staffing is in the ballpark.

Perhaps more controversial than attempts to use overall staffing ratios as prescriptive devices have been attempts to prescribe minimum staffing on individual fire apparatus. The International Association of Fire Fighters (IAFF; 1993), for example, strongly advocates minimum response strength of four firefighters per engine company. IAFF insists that those firefighters should arrive at the fire scene together on their engine. In contrast, NFPA (1992), which also encourages a minimum of four firefighters per engine prior to commencing fire-fighting operations, does not insist that they all arrive on the same vehicle.

> While members can be assigned and arrive at the scene of an incident in many different ways, it is strongly recommended that interior fire fighting operations not be conducted without an adequate number of qualified fire fighters operating in companies under the supervision of company officers.
> It is recommended that a minimum acceptable fire company staffing level should be 4 members responding on *or arriving with* each engine and each ladder company responding to any type of fire. The minimum acceptable staffing level for companies responding in high-risk areas should be 5 members responding *or arriving with* each engine company and 6 members responding *or arriving with* each ladder company. (NFPA, 1992, p. 40, emphasis added; reprinted with permission from *NFPA 1500: Fire Department Occupational Safety and Health Program,* Copyright © 1992, National Fire Protection Association, Quincy, MA 02269. This reprinted material is not the complete and official position of the National Fire Protection Association on the referenced subject, which is represented only by the standard in its entirety.)

Although the difference between the IAFF and NFPA positions may seem subtle, it is not. The NFPA position allows much greater flexibility in the deployment of firefighters when not actually engaged in fire fighting.

TABLE 12.4 Minimum Staffing and Average Response Times Among Large Cities

City by Rank	Minimum Staffing		Response Times
	Engine	Truck	
Phoenix	4	4	3.58
Austin	3	3	3.90
Philadelphia	4	5	4.00
Dallas	4	4	4.16
San Jose	4	5	4.20
El Paso	3	3	4.21
San Antonio	4	4	4.42
Fort Worth	3	4	4.47
Baltimore	3	4	4.50
Detroit	4	3	5.00
Houston	4	3	6.00
San Diego	4	4	< 5.00 (90%)

SOURCE: City of Dallas (Texas) Fire Department, *Comparative Analysis With Index Cities,* April 20, 1992, p. 7.

Individual cities, as well as ICMA, have opposed attempts to prescribe minimum staffing levels and to dictate deployment modes as unwarranted intrusions on local prerogatives. Although some communities have adopted response practices consistent with NFPA or even the more restrictive IAFF recommendations, surveys indicate that others—even fairly large cities such as Austin, El Paso, Fort Worth, and Baltimore—typically respond with fewer than the prescribed number of firefighters (Table 12.4). ICMA reports that in 1994, barely more than half of all cities required a minimum number of crew members per apparatus ("Fast Facts," 1995). For pumpers, the average minimum crew was 3.0; for ladder units, 2.8; and for other units, 2.3 crew members.

HAZARDOUS DUTIES

Spokespersons for proposals to increase the pay of firefighters often mention the hazardous nature of that occupation, sometimes contending that it is America's most dangerous work. Although a manager or mayor may choose to dispute the "most dangerous" label (Table 12.5), there is no disputing that fire fighting is hazardous work with considerable risk of injury or even death. Nevertheless, a reasonable command of relevant statistics on firefighter deaths and injuries will equip city officials not only to rebut any exaggerated claims but also to place local experience in a national context. If local firefighters' deaths and injuries exceed national norms, a careful review of the community's training program, firefighter fitness levels, and fireground procedures may be in order.

Only about half of all firefighter deaths occur while actually fighting fires (Table 12.6). Others occur en route to the fire scene or on the way back to the station, during training, or while performing other duties. Not surprisingly, death rates tend

TABLE 12.5 Death Rates in Fire Fighting and Other Hazardous Occupations

Occupation	Deaths per 100,000 Employees
Agriculture (including forestry and fishing)	44.0
Mining and quarrying (including oil and gas extraction)	43.0
Construction	31.0
Transportation and public utilities	22.0
Fire fighting (career-municipal)	13.8

SOURCE: Joe Erwin, "Firefighters See Red: Crew Size and Deaths or Injuries Are Debated," *Public Management, 75* (January 1993), p. 3, citing statistics from National Fire Protection Association and *Accident Facts,* 1992 edition, National Safety Council, Chicago, Illinois. Reprinted by permission of ICMA.

to be greater among older firefighters—reaching rates of 2.6 per 10,000 firefighters among those in their 50s and 4.7 per 10,000 among firefighters 60 years of age or older, compared with fewer than one death per 10,000 among firefighters less than 40 years of age (Peterson, 1991, p. 9.5).

Fire departments serving larger communities understandably experience greater numbers of fires and greater numbers of firefighter injuries in absolute terms, but is the job of fire fighting really more hazardous in large cities than in smaller communities? Statistics indicate that the answer is "yes." When measured as number of fireground injuries per 100 fires, larger communities experience a rate approximately 50% greater than do small towns (Table 12.7). Even that ratio, however, tends to understate the injury differential. Because big-city firefighters respond to more fires, they are more exposed to the risk of injury than are their small-town counterparts. Each year, 17 of every 100 big-city firefighters experience a fireground injury, compared with 1.3 of every 100 firefighters in communities with populations less than 2,500.

EQUIPMENT LEVELS

As every city official who has budgeted for major fire equipment already knows, fire engines are expensive. Unquestionably, fire-fighting efforts can be severely

TABLE 12.6 Firefighter Deaths, 1989, by Type of Duty

Cause	Number of Deaths	Percentage
Firefighting	58	50.0
Non-fire incident	6	5.2
Responding/returning from alarm	28	24.1
Other on-duty	12	10.3
Training	12	10.3

SOURCE: William Peterson, "Fire Department Occupational Safety and Health." In Arthur E. Cote and Jim L. Linville (Eds.), *Fire Protection Handbook,* 17th edition (Quincy, MA: National Fire Protection Association, 1991), p. 9.4. Reprinted with permission from *Fire Protection Handbook,* 17th edition, Copyright © 1991, National Fire Protection Association, Quincy, MA 02269.

TABLE 12.7 National Statistics on Reported Fires and Fireground Injury Rates

Population of Community Protected	Average Number of Reported Fires					Number of Fireground Injuries per 100 Fires					Number of Fireground Injuries per 100 Firefighters
	National Average	Northeast	North Central	South	West	National Average	North-east	North Central	South	West	National Average
500,000 to 999,999	5,718.0	*	*	6,252.2	5,272.2	3.0	*	*	2.4	3.7	17.1
250,000 to 499,999	2,790.0	*	2,973.0	2,992.2	2,121.3	3.3	*	4.0	2.6	2.4	15.5
100,000 to 249,999	1,178.9	1,982.5	1,246.8	1,127.7	864.1	3.3	7.1	2.7	2.0	2.3	14.1
50,000 to 99,999	442.6	630.9	388.7	486.2	356.1	3.0	4.7	4.0	1.7	2.0	11.6
25,000 to 49,999	222.4	205.9	195.3	256.2	243.8	2.2	3.6	2.7	1.7	1.5	8.9
10,000 to 24,999	117.6	114.8	101.1	140.5	124.1	1.9	3.0	2.0	1.4	1.2	5.7
5,000 to 9,999	56.3	52.3	51.9	61.9	64.9	1.8	2.9	1.7	1.6	1.2	3.2
2,500 to 4,999	32.9	33.9	28.0	40.0	36.5	2.4	3.3	2.5	1.3	3.0	2.9
Under 2,500	14.7	13.7	12.8	21.6	13.2	2.0	2.2	1.6	1.9	1.5	1.3

SOURCE: NFPA's Survey of Fire Departments for U.S. Fire Experience (1989), as reported in William Peterson, "Fire Department Occupational Safety and Health." In Arthur E. Cote and Jim L. Linville (Eds.), *Fire Protection Handbook*, 17th edition (Quincy, MA: National Fire Protection Association, 1991). p. 9.8. Reprinted with permission from *Fire Protection Handbook*, 17th edition, Copyright © 1991, National Fire Protection Association, Quincy, MA 02269.
NOTE: * denotes insufficient data.

TABLE 12.8 Average Apparatus and Station Rates, 1989-1991

Population Protected	Pumpers per 1,000 People	Aerial Apparatus per 1,000 People	Stations per 1,000 People
1,000,000 or more	0.033	0.017	0.036
500,000 to 999,999	0.066	0.022	0.054
250,000 to 499,999	0.064	0.018	0.061
100,000 to 249,999	0.075	0.018	0.068
50,000 to 99,999	0.086	0.022	0.075
25,000 to 49,999	0.123	0.029	0.089
10,000 to 24,999	0.208	0.041	0.122
5,000 to 9,999	0.381	0.039	0.211
2,500 to 4,999	0.617	0.037	0.366
Under 2,500	1.388	0.031	0.988
National average	0.274	0.028	0.186

SOURCE: NFPA Fire Service Survey, 1989-1991, as reported in Michael J. Karter, Jr., "NFPA Surveys U.S. Fire Departments." Reprinted with permission from *NFPA Journal*®, 1993 (July/August, Vol. 87, No. 4), Copyright © 1993, National Fire Protection Association, Quincy, MA 02269.
NOTE: These results reflect average apparatus and station rates per 1,000 people by population protected among reporting jurisdictions. The NFPA emphasizes that they do not reflect recommended rates or some defined fire protection standard.
Note that pumpers reported above had a capability of 500 gpm or greater. Many departments reported other suppression vehicles, including apparatus with pumps less than 500 gpm, hose wagons, brushfire vehicles, tankers, etc.

handicapped—sometimes with tragic consequences—by inadequate equipment. Nevertheless, few communities possess the financial resources to meet all their fire department's desires for the latest and most sophisticated fire-fighting equipment. City officials struggling to find the appropriate balance between their responsibility for ensuring public safety and their obligation for financial prudence may find helpful information in the statistical norms for fire equipment inventories in communities of various sizes.

Obviously, larger communities own more pumpers and aerial equipment and operate more fire stations than do smaller communities, but their *rates* of pumpers, aerial apparatus, and stations per 1,000 population are not as great (Table 12.8). Although considerable variation is evident, the national average is approximately one pumper per 3,700 population, one aerial apparatus per 35,700 population, and one fire station per 5,400 population.

Most cities with more than 100,000 people possess more than 10 pumpers (Table 12.9). Cities with populations of 10,000 to 50,000 tend to have 3 or 4 pumpers, whereas cities of less than 5,000 people are lucky to have more than 2. Aerial equipment is common among cities with populations greater than 100,000, fairly common among cities with more than 25,000 people, and relatively rare among cities of less than 10,000 people (Table 12.10).

TABLE 12.9 Pumpers (500 gpm or greater) by Community Size, 1989-1991

Population Protected	*Percentage of U.S. Fire Departments With*				
	1-5 Pumpers (%)	*6-9 Pumpers (%)*	*10-19 Pumpers (%)*	*20-39 Pumpers (%)*	*40 or More Pumpers (%)*
1,000,000 or more	0.0	0.0	0.0	0.0	100.0
500,000 to 999,999	0.0	0.0	0.0	59.1	40.9
250,000 to 499,999	0.0	2.2	39.1	54.4	4.3
100,000 to 249,999	12.8	35.3	44.3	6.8	0.8
50,000 to 99,999	56.1	34.5	9.4	0.0	0.0

Population Protected	*Percentage of U.S. Fire Departments With*				
	No Pumpers (%)	*1 Pumper (%)*	*2 Pumpers (%)*	*3-4 Pumpers (%)*	*5 or More Pumpers (%)*
25,000 to 49,999	0.4	1.6	15.4	53.9	28.7
10,000 to 24,999	0.3	5.0	28.4	54.5	11.8
5,000 to 9,999	1.1	10.3	41.9	41.4	5.3
2,500 to 4,999	2.1	19.9	48.6	28.5	0.9
Under 2,500	7.2	39.1	42.4	10.7	0.6

SOURCE: 1989-1991 NFPA Fire Service Survey, as reported in Michael J. Karter, Jr., *U.S. Fire Department Profile Through 1991* (Quincy, MA: National Fire Protection Association, October 1992), p. 19. Reprinted with permission from National Fire Protection Association, Quincy, MA 02269.
NOTE: Pumpers reported above had a capability of 500 gpm or greater. Many departments reported other fire suppression vehicles including apparatus with pumps less than 500 gpm, hose wagons, brushfire vehicles, tankers, etc.

TABLE 12.10 Aerial Apparatus by Community Size, 1989-1991

Population Protected	*Percentage of U.S. Fire Departments With*				
	No Aerial Apparatus (%)	*1-5 Aerial Apparatus (%)*	*6-9 Aerial Apparatus (%)*	*10-19 Aerial Apparatus (%)*	*20 or More Aerial Apparatus (%)*
1,000,000 or more	0.0	0.0	0.0	33.3	66.7
500,000 to 999,999	0.0	4.5	4.5	68.3	22.7
250,000 to 499,999	4.3	47.9	21.7	23.9	2.2

Population Protected	*Percentage of U.S. Fire Departments With*				
	No Aerial Apparatus (%)	*1 Aerial Apparatus (%)*	*2 Aerial Apparatus (%)*	*3-4 Aerial Apparatus (%)*	*5 or more Aerial Apparatus (%)*
100,000 to 249,999	8.3	18.8	24.8	30.1	18.0
50,000 to 99,999	14.2	42.9	29.3	13.1	0.5
25,000 to 49,999	20.4	63.4	14.9	1.2	0.1
10,000 to 24,999	41.2	54.8	3.6	0.4	0.0
5,000 to 9,999	73.9	25.3	0.8	0.0	0.0
2,500 to 4,999	88.7	10.7	0.6	0.0	0.0
Under 2,500	96.5	3.2	0.3	0.0	0.0

SOURCE: 1989-1991 NFPA Fire Service Survey, as reported in Michael J. Karter, Jr., *U.S. Fire Department Profile Through 1991* (Quincy, MA: National Fire Protection Association, October 1992), p. 20. Reprinted with permission from National Fire Protection Association, Quincy, MA 02269.

Fire Service Performance Measures

Statistics on staffing levels, firefighter deaths and injuries, and numbers of stations and heavy fire-fighting equipment can be useful to city officials as they grapple with budgetary and compensation issues, but those statistics are not benchmarks of performance. National or regional norms, perhaps categorized by community size or length of workweek, may inform officials whether local practices and characteristics lie within or outside normal bounds, but they do not address performance.

Benchmarks of performance provide external pegs against which local service measures may be compared. Ideally, the performance measures of a fire department include more than simple workload indicators such as the number of alarms, number of inspections, and hours of training. Although valuable, workload indicators only report what was done. They fail to reveal how efficiently the work was performed, its quality, or its effectiveness in achieving desired results. A set of measures offered by Harry Hatry and associates (1992) at the Urban Institute and provided in Table 12.11 goes well beyond workload to identify fire service effectiveness. Performance elements addressed in that set include aspects of prevention as well as various facets of suppression, including prompt response and fire containment. Other measures are included as fire service indicators in the Governmental Accounting Standards Board's (GASB) recommended indicators of service efforts and accomplishments (SEA; Table 12.12).

Intercity Comparison

Improvements in a municipal fire department's performance may be judged by comparing performance indicators from one period to another. Although such a review could reveal improvements, it would not address the adequacy of performance in a broader sense. Only by comparing local performance with some external benchmark can the adequacy of local operations truly be assessed. As noted by Parry, Sharp, Vreeland, and Wallace (1991) in GASB's volume on SEA indicators for fire department programs, the consideration of intercity norms can be useful.

> It would seem that recommended disclosures should focus on some intercity performance norms. A tie-in to average response time and average fire loss per capita experienced by peer cities, certification programs, and insurance grading systems provide benchmarks that place performance in context. (p. 24; reprinted by permission)

Workload measures from other municipal fire departments cannot perform that benchmarking function. Knowledge that the fire department in Boise, Idaho, responded to 7,611 emergency calls in 1991 is not especially useful to the fire chief or other municipal officials in Muleshoe, Texas, or Oliver Springs, Tennessee—or

(text continued on p. 109)

TABLE 12.11 Effectiveness Measures for Fire Services

Overall objective: To minimize losses to persons and property by helping to prevent fires from occurring and to reduce losses from fires that occur.

Objective	Quality Characteristic	Specific Measure	Data Collection Procedure
Overall loss minimization	Civilian casualties	• Number and rate of civilian injuries and deaths per 100,000 population	Incident reports, census
	Firefighter casualties	• Number of firefighter injuries and deaths per 100 firefighters and per 1,000 fires	Casualty reports
	Property loss	• Direct dollar loss from fires per $1,000 property served and per 100,000 population	Fire department or insurance company estimates
Prevention	Reported fire incidence	• Number of reported fires per 1,000 population, total and by type of residential occupancy[a]	Incident reports
	Reported building fire incidence	• Number of building fires per 1,000 occupancies, by type of occupancy and fire size	Incident reports, planning department records
	Reported plus unreported building fire incidence rate	• Number of unreported plus reported building fires per 1,000 households (or businesses)	Household and/or business survey
	Preventability of fires	• Percentage and rate of fires that are relatively preventable by inspection or education[b]	Incident report (special addition)
	Inspection effectiveness	• Rate of fires in inspected versus uninspected (or frequently inspected versus infrequently inspected) occupancies of a given type	Fire inspection files linked to incident reports
	Deterrence effectiveness for arson	• Number of incendiary and suspicious fires per 1,000 population	Incident reports, police and court records (for arson motive)
	Apprehension effectiveness for fire-related crimes	• Clearance and conviction rates for incendiary and suspicious fires	Fire incident and inspection data linked to police arrest records and fire department case records
	Juvenile fire setter recidivism	• Percentage of juveniles sent to treatment programs who set fires again within two years (recidivism rate)	Juvenile treatment program records

(continued)

TABLE 12.11 *Continued*

Objective	Quality Characteristic	Specific Measure	Data Collection Procedure
	Inspection outreach	• Percentage of occupancies inspected within x months	Inspection records
	Inspection violation "clearances"	• Percentage of building violations "cleared" by removal of hazard or imposition of penalty within x days	Inspection records
	Detector and sprinkler usage/maintenance	• Percentage of homes or businesses with (a) sprinklers, (b) working smoke detectors	Household survey, inspection records
	Public education outreach	• Percentage of citizens reached by public fire education	Household survey, public education contacts
Suppression	Firefighting effectiveness —dollar loss	• Average dollar loss per fire not out on arrival, by size on arrival and type of occupancy	Incident reports (including estimates of size on arrival)
	Firefighting effectiveness—spread	• Number or percentage of fires (not out on arrival of first unit) in which spread after arrival is limited to x square feet or z steps of the extent-of-flame scale	Incident report (including estimate of size on arrival)
	Firefighting effectiveness—time	• Time to control or extinguish the fire	Incident reports or dispatch center
	Speed of providing service	• Percentage of response times that are less (or more) than x minutes	Incident reports or dispatch center
		• Average response time	Incident reports or dispatch center
	Rescue effectiveness	• Number of "saves" versus number of casualties	New data element for incident report
	False alarms	• Number of false alarms	Incident reports
Overall	Citizen satisfaction	• Percentage of population rating fire protection services as satisfactory	Household or "client" survey

SOURCE: Harry P. Hatry, Louis H. Blair, Donald M. Fisk, John M. Greiner, John R. Hall, Jr., and Philip S. Schaenman, *How Effective Are Your Community Services? Procedures for Measuring Their Quality,* 2nd edition (Washington, DC: Urban Institute and International City/County Management Association, 1992), pp. 94-95. Reprinted by permission.
a. The term *occupancy* refers to a piece of property in terms of its use, for example, detached houses containing only one household, apartments, drug stores, or warehouses. There may be more than one type of occupancy in one building, and some occupancies—for example, garbage dumps and piers—are not buildings at all.
b. For example, fires started by flammables stored near ignition source, or in buildings with fire code violations relevant to the fire start, would be "relatively preventable." Fires started by hidden equipment defects not common to that equipment would be relatively "unpreventable."

TABLE 12.12 Service Efforts and Accomplishments Indicators Recommended for Fire Services

Overall Performance	Fire Suppression	Fire Prevention
Input Indicators		
Total Operating Expenditures	Personnel	Personnel
Total Capital Expenditures	Full-Time Personnel	Full-Time Personnel
Personnel	Part-Time and	Part-Time and
Full-Time Personnel	Volunteer Personnel	Volunteer Personnel
Part-Time and Volunteer	Total Man-Hours Worked	Total Hours Worked
Personnel	Total Operating Expenditures	Total Operating Expenditures
Total Hours Worked	Total Capital Expenditures	Total Capital Expenditures
	Hours in Training Programs	
	Percentage of Firefighters	
	Reaching an NFPA-Recommended	Certification
	Level	
Output Indicators		
Population Served	Number of Fire Calls Answered	Number of Inspections
Residential		Number of Education
Workforce		Programs Offered
Tourist, Average Daily		Number of Fire Investigations
Property Value Protected		Performed
Residential		
Commercial		
Public Property		

(continued)

to officials in most other communities across the nation. On the other hand, indicators that are standardized or measure service quality—for instance, indicators reporting that Boise's average response time was 3.5 minutes or that the fire department in Wichita, Kansas, was able to confine 91% of all fires to the room of origin—may usefully serve as benchmarks.

Fire Insurance Ratings

In almost all of the nation's 50 states, the Insurance Services Office (ISO) rates the fire readiness of individual communities as an informational service to potential insurers. Alternate systems performing the same function—for example, Texas's "key rate" system—may be found in the few states that are exceptions.

ISO rates are based on the adequacy of a community's water supply, its fire department, and its fire alarm system. Public protection classifications range from the most desirable rate of 1 to the least desirable of 10. Water supply considerations focus on hydrant availability, hydrant condition, and the availability of adequate supplies of water at sufficient pressure. Fire department considerations focus on staffing levels and geographic distribution of stations, training, and equipment. Fire alarm considerations, weighted at 10% compared with 40% for water supply

TABLE 12.12 *Continued*

Overall Performance	Fire Suppression	Fire Prevention
Outcome Indicators		
Percentage of Citizens Rating Performance Satisfactory	Water Supply	Number of Fires (Reported and Unreported)
ISO Fire Insurance Rating	Minimum Water Volume	Percentage of Fires Preventable
Total Dollars in Fire Losses	Available	by Inspection or Education
Total Fire-Related Deaths	Minimum Water Flow	Number of Fires of
Total Fire-Related Injuries	Available	Suspicious Origin
	Population With Access to	Fires in Inspected/
	Adequate Water Supply	Uninspected Buildings
	Response Time	Industrial
	Average Response Time	Other
	Percentage of Responses in	Citizens Participating in or
	under *x* Minutes	Aware of Education Programs
	Average Time to Control Fires	
	Single-Alarm, Residential	
	Single-Alarm, Industry	
	Two-Alarm, Industry	
	Percentage of Fires Spread Limited	
	to *x*% Square Feet on Arrival	
	Single-Alarm, Residential	
	Single-Alarm, Industry	
	Two-Alarm, Industry	
Efficiency Indicators		
	Operating Expenditures per Capita	Expenditures per Capita
	Capital Expenditures per Capita	Expenditures per $100,000
	Operating Expenditures per $100,000	of Property Protected
	of Property Protected	
	Capital Expenditures per $100,000	
	of Property Protected	

SOURCE: Robert W. Parry, Jr., Florence Cowan Sharp, Jannet Vreeland, and Wanda A. Wallace, *Fire Department Programs: Service Efforts and Accomplishments Reporting—Its Time Has Come* (Norwalk, CT: Governmental Accounting Standards Board, 1991), pp. 36-43. Reprinted by permission.

and 50% for fire department, focus primarily on dispatch operations and the availability of sufficient staff and equipment (Granito, 1991). Selected standards for ISO ratings may be found in Table 12.13.

A fire insurance rate has the advantage of conveying a great deal of information by means of a single indicator. Often, it is useful to compare staffing levels, costs, and deployment patterns with other communities possessing the same rating. If officials attempt to judge different fire departments on the strengths on their fire ratings, however, it is important to note two important facts. First, 40% of a community's rating is based on the adequacy of its water supply, something typically beyond the control of the fire department. Second, there appears to be a fairly strong correlation between favorable fire insurance rates and community size. Because large cities are more likely to possess favorable fire insurance ratings than are small towns, it may be most useful to select as benchmark partners other communities of similar population that have equal or better fire insurance rates.

Several municipalities supplying documents used in the development of this book reported their fire insurance rate as an indicator of performance in fire services. Some, including a few relatively small communities that defy the general trend, reported favorable public protection classifications. For example, the most favorable rating of 1 was reported by Coral Gables, Florida. A rating of 2 was reported by Greenville, South Carolina; Indianapolis, Indiana; Tallahassee, Florida; and Waukesha, Wisconsin. Class 3 ratings were reported by Boise, Idaho; Fairfield, California; Nashville-Davidson County, Tennessee (inside its urban service district); Oak Ridge, Tennessee; Sioux Falls, South Dakota; Smyrna, Georgia; and Vancouver, Washington.

Fire Incidence Rate

A fire department with a narrow view of its mission may expect to suppress fires within a defined geographic area—and to do nothing more. A broader view of fire department mission includes responsibility for minimizing loss of lives and property and incorporates a fire *prevention* role. Departments that subscribe to this broader view typically undertake more ambitious inspection programs in an effort to influence both the rate and severity of fire incidents. The effectiveness of such a strategy has been confirmed by NFPA studies indicating that average fire incidence rates in communities in which almost all public buildings are inspected annually are approximately one half the rate of communities with less ambitious inspection programs (Hall, Koss, Schainblatt, Karter, & McNerney, 1979).

An aggressive fire department with a broad view of its mission, then, hopes not only to suppress promptly those fires that do occur but also to prevent fires. Fire incidence rates record the success—or more accurately, the lack of success—in achieving that latter objective. City officials may wish to track their progress in reducing fire incidents from year to year; they may furthermore wish to compare their own community's figures with the fire incidence rates of other cities. Although both comparisons may provide useful information, special care may be required to interpret intercity comparisons properly. Gloating over possession of more favorable fire incidence rates than cities with substantially older housing stock and higher numbers of abandoned or poorly maintained manufacturing facilities not only may be unseemly but also may reflect an inaccurate assessment of relative fire department performance. With that note of caution in mind, a review of fire incidence rates in other communities may nevertheless provide local officials a context in which to judge their own numbers—giving them a sense of where they stand.

Among the documents examined in the preparation of this chapter, fire incidence rates were reported in three ways: (a) fire incidents per 1,000 structures, (b) structural fires per 1,000 population, and (c) fire incidents per 1,000 population (Table 12.14). Cities reporting those statistics managed to hold the rate of residen-

TABLE 12.13 Selected Standards for Fire Insurance Ratings

Basic Equipment for Class 9

- At least one apparatus with pump capacity of 50 gpm or more at 150 psi and at least a 300-gallon water tank

Basic Equipment for Class 8 or Better

- Apparatus with pump capacity of 250 gpm or more at 150 psi
- Water system capable of delivering 250 gpm for 2 hours, plus consumption at the maximum daily rate at a fire location

Fire Flow at Residential Structures

- For 1- and 2-family dwellings not exceeding 2 stories in height, the following Needed Fire Flows shall be used:

Distance Between Buildings	Needed Fire Flow
Over 100′	500 gpm
31-100′	750
11-30′	1000
10′ or less	1500

- Other habitational buildings, up to 3500 gpm maximum

Reserved Telephone Lines at Communication Center

Population Served	Number of Reserved Lines	
	Fire	Business
Up to 40,000	1	1
40,001-125,000	2	2
125,001-300,000	3	3
Over 300,000	4	3

When emergency calls for other than fire are received over the fire number, double the number of needed reserved fire lines indicated above. Automatic telephone dialing equipment used to report alarms from private fire detection systems should have an emergency line separate from the normal fire and business numbers.

(continued)

tial fires to less than 10 per 1,000 occupancies. They managed to hold the rate of structural fires to less than 4 per 1,000 population. And fire incidents of all types occurred at a rate of less than 10 per 1,000 population. As a further reference point, the LCC (1994) considers fire incidence rates below 9.8 per 1,000 population to be characteristic of high-service-level cities, rates above 11.8 to reflect low-service levels, and rates in between to indicate medium fire protection service (p. 24).

False Alarms

Not every alarm summons firefighters to an actual fire incident. Nationally, about $1/10$ of all emergency calls received by fire departments, including calls for

TABLE 12.13 *Continued*

Number of Engine Companies

Needed Fire Flow (gpm)	Number of Engine Companies Needed	Number of Ladder Companies Needed
500-1,000	1	—
1,250-2,500	2	—
3,000-3,500	3	—
4,000-4,500	4	1
5,000-5,500	5	2
6,000-6,500	6	2
7,000-7,500	7	3
8,000-8,500	8	3
9,000-9,500	9	4
10,000	10	4
11,000	11	4
12,000	12	5

Hydrant Distribution

- Full credit awarded if hydrant is no more than 300 feet away; partial credit for 301 to 600 feet away; less credit for 601 to 1,000 feet away
- Built-on areas not within 1,000 feet of a recognized water system may still be eligible for class 9

SOURCE: Includes copyrighted material of Insurance Services Office, Inc., with its permission. Excerpted from *Fire Suppression Rating Schedule,* Edition 6-80. Copyright © Insurance Services Office, Inc., 1980.

medical aid, are false alarms (Hall & Cote, 1991, p. 1.22). Some cities are fortunate to have false alarm rates well below the national average (Table 12.15), but even those communities devote considerable resources to needless fire runs and impose unnecessary risks on firefighters and others sharing city streets. Every city that experiences *any* false alarms has a false alarm problem; comparison with other communities and the national average permits them to know how big that problem is.

Response Time

Quick response is instrumental to fire rescue and effective fire suppression. Although prompt arrival is no more important than competent performance at the scene, studies have shown a positive correlation between quicker response times and lower loss of life and property (Gordon, Drozda, & Stacey, 1969; Rider, 1979). Quick response counts. In fire emergencies, it counts a great deal.

TABLE 12.14 Fire Incidence Rates: Selected Cities

FIRE INCIDENTS PER 1,000 STRUCTURES

Gresham, OR
Target: Not more than 6.5 residential fire incidents per 1,000 occupancies (0.65%); not more than 5.5 commercial fire incidents per 1,000 occupancies (0.55%)

Savannah, GA
9.0 dwelling fires per 1,000 structures (0.90%) (1993)

Duncanville, TX
9.6 residential fire incidents per 1,000 structures (0.96%); 29.3 nonresidential fire incidents per 1,000 structures (2.93%) (1990)

STRUCTURAL FIRES PER 1,000 POPULATION

Sunnyvale, CA
Target: 1.43 (1994)

Sacramento, CA
1.6 (1992)[a]

Seattle, WA
1.8 (1991)[a]

Greenville, SC
1.9 (1991)

Charlotte, NC
2.4 (1992)[a]

Portland, OR
3.3 (1989); 3.0 (1990); 2.9 (1991); 2.5 (1992)

Denver, CO
2.6 (1991)[a]

Kansas City, MO
2.6 (1992)[a]

Cincinnati, OH
3.7 (1991)[a]

FIRE INCIDENTS PER 1,000 POPULATION

Iowa City, IA
Fires per 1,000 population: 6.8 (1990); 4.1 (1991)

Fayetteville, AR
4.7 working fires per 1,000 population (1991)

Chandler, AZ
5.74 fire incidents per 1,000 population (1991)

Portland, OR
Fire incidents per 1,000 population: 7.8 (1989); 7.0 (1990); 6.4 (1991); 6.9 (1992)

Cincinnati, OH
Fire incidents per 1,000 population: 9.15 (1987); 10.05 (1988); 8.5 (1989); 7.2 (1990); 7.8 (1991)

Corpus Christi, TX
8.1 fires per 1,000 population (1991)

Peoria, AZ
8.14 fires per 1,000 population (1991)

Greenville, SC
8.3 fires per 1,000 population (1990)

a. As reported in City of Portland (Oregon), *Service Efforts and Accomplishments: 1991-92* (Office of the City Auditor, January 1993), pp. B-2, B-3.

Response time consists of two major components: (a) the time between a citizen's call for help and the dispatch of an emergency unit and (b) the time between dispatch and arrival.[1] Excessive speed by heavy fire equipment is a poor way to reduce response time because it imposes undue risk on the lives of firefighters, motorists, and pedestrians. Far better avenues for reducing response time lie in minimizing the time consumed in the first component, in reducing the time required to get out of the station, and also in selecting station locations to minimize travel distances.

Although the LCC (1994, p. 23) prescribes a 5-minute response for high-service-level departments and a regional standard in Pennsylvania calls for first alarm responses within 8 minutes (SPRPC, 1990, p. VI-5), an average response time of 4 minutes to fire emergencies was the median among documents examined for this chapter (Table 12.16). Although a few departments reported average response

TABLE 12.15 False Alarms

National Statistics

NFPA survey[a]

10.6% of all fire department emergency calls (including medical aid) in 1980-1988 were false alarms

Local Statistics

False Alarms as a Percentage of All Alarms Received	Ratio of Actual Fires to False Alarms	False Alarms per 1,000 Population
Tallahassee, FL	**Savannah, GA**	**Loveland, CO**
4.0% (1991)	1.7:1 (1988);	2.85 (1989);
Southfield, MI	1.6:1 (1989);	3.08 (1990);
6% (1991)	1.6:1 (1990);	0.77 (1991)
Longview, TX	1.5:1 (1991);	**Tallahassee, FL**
11.0% (1991)	1.6:1 (1992, est.)	2.2 (1991)
	Southfield, MI	
	1.2:1 (1991)	

a. National Fire Protection Association survey cited in John R. Hall, Jr., and Arthur E. Cote, "America's Fire Problem and Fire Protection," p. 1.22. In Arthur E. Cote and Jim L. Linville (Eds.), *Fire Protection Handbook,* 17th edition (Quincy, MA: NFPA, 1991).

times of less than 3 minutes, averages between 3 minutes and 6 minutes were much more common among this group of cities.

Many fire departments respond not only to fire emergencies but to other calls as well. Some maintain separate time statistics for other categories of emergency, often reporting somewhat slower response time averages. For example, Gresham, Oregon, reported an average response time of 20 minutes by its hazardous materials unit. Similarly, the fire department of Houston, Texas, attempts to respond to at least 95% of all hazardous material incidents within 20 minutes.

Controlling the Fire

The prompt arrival of a highly trained and well-equipped fire company is the desire of every fire chief—and the hope of every emergency caller. Proper equipment and staffing levels remain a topic of debate, with fire service organizations and fire unions calling for numbers that exceed the equipment and staffing levels found in many—and perhaps even most—cities. One such prescription, appearing in an NFPA publication (Granito, 1991), is provided in Table 12.17.

Firefighters and equipment are important suppression resources, but in and of themselves, tallies showing the number of firefighters in a community and the number of fire trucks offer a poor gauge of fire service performance. Suppression effectiveness is a matter of how those firefighters and their equipment perform in

TABLE 12.16 Average Response Times to Fire Emergencies

City	Response Time	City	Response Time
Winston-Salem, NC	Target: 4 minutes or less Actual: 2.24-minute average; percentage of responses within 4 minutes: 92% (1992)	Phoenix, AZ	Target: Respond to at least 25% of all fire and EMS calls within 3 minutes Actual: 16.0% (1988); 15.25% (1989); 15.5% (1990)
Wilmington, DE	Target: Average of 2 minutes or less Actual: 2-minute 21-second average (1992)		Target: 4 minutes or less Actual: 3-4 minute average
Indianapolis, IN	2.4-minute average (1991)		Average response time for fire and emergency medical calls: 3.97-minute average (1989);
Greenville, SC	2.66-minute average (1990) 2.50-minute average (1991)		3.82-minute average (1990); 3.8-minute average (1993)
Raleigh, NC	Target: 4 minutes or less Actual: 2.98-minute average (1991)	Richmond, VA	3-minute 52-second average (Jan.-June 1991)
Boston, MA	94.7% within 3 minutes; 3.2% between 3 and 4 minutes; 1.1%, 4 minutes or longer (1992)	Wichita, KS	3-minute 52-second average (Jan.-June 1992)
Duncanville, TX	3-minute average (1990)		3.9-minute average (1990) 3.9-minute average (1991)
Ocala, FL	3-minute average (1992)	Largo, FL	4-minute average (1991)
Fayetteville, AR	3.1-minute average (1991) 3.2-minute average (1992)	Lubbock, TX	4.0-minute average; 28.2% taking 5 to 8 minutes; 4.9% taking more than 8 minutes (1991)
Iowa City, IA	3.33-minute average (1990)	Nashville-Davidson County, TN	4-minute average inside urban service district
Dayton, OH	3.44-minute average		12-minute average outside urban service district (1991)
Boise, ID	3.5-minute average (1991)		
Vancouver, WA	Target: 4 minutes or less Actual: 3-4 minute average	Smyrna, GA	4-minute average (1992)
		Tallahassee, FL	4-minute average (1991)
Oak Ridge, TN	4-minute average (1991) 3.6-minute average (1992)	Portland, OR	74.1% in 4 minutes or less (1991)
		Reno, NV	55% within 4 minutes or less (1990)
Gresham, OR	Target: 3.65-minute average	Amarillo, TX	4-minute average from dispatch to arrival (1991)
Hurst, TX	Target: Less than 4 minutes Actual: 3.8-minute average (1990)	Peoria, AZ	4-minute 1-second average (1991)

City	Response Time
Santa Ana, CA	5.1-minute average response time; 68.1% within 5 minutes; 83.8% within 6 minutes; 91.3% within 7 minutes (1987)
Huntington Beach, CA	75% within 5 minutes; 98% within 10 minutes (1988)
Corvallis, OR	Target: Within 5 minutes from dispatch (within city)
Orlando, FL	80% within 5 minutes from dispatch (1991)
Chesapeake, VA	5.6-minute average (residential); 5.2-minute average (commercial) (1991)
Sunnyvale, CA	Targets: (a) 5.6-minute average from receipt of call; (b) 4.6-minute average from dispatch; (c) 90% within 6 minutes of call; (d) 90% within 5 minutes of dispatch (1994)
San Jose, CA	Targets: (a) 90% of dispatched units en route within 2 minutes; (b) travel time of 4 minutes or less for at least 70% of first-arriving units; (c) travel time of 6 minutes or less for at least 90% of first-arriving units. Actual: (a) 90%; (b) 75%; (c) 92% (1989)
Grand Prairie, TX	6.6-minute average (YTD 12/31/92)
Houston, TX	Target: Respond to at least 95% of fire incidents within 10 minutes (1993)
Shreveport, LA	4.05-minute average (1991)
Charlotte, NC	4-minute 4-second average; 99% within 9 minutes (5 months, 1994)
Oklahoma City, OK	4.27-minute average (1991); 4.12-minute average (1992)
Irving, TX	4.22-minute average (1991)
Sacramento, CA	4.34-minute average (1991)
Alexandria, VA	4.01-minute average (1991); 3.87-minute average (1992); 4.43-minute average (1993)
Tucson, AZ	Target: 4.6-minute average
Blacksburg, VA	4.63 minutes, in town; 7.49 minutes, outlying county; 4.58 minutes, Virginia Tech campus (1994)
Fort Worth, TX	4.65-minute average (1991)
Burbank, CA	4.71-minute average (1989)
New York City	4-minute 51-second average (Manhattan); 4-minute 29-second average (Bronx); 5-minute 16-second average (Staten Island); 4-minute 14-second average (Brooklyn); 5-minute 3-second average (Queens) (1992)
College Station, TX	Target: 5 minutes or less
Milwaukee, WI	94% within 5 minutes (1994)
Corpus Christi, TX	5-minute 1-second average; 92% within 5 minutes (1991)

TABLE 12.17 Recommended Equipment and Fire Force for Interior Attack

High-Hazard Occupancies (Schools, hospitals, nursing homes, explosive plants, refineries, high-rise buildings, and other high life hazard or large fire potential occupancies):

At least 4 pumpers, 2 ladder trucks (or combination apparatus with equivalent capabilities), 2 chief officers, and other specialized apparatus as may be needed to cope with the combustible involved; not less than 24 firefighters and 2 chief officers.

Medium-Hazard Occupancies (Apartments, offices, mercantile and industrial occupancies not normally requiring extensive rescue or fire-fighting forces):

At least 3 pumpers, 1 ladder truck (or combination apparatus with equivalent capabilities), 1 chief officer, and other specialized apparatus as may be needed or available; not less than 16 firefighters and 1 chief officer.

Low-Hazard Occupancies (One-, two-, or three-family dwellings and scattered small businesses and industrial occupancies):

At least 2 pumpers, 1 ladder truck (or combination apparatus with equivalent capabilities), 1 chief officer, and other specialized apparatus as may be needed or available; not less than 12 firefighters and 1 chief officer.

Rural Operations (Scattered dwellings, small businesses, and farm buildings):

At least 1 pumper with a large water tank (500 gal [1.9 m^3] or more), one mobile water supply apparatus (1000 gal [3.78 m^3] or larger), and such other specialized apparatus as may be necessary to perform effective initial fire-fighting operations; at least 12 firefighters and 1 chief officer.

Additional Alarms:

At least the equivalent of that required for Rural Operations for second alarms; equipment as may be needed according to the type of emergency and capabilities of the fire department. This may involve the immediate use of mutual aid companies until local forces can be supplemented with additional off-duty personnel. In some communities, single units are "special called" when needed, without always resorting to a multiple alarm. Additional units also may be needed to fill at least some empty fire stations.

SOURCE: John Granito, "Evaluation and Planning of Public Fire Protection." In Arthur E. Cote and Jim L. Linville (Eds.), *Fire Protection Handbook,* 17th edition (Quincy, MA: National Fire Protection Association, 1991), p. 10.41. Reprinted with permission from *Fire Protection Handbook,* 17th edition, Copyright © 1991, National Fire Protection Association, Quincy, MA 02269.

the heat of an emergency. Some cities have begun to use better and more direct gauges, measuring and reporting suppression effectiveness through indicators of fire spread and control time (Table 12.18). National statistics suggest that almost three fourths of all structural fires are contained through fire suppression efforts to the room of the fire's origin (Hall & Cote, 1991, p. 1.15). Most of the examined documents that report fire spread indicated even higher percentages contained at that level. Statistics on control time exhibited considerable variation, with most reported averages falling in the 15- to 30-minute range.

Other Duties

Fire departments perform a variety of duties in support of their primary mission. For some of those duties, relevant standards have been developed, or general norms are evident. A few pertaining to the functions of record keeping and the inspection and servicing of equipment will be mentioned here.

TABLE 12.18 Under Control: Performance Targets and Actual Experience of Selected Cities

Controlling Fire Spread

Corvallis, OR	Target: Control structural fires and fires involving real property within the level of involvement found on arrival Actual: 95% (1991); 100% (1992)
Oak Ridge, TN	Percentage of building fires confined to room of origin: 89% (1991); 98% (1992)
Wichita, KS	Among fires not out on arrival, percentage confined to room of origin: 92% (1990); 91% (1991)
Winston-Salem, NC	87% of fires contained within room(s) involved on arrival (1992)
Lubbock, TX	83.8% of structural fires confined to room of origin (1991)
Wilmington, DE	Target: At least 92% of fires confined to room of origin Actual: 74% (1992)
Greenville, SC	98% of fires confined to one increment of flame spread (1990)
Vancouver, WA	Targets: Extinguish 95% of all residential fires and 90% of all commercial fires before total destruction occurs
Fayetteville, AR	Percentage of working fires stopped short of total loss: 89.8% (1991); 90.2% (1992)
National Statistics	73% of structural fires confined to room of origin; 4.4% confined to floor but not room of origin; 22.6% spread beyond floor of origin (1983-1987)[a]

Control Time

Oak Ridge, TN	Control time following arrival: 2.9 minute average (1991); 2.4 minute average (1992)
Alexandria, VA	Average time spent at scene of incident: 15.9 minutes (1989); 16.6 minutes (1990); 16.7 minutes (1991); 16.7 minutes (1992); 16.5 minutes (1993)
San Jose, CA	Target: Control at least 75% of emergency fire incidents within 15 minutes of arrival; at least 95% within 1 hour Actual: 79% controlled within 15 minutes; 97% within 1 hour (1989)
Indianapolis, IN	Average time to control a residential fire: 20 minutes
Fayetteville, AR	Percentage of working fires controlled in 45 minutes or less: 91.8% (1991) Structural fire control time average: Less than 30 minutes (1992)
Shreveport, LA	Average time to control (all incidents): 33.31 minutes (estimate, 1993)

a. John R. Hall, Jr., and Arthur E. Cote, "America's Fire Problem and Fire Protection," p. 1.15. In Arthur E. Cote and Jim L. Linville (Eds.), *Fire Protection Handbook* (Quincy, MA: NFPA, 1991), citing NFIRS and NFPA survey.

Perhaps every fire department keeps some type of records of fire incidents. In many cases, however, those records fall short of even the bare essentials needed for minimal analysis. Basic fire incidence data should include eight fundamental elements:

1. time of dispatch,
2. address,

3. identifications and times of responding units,
4. arrival times at the scene,
5. departure times from the scene,
6. radio traffic tape logs,
7. telephone tape logs, and
8. times of significant events at the scene. (Stauffer, 1991, p. 4.10)

Key focal points of maintenance efforts in most fire departments are fire vehicles, pumps, and hydrants. Although principal maintenance of fire vehicles often is performed by mechanics outside the fire department, fire officials can keep tabs on the increasing age of fire vehicles with an eye toward eventual replacement. If properly maintained, most fire trucks can sustain a lengthy life of service. Generally, however, fire officials prefer their inventory to include some fairly new units of utmost reliability. Although the fire department in Portland, Oregon, projects the useful life of an engine to be 20 years, followed by 5 years in a reserve capacity, the average age should be less to avoid high maintenance costs and problems of low reliability. The average age of Portland's fire engines climbed to more than 10 years at various times in the early and mid-1980s, before declining to a more satisfactory average age of about 6 years in 1991. In New York City, the average age of pumpers was 90 months (7.5 years), and the average age of ladder trucks was 77 months (6.4 years) in 1992. The city of Boston attempts to hold its average age of fire equipment below 7 years.

Periodic testing of pumps to ensure their readiness is an important precaution. Such testing should include all the prescribed steps in Table 12.19. The standards noted within those steps may be considered benchmarks.

The middle of an emergency is no time to discover a faulty fire hydrant. Periodic inspections at reasonable intervals can minimize that occurrence, and such inspection need not consume a great amount of time. Fort Collins, Colorado, and Orlando, Florida, attempt to test and service hydrants at least once every 4 years; Oak Ridge, Tennessee, every 3 years. Raleigh, North Carolina, reports that an average hydrant inspection requires 0.14 hours (8.4 minutes). Properly marked hydrants reveal their water flow by the color of their bonnet and nozzle cap (Table 12.20).

Odds and Ends

Little known facts, figures, and miscellaneous tidbits of information sometimes come in handy. Although most of the performance dimensions mentioned in Table 12.21 were reported by too few cities to depict any trends or norms, they may nevertheless prove helpful to municipal officials.

One item that *can* be reported as a norm is the national average of residences with smoke detectors, which stood at 81% in 1988 (Hall & Cote, 1991, p. 1.16). Communities such as Oak Ridge, Tennessee, which have made the installation of

TABLE 12.19 Service Tests of Fire Department Pumps

NFPA 1911, *Standard for Service Tests of Pumps on Fire Department Apparatus,* outlines the procedures for the service test. The test should cover the following items:

1. A test pumping from draft for 20 minutes at 100% of rated capacity, at 150 psi (1039 kPa) net pump pressure; 10 minutes at 70% of rated capacity at 200 psi (1379 kPa) net pump pressure; and 10 minutes at 50% of rated capacity at 250 psi (1724 kPa) net pump pressure.

2. An engine speed test to determine if the engine is capable of reaching its no-load governed speed.

3. A vacuum test to ensure that the pump and the attached piping are still tight.

4. A pressure-control test to ensure that the pressure-control device can control the discharge within the specified limits.

5. A check of the accuracy of the gauges and flow meters.

The purpose of the test is to ensure that the pump is generally in good condition, that the pump casing and various fittings are tight, and that the transfer valve is operating properly, if the pump is a series parallel type. If rated capacity cannot be obtained at 150 psi (1034 kPa) and the pump is not cavitating, or it appears that the pump, engine, pump accessories, or other parts of the power train and pumping equipment are not in good condition, the apparatus manufacturer, an authorized representative, or a competent mechanic should be contacted for advice so the condition can be corrected.

SOURCE: Excerpted from A. K. Rosenhan, "Fire Department Apparatus, Equipment, and Protective Clothing." In Arthur E. Cote and Jim L. Linville (Eds.), *Fire Protection Handbook,* 17th edition (Quincy, MA: National Fire Protection Association, 1991), p. 9.93. Reprinted with permission from NFPA 1911, *Service Tests of Pumps on Fire Department Apparatus,* Copyright © 1991, National Fire Protection Association, Quincy, MA 02269. This reprinted material is not the complete and official position of the National Fire Protection Association on the referenced subject, which is represented only by the standard in its entirety.
NOTE: psi = pounds per square inch; kPa = kilopascals.

residential smoke detectors a major thrust of their fire safety programs, may find the national average to be a useful benchmark. By 1992, 94% of all Oak Ridge residences were equipped with smoke detectors. Individual cities' performance targets and experience in response to extrication situations and fire safety complaints are similarly instructive.

Harried municipal executives, besieged by local fire personnel claiming to be overworked, may be especially interested in a statistic that may set the standard for workload. Incredible but true, engine companies in New York City responded to an average of 2,814 calls in 1992. Now, that's busy![2]

TABLE 12.20 Uniform Marking of Fire Hydrants

Hydrant Classification	*Hydrant Flow*	*Color of Bonnets and Nozzle Caps*
AA	1500 gpm or greater	Light blue
A	1000 to 1499 gpm	Green
B	500 to 999 gpm	Orange
C	Less than 500 gpm	Red

SOURCE: Gerald Schultz, "Water Distribution Systems." In Arthur E. Cote and Jim L. Linville (Eds.), *Fire Protection Handbook,* 17th edition (Quincy, MA: National Fire Protection Association, 1991), p. 5.59. Reprinted with permission from *Fire Protection Handbook,* 17th edition, Copyright © 1991, National Fire Protection Association, Quincy, MA 02269.

TABLE 12.21 Fire Service Odds and Ends

RESIDENCES WITH SMOKE DETECTORS

Based on Harris survey[a]
77% nationally (1986); 82% nationally (1987); 81% nationally (1988)

Oak Ridge, TN
94% (1992)

BUILT-IN PROTECTION IN COMMERCIAL STRUCTURES

Oak Ridge, TN
96% of all newly constructed or substantially renovated commercial structures are equipped with automatic fire suppression or alarm systems (1992)

PROMPT RESPONSE TO FIRE SAFETY COMPLAINTS

St. Petersburg, FL
Target: Within 16 working hours

EXTRICATION

Corvallis, OR
Target: Rescue and extricate persons trapped by motor vehicle accidents, industrial mishaps, or other emergencies within 15 minutes of arrival
Actual: 100% (1991); 100% (1992)

RESPONSES PER COMPANY

San Antonio, TX
Average annual responses per pumper company: 935 (estimate, 1992)

New York City
Average annual responses per engine company: 2,685 (1991); 2,814 (1992)
Average annual responses per ladder company: 2,887 (1991); 2,922 (1992)

a. Cited in John R. Hall and Arthur E. Cote, "America's Fire Problem and Fire Protection." In Arthur E. Cote and Jim L. Linville (Eds.), *Fire Protection Handbook,* 17th edition (Quincy, MA: National Fire Protection Association, 1991), p. 1.16.

Notes

1. Although the call-to-dispatch and dispatch-to-arrival components both should be included in reported response times (and both are clearly part of response time from the perspective of a citizen awaiting arrival of an emergency unit), some departments may report only the second component—the time between dispatch of an emergency unit and its arrival at the scene. If a municipal official suspects that the reported response time of a comparison city is unrealistically fast, further investigation may reveal a different definition of "response time" in that community.

2. As impressive as the New York City figure is, some other cities may be able to challenge the heavy workload of engine companies in that city. Phoenix, Arizona, for example, reported that in the 1989-1990 fiscal year, 56 of its units responded to more than 2,000 calls, 28 responded to more than 3,000, and 5 units responded to more than 4,000 calls (City of Phoenix Auditor Department, 1991, p. 10).

13

Library

In 1943, Clarence Ridley and Herbert Simon (1943) praised the advances being made in measuring the "intangibles" of library service (p. xiv). They encouraged the leaders of other municipal functions to take similar strides. Now, more than 50 years later, library services continue to maintain a performance measurement edge over many other municipal services in the scope and sophistication of widely reported indicators and in the availability of detailed statistics identifying national performance norms. A citizen or municipal official wishing to assess the adequacy of the local public library has a reasonably rich array of potential benchmarks on which to draw.

History of Library Standards[1]

Until recently, the Public Library Association (PLA) and its parent organization, the American Library Association (ALA), had maintained a long-standing role in the development of standards for public libraries. Those standards changed through the years, shifting from initial efforts to prescribe appropriate levels of financial support and staff credentials to the formulation of qualitative library goals and, eventually, quantitative standards pertaining to library collection, facilities, services, and performance.

Lists of recommended books have guided librarians through the years in the development of local collections. One buying guide published by ALA in 1913 recommended subscriptions to periodicals such as *McClure's Magazine* and *National Geographic* and the purchase of books from an array of titles that included the now quaint-seeming *Worry, the Disease of the Age* and *Woman's Part in Government Whether She Votes or Not* at $1.35 and $1.50, respectively (Brown & Webster, 1913).

In 1921, ALA recommended $1 per capita as the minimum reasonable support for a public library (Withers, 1974, p. 320). Through the years, the recommended minimum allocation was ratcheted upward until in 1956, per capita expenditure prescriptions were dropped altogether, partially in acknowledgment of the uneven effects of library size and mission on resource requirements and partially in recognition of the imperfect relationship between expenditure level and service quality (pp. 322-323).

Other standards introduced by ALA in 1933 offered broad scope but possessed only limited usefulness for the purpose of evaluating a library's performance. Phrases such as "reasonably adequate library service," maintaining a collection "adequate to the needs of the community," and providing "a professional staff of high quality and adequate number" (Withers, 1974, pp. 320-321) dotted those standards and offered an imprecise yardstick against which to measure performance. Perhaps even more important to library practitioners and patrons, these qualitative "standards" offered little leverage for prying resources from the city treasury. In short, they failed to arm library directors with a persuasive means of demonstrating to budget makers local shortcomings in facilities, services, and funding. A series of quantitative standards began to emerge at midcentury which many library directors proved more adept at applying to that purpose.

The introduction of standards that prescribed acceptable travel time for library patrons, library operating hours, collection size and quality, collection maintenance practices, library staffing levels, and physical characteristics of the library facility (Table 13.1) served the cause of librarians whose operations fell below the prescribed numbers and who wished to use the standards as a peg for demonstrating and remedying local deficiencies. Such standards had several advantages—not the least of which was ease of application. It was a simple matter to compare local collection, staffing, and facility characteristics with national prescriptions and identify areas of conformance or deficiency.

Nevertheless, quantitative standards proved to be controversial. In part, they were criticized as the musings of professionals, goals established largely without the benefit of careful research. In part, they were criticized for failing to fully recognize the ramifications of service population differences and variations in service emphasis from one community library to another. Furthermore, quantitative standards were criticized for undercutting the efforts of top-quality libraries that felt penalized for their successes when budget officials and governing bodies cited above-standard performance as the rationale for diverting resources

TABLE 13.1 Selected Midcentury Public Library Standards

- Maximum acceptable travel time to library:
  15 minutes, urban; 30 minutes, rural

- Acceptable range of operating hours:
  Community libraries serving 10,000 to 25,000 population, 45 to 66 hours per week; community libraries serving larger populations, 66 to 72 hours per week; central resource libraries, at least 66 hours per week

- Systematic "weeding" of unneeded items from collection:
  Average annual withdrawals of at least 5%

- Minimum percentage of requested adult nonfiction titles included in local collection:
  35%-50% for libraries serving a population of less than 10,000; 50%-65%, 10,000-24,999 population; 65%-80%, 25,000-49,999; 80%-95%, 50,000-99,999

- Minimum percentage of requested juvenile titles included in local collection:
  80%

- Maximum acceptable waiting time for reserved materials:
  6 weeks

- Minimum collection size:
  2 to 4 volumes per capita

- Minimum annual additions and replacements:
  One sixth of a volume per capita for libraries serving a population of 500,000 or less; one eighth of a volume per capita for larger libraries

- Minimum number of periodical subscriptions:
  1 per 250 population

- Minimum staff size:
  1 staff member per 2,000 population

- Library lighting intensity:
  50 foot-candles

- Minimum library floor space:
  For communities of 25,000 to 49,999 population, 5,000 square feet plus 1 additional square foot for every 10 books beyond a 50,000-volume collection; for smaller communities, 1,500 square feet plus 1 additional square foot for every 10 books beyond a 15,000-volume collection

- Minimum shelf space:
  For communities of 25,000 to 49,999 population, 6,300 linear feet plus 1 additional foot for every 8 books beyond a 50,000-volume collection; for smaller communities, 1,875 linear feet plus 1 additional foot for every 8 books beyond a 15,000-volume collection

---

SOURCES: American Library Association, *Minimum Standards for Public Library Systems* (Chicago: ALA, 1967); American Library Association, *Interim Standards for Small Public Libraries* (Chicago: ALA, 1962); F. N. Withers, *Standards for Library Service: An International Survey* (Paris: The UNESCO Press, 1974), pp. 320-338. Reproduced by permission of the American Library Association and UNESCO; copyright © American Library Association, 1967 and 1962; copyright © UNESCO, 1974.

to "needier" purposes (Baker & Lancaster, 1991, pp. 321-334; Withers, 1974, pp. 320-338).

In 1986, the International Federation of Library Associations and Institutions (IFLA) published a summary of the public library standards it had been recommending as minimal guidelines since the 1970s. In its *Guidelines for Public Libraries,* the IFLA cautioned that some of its standards were "not likely to be

TABLE 13.2 Selected Standards for Public Libraries: International Federation of Library Associations and Institutions

Minimum Population Thresholds	
Preferred minimum population	150,000
Minimum population normally regarded as viable	50,000
Absolute minimum population	3,000
Minimum Service Hours	
Main library (urban)	60 hours per week
Minimum Book Stocks	
In smallest libraries	3 volumes per capita
In medium and larger libraries	2 volumes per inhabitant
[When children 14 years of age or younger constitute 25%-30% of the population, children's books should comprise one third of the total.]	
Reference books	Up to 10% of total stock (at least 100 volumes)
Annual Additions to Library Collection	
In smallest libraries	300 volumes per 1,000 population
In medium and larger libraries	250 volumes per 1,000 population
Children's books	One third of additions if children (14 or younger) comprise 25%-30% of population
Reference books in libraries serving populations of more than 50,000	10% of all additions
Periodical Holdings, Including Newspaper	
In libraries serving up to 5,000 people	50
In libraries serving more than 5,000	10 per 1,000 population
[These figures include multiple copies, foreign language periodicals, and periodicals for children.]	

universally relevant" (p. 61) but proceeded nevertheless to offer guidance on such diverse topics as desirable minimum service population, operating hours, collection size, staff, and library distance from patrons (Table 13.2).

Even as the IFLA was moving to consolidate its quantitative standards, the PLA was moving in the opposite direction. By the late 1980s, the PLA had abandoned its quantitative standards for public libraries—in contrast not only to IFLA but also to the continued use of quantitative standards by other ALA affiliates for various types of nonpublic libraries. The PLA's departure from prescriptive numerical standards has been applauded by some librarians but has prompted concern among others who miss the evaluative usefulness of national standards and who "still want published quantitative figures to help them argue for additional resources when their services are less than optimum" (Baker & Lancaster, 1991, p. 328). Many state-level associations—at least 40 by one recent count (Weech, 1988)—have filled the void by offering their own prescriptive, quantitative standards for public libraries.

TABLE 13.2 *Continued*

Records and Taped Recordings of All Kinds

Stock for use within library	2,000
Annual additions	300
Records for home use	No recommendation

Foreign Language Collections

For foreign-language minorities of less than 500 persons	100 volumes
For foreign-language minorities between 500-2,000 persons	1 book per 5 persons
For foreign-language minorities greater than 2,000 persons	1 book per 10 persons
Foreign-language periodicals	1 per 500 foreign-language minorities

Staff

Total nonmanual staff (professional, clerical, administrative):

In smallest libraries	1 full-time qualified librarian with clerical assistance
In medium and larger libraries	1 per 2,000 population
In very large libraries	1 per 2,500 population

Qualified librarians

In compact urban libraries	33% of total staff
In less compact libraries with many branches	40% of total staff

[In a large library system, one third of professional librarians should be specialized in children's work.]

Facility Location

Branch library within 1 mile of most residents

Main library within 2.5 miles of most residents

SOURCE: Adapted from *Guidelines for Public Libraries* (IFLA Publication 36). München, Germany: K. G. Saur for the International Federation of Library Associations and Institutions, 1986, pp. 61-63. Reprinted with the permission of K. G. Saur Verlag (Munich, Germany).

Performance Measures

Recommended sets of performance measures for public libraries typically include several indicators pertaining to library characteristics and services, often paralleling IFLA or earlier PLA standards. Some, such as the set of library measures recommended by Hatry and associates (1992), advocate extensive use of direct feedback from library patrons and other community residents (Table 13.3).

Other sets of recommended measures rely less on survey responses and more on library use, access, and performance statistics. Among the leading measures of library use, material use, materials access, reference services, and library programming recommended by PLA are the following:

TABLE 13.3 Measures of Effectiveness for Library Services

Overall objective: To present library services that provide the greatest possible satisfaction to citizens and are comprehensive, timely, helpful, and readily available in facilities that are attractive, accessible, and convenient.

Objective	Quality Characteristic	Specific Measure[a]	Data Collection Procedure
Overall citizen satisfaction	Citizen satisfaction	• Percentage of citizens who rate library service as satisfactory	Household survey
	User satisfaction	• Percentage of users who rate library service as satisfactory	Library user survey or household survey
	Use— household	• Percentage of households using (or not using) a public library a minimum number of times, such as four times over past twelve months, with reasons for nonuse	Household survey
	Registration— household	• Percentage of households with (or without) a valid registration card	Library tabulations or household survey
	Use—visitation	• Number of visits to library facilities per capita (or per household)	Library tabulations of number of users with population estimates from city planning department
	Use—circulation	• Circulation per capita by type of material—e.g., fiction, nonfiction, records, film—divided between internal and external use	External circulation figures, kept by most public libraries; internal figures, estimated from periodic sampling
Comprehensiveness/ timeliness/availability of holdings and service	Diversity and availability of holdings	• Percentage of user searches—title, subject/author, browser— that are successful	Library tabulations from special user survey
		• Percentage of users who rate materials as satisfactory	Library user survey or household survey
		• Percentage of nonusers who cite poor materials as a reason for nonuse	Household survey
	Speed of service	• Percentage of requests available within 7, 14, and 30 days or longer	Library tabulations of delivery dates
		• Percentage of users who rate speed of service (e.g., book retrieval and checkout) as satisfactory	Library user survey or household survey

TABLE 13.3 *Continued*

Objective	Quality Characteristic	Specific Measure[a]	Data Collection Procedure
Helpfulness of staff	Helpfulness—courtesy of staff	• Percentage of library users who rate helpfulness and general attitude of library staff as satisfactory	Library user survey or household survey
	Helpfulness-quality of reference service	• Percentage of users of library reference services who rate them as satisfactory	Library user survey
		• Percentage of reference transactions successfully completed	Library tabulations of reference responses
Attractiveness, accessibility, and convenience of facility	User satisfaction with comfort, crowdedness, noise, etc.	• Percentage of library users who rate the comfort, crowdedness, noise, cleanliness, and temperature/ventilation as satisfactory	Library user survey or household survey
	Nonuser satisfaction	• Percentage of nonusers who cite lack of comfort, crowdedness, noise, cleanliness, or temperature/ventilation as reasons for nonuse	Household survey
	Physical accessibility, convenience	• Percentage of users who rate convenience as satisfactory	Library user survey or household survey
		• Percentage of nonuser households who give poor physical accessibility as a reason for nonuse	Household survey
		• Percentage of citizens who live within a specific travel time (such as 15 minutes) of a public library	Counts from mapping latest census tract population figures against location of facilities with appropriate travel time radius drawn around each facility
	Hours of operation	• Percentage of users who rate hours of operation as satisfactory	Library user survey or household survey
		• Percentage of nonuser households who give poor hours as a reason for nonuse	Household survey

SOURCE: Harry P. Hatry, Louis H. Blair, Donald M. Fisk, John M. Greiner, John R. Hall, Jr., and Philip S. Schaenman, *How Effective Are Your Community Services? Procedures for Measuring Their Quality,* 2nd edition (Washington, DC: Urban Institute and International City/County Management Association, 1992), pp. 58-59. Reprinted by permission.
a. Officials who wish to focus on amount of local dissatisfaction may substitute "unsatisfactory" for the term "satisfactory" in many of these measures.

Library Use

• *Annual Library Visits per Capita* is the average number of library visits during the year per person in the area served. It reflects the library's walk in use, adjusted for the population served.

• *Registration as a Percentage of Population* is the proportion of the people in the area served who are currently registered as library users, reflecting the proportion of the people who are potential library users who have indicated an intention to use the library.

Material Use

• *Circulation per Capita* is the annual circulation outside the library of materials of all types per person in the legal service area.

• *In-Library Materials Use per Capita* is the annual number of materials of all types used within the library per person in the area served.

• *Turnover Rate* measures the intensity of use of the collection. It is the average annual circulation per physical item held.

Materials Access

• *Title Fill Rate* is the proportion of specific titles sought that were found during the user's visit. It is not the proportion of users who were successful, because one user may have looked for more than one title; it is the proportion of the searches that were successful.

• *Subject and Author Fill Rate* is the proportion of searches for materials on a subject or by an author that were filled during the user's visit.

• *Browsers' Fill Rate* is the proportion of users who were browsing, rather than looking for something specific, who found something useful.

• *Document Delivery* measures the time that a user waits for materials not immediately available, including reserve and interlibrary loans. It is expressed as the percent of requests filled within 7, 14, 30 days, and over 30 days.

Reference Services

• *Reference Transactions per Capita* is the annual number of reference questions asked per person in the area served.

• *Reference Completion Rate* is the proportion of reference questions answered satisfactorily.

Programming

• *Program Attendance per Capita* is the annual number of people attending programs per person in the area served. (Van House, Lynch, McClure, Zweizig, & Rodger, 1987, pp. 37-72; excerpted by permission of the American Library Association)

Benchmarks

Persons wishing to assess the adequacy of local public library facilities, collection, staff, and performance may legitimately relate current local marks to PLA standards of the past (see Table 13.1), as long as the date of those guidelines and the revised policy of PLA is acknowledged. Potential benchmarks with fewer encum-

TABLE 13.4 Selected Standards for Public Library Service in Ohio

- Achieve a title fill rate of 75% or greater.
- Achieve author and subject fill rate of 90% or greater.
- Provide within 7 days at least 40% of requested materials that are not immediately available; at least 80 percent within 30 days.
- Provide an up-to-date collection with at least 25% of holdings published within the last 5 years.
- Achieve a turnover rate of 5 or greater (circulation plus in-house use divided by holdings).
- Provide prompt and correct responses to reference inquiries, with 80% answered by the end of the business day.
- Provide sufficient photocopy equipment to ensure that no patron need wait longer than 10 minutes for such equipment.
- Provide convenient weekend and evening operating hours, with not more than 65% of operating hours coming during the daytime on weekdays.
- Provide sufficient parking to ensure the availability of at least one free parking space within one tenth of a mile during at least 80% of library operating hours.
- Operate the library under the management of a professional librarian who is a graduate of an ALA-accredited library program.

SOURCE: Excerpted from *Standards for Public Library Service in Ohio: 1992 Revision and Survey of Compliance* (Columbus: Ohio Library Council, 1995). Used by permission.

brances include the more recent IFLA standards (see Table 13.2), standards adopted by library associations at the state level, current PLA statistics on public library performance indicators, statistics compiled by the U.S. Department of Education, and standardized targets and performance indicators from individual municipalities.

The sets of standards developed by state-level library associations typically address a broad array of library characteristics and performance dimensions. The standards adopted in Ohio, for instance, prescribe the desired turnover rate for library holdings, fill rates, promptness and accuracy of responses to reference inquiries, desired currency of holdings, operating hours, and adequacy of parking (Table 13.4). Ohio's standards also include a list of minimum contents of an up-to-date reference collection, a list that differs little from those of several other states (Table 13.5).

The standards specified for public libraries in North Carolina similarly encompass a broad array of library characteristics (Table 13.6). Although some features of the North Carolina standards resemble those in Ohio, a principal distinction is North Carolina's practice of identifying three levels within each standard—one level regarded as "an absolute minimum level," a second level representing "a significant degree of progress" (labeled here as "good"), and a third level denoting "a high level of accomplishment" (labeled here as "excellent"; *Standards for North Carolina Public Libraries,* 1988, p. 3).

The benchmarking contributions of PLA and the U.S. Department of Education come in the form of descriptive statistics. Both collect information on public libraries and calculate relevant statistics for libraries on the basis of service

TABLE 13.5 Minimum Contents of an Up-to-Date Reference Collection: Ohio Standards

A public library in Ohio is considered to possess an up-to-date reference collection if it possesses at least one copy of each of the following reference tools:

Reference Item	Copyright Date
General encyclopedia	Copyright date within last 3 years
Dictionary (unabridged)	Most recent edition
Dictionary (abridged)	Copyright date within last 3 years
Thesaurus	Copyright date within last 10 years
Book of quotations	Copyright date within last 5 years
Biographical dictionary	Copyright date within last 5 years
General periodical index	Current year
Almanac	Current year
Statistical abstract	Current year
Atlas	Copyright date within last 3 years
Books in Print	Latest edition
Local phone book	Latest version
Phone book for nearest metro area	Latest version

SOURCE: *Standards for Public Library Service in Ohio: 1992 Revision and Survey of Compliance* (Columbus: Ohio Library Council, 1995), p. 15. Reprinted by permission.
NOTE: In 1986, fewer than 31% of 172 responding Ohio public libraries reported owning all the above reference items. By 1992, 68% of 138 respondents reported having achieved this standard.

population clusters. The remaining information in this chapter will draw extensively on those statistics as well as standardized targets and performance indicators from individual municipalities.

Staffing Norms and Other Input "Standards": Handle With Care

As with staffing prescriptions for other municipal functions, guidelines calling for a given number of librarians per 1,000 population are input standards rather than standards by which *performance* may be assessed. A library that discovers innovative ways to achieve its objectives with fewer than the customary number of employees would be penalized inappropriately by the blind application of staffing standards. Nevertheless, national norms and staffing indicators from other jurisdictions do offer supplemental information of value in a more comprehensive assessment of performance by placing in context local contentions of overstaffing or understaffing.

The general staffing pattern among public libraries in 1990 was approximately one half full-time equivalent (FTE) staff member per 1,000 population, except among the smallest libraries, in which the staffing ratio was somewhat higher (Table 13.7). Relatively little variation was found for average staffing ratios from one population cluster to another. Much greater variation was evident in the

TABLE 13.6 Selected Standards for North Carolina Public Libraries

	Standard		
	Minimum	*Good*	*Excellent*
Hours and days of operation per week	62/6	66/6	70/7
Title fill rate	40%	50%	60%
Author/subject fill rate	60%	70%	80%
Browsing fill rate	70%	80%	90%
Patron wait for library-owned material not immediately available			
• within 7 days	50%	60%	70%
• within 30 days	70%	80%	90%
Circulation per capita	3	5	7
Collection turnover rate	3	4	5
Reference transactions per capita	0.25	0.50	1
Reference fill rate (within 24 hours)	50%	70%	90%
Program attendance as percentage of population	25%	50%	100%
Library visits per capita	1	2	3
Registered patrons as percentage of population	20%	30%	40%
Books per capita	1	2	3
Percentage of collection published in last 5 years	15%	20%	25%
Percentage of collection "weeded" annually	3%	4%	5%
Periodical titles per 1,000 population	2	3	4
Ratio of FTE staff:population	1:4,000	1:3,000	1:2,000
Percentage of operating budget allocated to materials acquisition	15%	20%	25%
Percentage of operating budget allocated to training/travel	0.5%	0.75%	1%

Facility standards:

- Every member of the service population should have access to a library located within 15 minutes driving time.
- Approximately 1.5 square feet of convenient parking should be available for each square foot of library building.
- A full-service library should have 0.5 square feet of building per capita or 10,000 square feet, whichever is greater.

SOURCE: *Standards for North Carolina Public Libraries* (North Carolina Library Association/Public Library Section and North Carolina Public Library Directors Association, 1988). Reprinted by permission.

documents of individual municipalities reporting circulation per FTE and reference questions per staff member (Table 13.8).

Library staff professionalism, as measured by the employment of a library director with a graduate degree from an ALA-accredited library science program, is correlated with jurisdiction size (Table 13.7). Two thirds of all public libraries were headed by a director possessing a master's degree from an ALA-accredited program in 1990; 90% of those serving a population between $\frac{1}{2}$ and 1 million reported that level of professionalism.

The U.S. Department of Education reports total operating expenditures of $4.1 billion for public libraries in 1990, with 63% devoted to staff expenses and 16% to library collections (Chute, 1992, p. iii). The staff expenditure percentage

TABLE 13.7 Library Staffing in 1990: Staffing by Service Population Categories

							Service Population						
	National Average	1 Million or More	500,000- 999,999	250,000- 499,999	100,000- 249,999	50,000- 99,999	25,000- 49,999	10,000- 24,999	5,000- 9,999	2,500- 4,999	1,000- 2,499	1-999
Paid FTE staff per 1,000 population	0.4	0.4	0.4	0.4	0.4	0.4	0.5	0.5	0.5	0.5	0.6	1.1
Percentage of paid FTE librarians with ALA-MLS[a]	67.6%	82.4%	90.3%	86.6%	81.7%	74.1%	66.8%	54.7%	29.6%	13.6%	5.7%	2.6%

SOURCE: Adrienne Chute, *Public Libraries in the U.S.: 1990* (U.S. Department of Education, Office of Educational Research and Improvement, National Center for Education Statistics, NCES 92-028, June 1992), p. 27.

a. ALA-MLS: A master's degree from a program accredited by the American Library Association (ALA).

TABLE 13.8 Workload-to-Staff Ratios in Selected Cities

Circulation per FTE

21,801 (1991)	Fayetteville, AR
50,162 (1991)	Corpus Christi, TX
67,048 (1991)	Denton, TX

Reference Questions per
Reference Staff Member (Annual)

5,155 (youth) to 12,442 (adult) (1991)	Boise, ID
8,738 (1991)	Tacoma, WA
12,054 (adult) (1991)	Denton, TX
33,077 (1991)	Corpus Christi, TX

reported for 1991 by PLA differs fairly substantially, but the reported percentage allocation for library materials is quite close to the Department of Education figure. PLA figures suggest that most libraries allocate 44% to 57% of their expenditures to salaries and 12% to 18% to materials (Table 13.9). As in the case of staffing ratios, such figures should be regarded as input allocation norms rather than as standards. They reflect the customary means (i.e., allocation patterns) to desired ends, rather than ends in themselves.

Adequacy of Collection

Although figures provided by the U.S. Department of Education and PLA for average library holdings differ somewhat from one another, they nevertheless

TABLE 13.9 Percentage of Library Operating Expenditures Devoted to Salaries and Materials: 1991

Library Service Population	N	Salaries as Percentage of Operating Expenditures			Materials as Percentage of Operating Expenditures		
		Lower Quartile (25%)	Median	Upper Quartile (75%)	Lower Quartile (25%)	Median	Upper Quartile (75%)
1 million or greater	21	43.5%	50.2%	56.1%	11.3%	14.1%	15.1%
500,000-999,999	51	46.1%	50.6%	54.6%	11.4%	14.8%	17.3%
250,000-499,999	82	47.6%	51.2%	55.2%	13.0%	14.6%	17.2%
100,000-249,999	202	47.4%	52.1%	57.8%	11.8%	14.2%	17.1%
50,000-99,999	72	47.3%	52.7%	56.9%	12.2%	14.1%	16.6%
25,000-49,999	71	48.8%	53.0%	58.5%	13.2%	15.5%	17.7%
10,000-24,999	32	48.8%	52.6%	56.4%	12.9%	16.3%	19.4%
5,000-9,999	19	41.2%	51.7%	57.6%	12.9%	15.1%	20.4%

SOURCE: Adapted from Public Library Association, *Statistical Report '92: Public Library Data Service* (Chicago: American Library Association, 1992), pp. 36-38. Used by permission of American Library Association.
NOTE: *N* = number of libraries reporting.

exhibit sufficient consistency to permit individual libraries to judge their collection size in a national context (Table 13.10). For example, whether the precise norm for collection size among public libraries serving populations of 100,000 to 249,999 people is the 2.06 books and serial volumes per capita reported by the Department of Education for 1990 or the 2.2 volumes per capita reported as the median in 1991 by PLA should not be of great concern to a library in that population range wishing to assess its collection size. If its holdings per capita are less than both figures or greater than both figures, the message is clear. If its per capita holdings fall between the PLA and Department of Education figures, it may consider its collection size to be consistent with national norms.

Although the national average for collection size may be the 2.46 books and serial volumes per capita reported by the Department of Education—or a slightly higher figure to reflect audio and video materials, films, and serial subscriptions that might also be included in a count of library holdings—it is important to note the variation across population clusters. Libraries serving larger populations typically have a larger total collection but a smaller number of volumes per capita than do libraries serving smaller populations. For that reason, benchmarks for collection size should be drawn from appropriate service population clusters.

The adequacy of a library's collection may be judged not only by its sheer numbers but also by its quality. One approach to a quality-of-collection assessment tests the local holdings to see how many entries from a renowned bibliography or book list are possessed by the library. ALA's regularly published "best books" lists may be used for that purpose, as can the *Fiction Catalog* and the *Public Library Catalog*. Additional sources may be found in the *Pacific Northwest Collection Assessment Manual* (1990). Information is readily available on the application of this assessment technique and the identification of applicable bibliographies (e.g., Baker & Lancaster, 1991; *Pacific Northwest Collection Assessment Manual*, 1990).

Other methods of evaluating the quality of a public library's collection typically rely on indicators reflecting the popularity of that collection among the library's service population or the success of patrons in securing the materials they seek. A library with greater-than-average collection turnover probably has a greater-than-average percentage of its collection in high-demand and well-received books and materials (Table 13.11). Circulation per capita and its variant, circulation per registered borrower, provide an even more revealing assessment of the adequacy of a library's collection as judged by community residents as a whole and by those who use or intend to use the library, respectively (Table 13.12). Not surprisingly, reported rates of circulation among card-carrying patrons are higher than among the potential service population as a whole. A word of caution is in order regarding the use of per patron figures. Assessments based on circulation per registered borrower are especially vulnerable to variations among libraries in the practice of periodically purging their patron lists. Libraries with "lean" patron lists are rewarded in this calculation.

Figures reported for circulation per capita by the U.S. Department of Education for 1990 and by PLA for 1991 differ minimally in some population categories and

TABLE 13.10 Public Library Collection Size

Library Service Population	U.S. Department of Education Survey, 1990												Public Library Association Survey, 1991 Holdings per Capita		
	N	Books and Serial Vols. per Capita	N	Audio Materials per 1,000 Pop.	N	Films per 1,000 Pop.	N	Video Materials per 1,000 Pop.	N	Serial Subscripts per 1,000 Pop.	N	Lower Quartile (25%)	Median	Upper Quartile (75%)
1 million or greater	20	2.13	20	72.66	19	2.60	19	8.24	20	4.69	20	1.6 vols.	2.2 vols.	2.6 vols.
500,000-999,999	50	2.14	43	93.40	44	2.71	44	13.54	49	5.62	50	1.7 vols.	2.4 vols.	3.2 vols.
250,000-499,999	83	2.09	82	50.97	79	3.43	81	11.08	82	4.86	80	1.7 vols.	2.0 vols.	2.8 vols.
100,000-249,999	288	2.06	279	67.69	269	2.83	276	13.72	289	6.16	200	1.6 vols.	2.2 vols.	3.0 vols.
50,000-99,999	492	2.19	483	72.66	451	4.06	465	16.92	490	5.79	74	2.3 vols.	3.0 vols.	4.2 vols.
25,000-49,999	845	2.57	834	83.49	747	2.59	775	22.26	841	7.14	74	2.2 vols.	3.3 vols.	4.7 vols.
10,000-24,999	1,612	3.06	1,599	88.47	1,433	3.06	1,483	23.38	1,607	13.22	31	2.3 vols.	3.2 vols.	4.5 vols.
5,000-9,999	1,431	3.72	1,410	88.97	1,266	3.00	1,324	28.32	1,430	10.92	20	2.4 vols.	5.1 vols.	6.2 vols.
2,500-4,999	1,363	4.71	1,338	92.08	1,203	4.41	1,238	30.85	1,361	13.48	11	3.5 vols.	4.9 vols.	7.8 vols.
1,000-2,499	1,613	6.70	1,612	103.46	1,460	3.37	1,470	36.71	1,620	18.55	—	—	—	—
1-999	881	11.82	875	146.56	838	4.68	830	58.93	882	29.60	—	—	—	—
National average	8,711	2.46	8,608	76.42	7,846	3.10	8,043	16.37	8,704	7.07	—	—	—	—

SOURCE: Adapted from Adrienne Chute, *Public Libraries in the U.S.: 1990* (U.S. Department of Education, Office of Educational Research and Improvement, National Center for Education Statistics, NCES 92-028, June 1992), pp. 67-68, and Public Library Association, *Statistical Report '92: Public Library Data Service* (Chicago: American Library Association, 1992), pp. 65-67, by permission.
NOTE: *N* = number of libraries reporting.

137

TABLE 13.11 Annual Collection Turnover: 1991

Library Service Population	N	Lower Quartile (25%)	Median	Upper Quartile (75%)
1 million or greater	20	1.25	1.76	2.17
500,000-999,999	50	2.04	2.98	3.90
250,000-499,999	77	1.91	2.85	3.86
100,000-249,999	199	1.96	2.84	3.61
50,000-99,999	74	2.00	2.87	3.52
25,000-49,999	72	1.92	2.46	3.05
10,000-24,999	31	1.80	2.54	3.23
5,000-9,999	18	1.44	1.87	2.36
2,500-4,999	11	0.78	1.86	2.08

SOURCE: Adapted from Public Library Association, *Statistical Report '92: Public Library Data Service* (Chicago: American Library Association), pp. 113-115. Used by permission.
NOTE: *N* = number of libraries reporting.

fairly substantially in others. As in the case of collection holdings per capita, jurisdictions comparing their own numbers with these figures will receive an unambiguous message if they exceed or fall short of both the PLA and Department of Education numbers in their own population cluster. They may consider their own mark to be in the average range if they fall between the reported norms.

The adequacy of a library's collection may also be gauged by fill rates—the extent of patrons' success in securing the materials of a specific or general nature that they are seeking (Tables 13.13 and 13.14). Libraries routinely approach or exceed a title fill rate of 60%, considered by North Carolina standards to be "a high level of accomplishment."

Effectiveness in Reaching Service Population

The public library's mission encompasses more than providing quality services to persons who seek out the library and avail themselves of its offerings. Typically, libraries also encourage as many persons within their service population as they can to become registered and active patrons. Registration of one half the service population is not uncommon (Table 13.15). Library visits average more than three per capita annually, according to Department of Education figures, with a somewhat lower number of visits per capita among libraries serving the largest populations and higher figures among smaller libraries (Table 13.16).

Adequacy of Services

Special features such as reference services and interlibrary loan capabilities can enhance the library's offerings. The number of reference transactions tends to be

TABLE 13.12 Circulation per Capita, In-Library Use of Materials per Capita, and Circulation per Registered Borrower

Library Service Population	U.S. Department of Education Survey, 1990		Public Library Association Survey, 1991											
			Circulation per Capita				In-Library Use per Capita				Circulation per Registered Borrower[a]			
	N	Circulation per Capita	N	Lower Quartile (25%)	Median	Upper Quartile (75%)	N	Lower Quartile (25%)	Median	Upper Quartile (75%)	N	Lower Quartile (25%)	Median	Upper Quartile (75%)
1 million or greater	20	3.52	21	2.8	3.9	4.6	10	n.a.	1.6	n.a.	20	8.3	11.1	13.4
500,000-999,999	50	6.13	51	3.9	6.6	9.7	25	2.1	3.6	4.5	43	9.6	13.1	15.9
250,000-499,999	83	5.12	79	4.1	5.9	8.0	39	1.0	2.8	4.3	61	10.8	14.3	17.4
100,000-249,999	290	5.44	201	4.0	6.2	8.3	94	1.1	2.5	4.4	155	10.3	13.8	18.0
50,000-99,999	493	5.36	74	5.3	7.8	10.9	35	2.0	3.4	6.7	63	11.6	15.6	19.6
25,000-49,999	844	6.19	72	5.5	7.9	11.7	26	1.8	4.0	5.7	59	11.1	15.0	18.5
10,000-24,999	1,613	6.82	32	6.3	9.3	11.1	12	0.6	3.9	5.1	25	10.8	15.8	18.1
5,000-9,999	1,435	7.09	18	5.7	10.8	11.7	7	n.a.	1.5	n.a.	16	10.8	15.7	19.9
2,500-4,999	1,364	7.09	—	—	—	—	—	—	—	—	—	—	—	—
1,000-2,499	1,608	7.72	—	—	—	—	—	—	—	—	—	—	—	—
1-999	862	10.04	—	—	—	—	—	—	—	—	—	—	—	—
National average	8,693	5.59	—	—	—	—	—	—	—	—	—	—	—	—

SOURCE: Adapted from Adrienne Chute, *Public Libraries in the U.S.: 1990* (U.S. Department of Education, Office of Educational Research and Improvement, National Center for Education Statistics, NCES 92-028, June 1992), p. 88, and Public Library Association, *Statistical Report '92: Public Library Data Service* (Chicago: American Library Association, 1992), pp. 65-67, 113-115, by permission.

a. These figures include only those libraries that report having purged their registration files at least once during the prior 3 years.

N = number of libraries reporting.

n.a. = not applicable (quartiles are not recorded when the number of reporting libraries is less than 11).

TABLE 13.13 Fill Rates: 1991

Library Service Population	Reference Completion Rate				Title Fill Rate				Subject/Author Fill Rate				Browser Fill Rate			
	N	Lower Quartile (25%)	Median	Upper Quartile (75%)	N	Lower Quartile (25%)	Median	Upper Quartile (75%)	N	Lower Quartile (25%)	Median	Upper Quartile (75%)	N	Lower Quartile (25%)	Median	Upper Quartile (75%)
1 million or greater	1	n.a.	91.0%	n.a.	2	n.a.	71.1%	n.a.	2	n.a.	80.9%	n.a.	2	n.a.	94.7%	n.a.
500,000-999,999	6	n.a.	90.5%	n.a.	7	n.a.	64.1%	n.a.	6	n.a.	72.3%	n.a.	6	n.a.	91.1%	n.a.
250,000-499,999	18	71.0%	85.9%	93.0%	15	63.0%	72.0%	79.0%	13	65.0%	80.0%	83.0%	13	89.0%	92.0%	95.1%
100,000-249,999	55	76.6%	86.0%	92.3%	40	64.2%	68.0%	71.0%	39	69.4%	76.8%	81.7%	39	89.0%	92.0%	94.0%
50,000-99,999	30	83.0%	88.0%	93.0%	28	63.7%	69.0%	73.0%	28	74.0%	77.0%	79.0%	28	91.0%	94.0%	95.0%
25,000-49,999	17	84.0%	91.0%	94.0%	14	59.0%	65.0%	73.6%	14	73.0%	79.0%	79.6%	13	90.0%	91.0%	92.6%
10,000-24,999	5	n.a.	78.0%	n.a.	3	n.a.	72.0%	n.a.	3	n.a.	75.0%	n.a.	3	n.a.	93.0%	n.a.

SOURCE: Adapted from Public Library Association. *Statistical Report '92: Public Library Data Service* (Chicago: American Library Association, 1992). pp. 113-115, by permission.

N = number of libraries reporting.

n.a. = not applicable (quartiles are not recorded when the number of reporting libraries is less than 11).

TABLE 13.14 Success Rates in Providing Needed Materials: Selected Standards, Targets, and Actual Experience

Chandler, AZ
57.4% of requests for specific materials (e.g., a particular book title) fulfilled (1991)

Lubbock, TX
Target: Achieve 70% material availability rate
Actual: 61% (1991)

Sunnyvale, CA
Targets: 90% success rate for patron browsing searches in adult collection; 67% success rate for patron title searches in adult collection (1994)

Ohio Library Council
Ohio standards: (a) title fill rate of 75% or greater; (b) author and subject fill rate of 90% or greater
Actual: (a) 75% or greater title fill rate reported by 26% of responding Ohio libraries in 1992; (b) 90% or greater author/subject fill rate reported by 16% of responding Ohio libraries in 1992

NC Library Association/Public Library Section, NC Public Library Directors Association
North Carolina standards: (a) title rate of 40% (minimum), 50% (good), 60% (excellent); (b) author/subject fill rate of 60% (minimum), 70% (good), 80% (excellent); (c) browsing fill rate of 70% (minimum), 80% (good), 90% (excellent)

disproportionately greater in libraries serving large populations than in their smaller population counterparts. The averages range from approximately one reference transaction for every two residents annually among the smallest libraries to more than triple that rate among libraries serving the largest populations (Table 13.17). Many libraries take pride in providing prompt and thorough answers to reference inquiries, typically striving to achieve a high percentage of same-day responses (Table 13.18).

A library may be the sender or the receiver in interlibrary loan transactions. In the case of sending and in the case of receiving, the national average for interlibrary loans is 0.02 per capita (Table 13.19). The averages of different population clusters

TABLE 13.15 Registrations as a Percentage of Library Service Population: 1991[a]

Library Service Population	N	Lower Quartile (25%)	Median	Upper Quartile (75%)
1 million or greater	20	22.6%	40.7%	47.1%
500,000-999,999	43	42.7%	53.9%	59.5%
250,000-499,999	64	32.3%	39.3%	48.5%
100,000-249,999	155	34.9%	47.4%	59.9%
50,000-99,999	63	40.3%	52.0%	66.7%
25,000-49,999	60	38.4%	58.2%	85.2%
10,000-24,999	25	47.2%	62.5%	78.8%
5,000-9,999	17	50.0%	57.1%	74.5%

SOURCE: Adapted from Public Library Association, *Statistical Report '92: Public Library Data Service* (Chicago: American Library Association, 1992), pp. 65-67. Used by permission.
a. These figures include only those libraries that report having purged their registration files at least once during the prior 3 years.
N = number of libraries reporting.

TABLE 13.16 Public Library Visits per Capita

Library Service Population	U.S. Department of Education Survey, 1990		Public Library Association Survey, 1991 Visits per Capita			
	N	Visits per Capita	N	Lower Quartile (25%)	Median	Upper Quartile (75%)
1 million or greater	13	2.26	15	2.5	2.8	3.3
500,000-999,999	26	2.73	32	2.0	3.6	4.7
250,000-499,999	56	2.64	57	2.7	3.5	4.7
100,000-249,999	221	2.70	148	2.3	3.7	5.0
50,000-99,999	344	3.28	58	3.4	5.4	8.0
25,000-49,999	533	3.80	51	3.7	5.6	9.4
10,000-24,999	1,029	4.20	22	4.2	5.2	8.5
5,000-9,999	992	4.15	10	n.a.	5.7	n.a.
2,500-4,999	1,013	3.94	9	n.a.	3.5	n.a.
1,000-2,499	1,185	4.45				
1-999	673	6.01				
National average	6,110	3.13				

SOURCE: Adapted from Adrienne Chute, *Public Libraries in the U.S.: 1990* (U.S. Department of Education, Office of Educational Research and Improvement, National Center for Education Statistics, NCES 92-028, June 1992), p. 83, and Public Library Association, *Statistical Report '92: Public Library Data Service* (Chicago: American Library Association, 1992), pp.113-115, by permission.
N = number of libraries reporting.
n.a. = not applicable (quartiles are not recorded when the number of reporting libraries is less than 11).

TABLE 13.17 Reference Transactions per Capita

Library Service Population	U.S. Department of Education Survey, 1990		Public Library Association Survey, 1991 Reference Transactions per Capita			
	N	Reference Transactions per Capita	N	Lower Quartile (25%)	Median	Upper Quartile (75%)
1 million or greater	18	1.81	21	0.6	1.6	2.0
500,000-999,999	39	1.12	50	0.8	1.2	1.9
250,000-499,999	72	0.90	74	0.6	1.1	1.4
100,000-249,999	274	0.70	191	0.4	0.8	1.2
50,000-99,999	440	0.71	69	0.5	1.1	1.9
25,000-49,999	724	0.61	62	0.4	0.7	1.2
10,000-24,999	1,365	0.59	29	0.3	0.7	1.2
5,000-9,999	1,177	0.89	14	0.2	0.5	0.5
2,500-4,999	1,158	0.56	7	n.a.	0.8	n.a.
1,000-2,499	1,302	0.60				
1-999	652	0.76				
National average	7,248	0.92				

SOURCE: Adapted from Adrienne Chute, *Public Libraries in the U.S.: 1990* (U.S. Department of Education, Office of Educational Research and Improvement, National Center for Education Statistics, NCES 92-028, June 1992), p. 83, and Public Library Association, *Statistical Report '92: Public Library Data Service* (Chicago: American Library Association, 1992), pp. 113-115, by permission.
N = number of libraries reporting.
n.a. = not applicable (quartiles are not recorded when the number of reporting libraries is less than 11).

TABLE 13.18 Prompt Response to Reference Inquiries: Selected Standards, Targets, and Actual Experience

Wichita, KS
Percentage of reference responses completed on day of inquiry: 92.4% (1990); 92.2% (1991)

Oakland, CA
Percentage answered promptly with no need for referral: 96% (1990); 92% (1991); 95% (1992)

Tacoma, WA
Percentage of telephone inquiries answered within 5 minutes: 95.2% (1991)

Sunnyvale, CA
Target: 95% answered on same day (1994)

Alexandria, VA
Percentage answered within 24 hours: 93.0% (1991); 92.5% (1992); 93.0% (1993) (Note: approximately 85% answered immediately)

Ohio Library Council
Ohio standard: At least 80% answered by end of business day
Actual: 46% of responding Ohio libraries reported meeting this standard in 1992

NC Library Association/Public Library Section, NC Public Library Directors Association
North Carolina standard: At least 50% of reference responses successfully completed within 24 hours (minimum standard); 70% (good); 90% (excellent)

exhibit much greater variation for interlibrary receipts (ranging from less than 0.01 among the largest libraries to 0.19 among the smallest) than for transactions involving sending materials.

Libraries that promptly secure materials that are not readily available to their patrons provide a higher quality of service than those that cannot or do not respond as quickly (Table 13.20). Most libraries manage to deliver a majority of requested

TABLE 13.19 Interlibrary Loans per Capita: 1990

Library Service Population	Interlibrary Loans per Capita		
	N	*Out*	*In*
1 million or more	19	0.01	< .01
500,000-999,999	45	0.01	0.01
250,000-499,999	81	0.03	0.01
100,000-249,999	276	0.02	0.02
50,000-99,999	458	0.02	0.02
25,000-49,999	769	0.02	0.03
10,000-24,999	1,462	0.03	0.05
5,000-9,999	1,301	0.02	0.06
2,500-4,999	1,225	0.02	0.09
1,000-2,499	1,447	0.02	0.11
1-999	816	0.02	0.19
National average	7,932	0.02	0.02

SOURCE: Adapted from Adrienne Chute, *Public Libraries in the U.S.: 1990* (U.S. Department of Education, Office of Educational Research and Improvement, National Center for Education Statistics, NCES 92-028, June 1992), p. 88.
N = number of libraries reporting.

TABLE 13.20 Document Delivery Rates: 1991

Library Service Population	N	Percentage Delivered Within 7 Days			N	Percentage Delivered Within 30 Days		
		Lower Quartile (25%)	Median	Upper Quartile (75%)		Lower Quartile (25%)	Median	Upper Quartile (75%)
1 million or greater	2	n.a.	58.0%	n.a.	2	n.a.	60.4%	n.a.
500,000-999,999	3	n.a.	30.0%	n.a.	3	n.a.	60.0%	n.a.
250,000-499,999	9	n.a.	55.0%	n.a.	9	n.a.	85.0%	n.a.
100,000-249,999	32	30.0%	49.7%	63.3%	32	65.0%	77.0%	87.0%
50,000-99,999	26	28.8%	46.0%	53.0%	27	56.0%	72.7%	86.0%
25,000-49,999	18	25.0%	36.0%	44.0%	19	67.0%	77.0%	85.0%

SOURCE: Adapted from Public Library Association, *Statistical Report '92: Public Library Data Service* (Chicago: American Library Association, 1992), pp. 113-115. Used by permission.
N = number of libraries reporting.
n.a. = not applicable (quartiles are not recorded when the number of reporting libraries is less than 11).

documents within 30 days. In only two population clusters, however, does the typical library deliver more than half of the requested documents within 7 days. Once again, variation in document delivery rates across service population clusters indicates that benchmarks should be drawn carefully from the appropriate population category. Standards and targets for prompt action on requested materials adopted by selected libraries testify to the importance of that service in some libraries (Table 13.21).

TABLE 13.21 Library Performance in Securing Material Not Immediately Available: Selected Standards, Targets, and Actual Experience

Fayetteville, AR
   78% of requests filled within 30 days (1991)

Ohio Library Council
   Ohio standards: (a) Provide within 7 days at least 40% of requested materials that are not immediately available; (b) at least 80% within 30 days
   Actual: (a) Rate of 40% within 7 days met by 35% of responding Ohio libraries in 1992; (b) rate of 80% within 30 days met by 42% of responding Ohio libraries in 1992

Houston, TX
   Targets: Fill requests from system libraries within 24 hours; achieve average within-region turnaround of 10 days and outside-region turnaround of 20 days

Amarillo, TX
   Targets: Average response time of 10 days for interlibrary loan items to other libraries; 20 days for items from other libraries; process requests within 24 hours

TABLE 13.22 Process Targets and Miscellaneous Library Benchmarks From Selected Cities

PROMPT RESHELVING OF RETURNED MATERIALS

Burbank, CA
 Target: Within 1 working day

Tacoma, WA
 Percentage within 18 hours: 60% (1991)

Oak Ridge, TN
 Target: Within 24 hours

Sunnyvale, CA
 Target: Within 24 hours

Iowa City, IA
 Target: Within 2 days
 Actual: 4 days (average, 1990); 1.5 days (average, 1991)

Fort Worth, TX
 Target: Within 2 days average

Lubbock, TX
 Percentage of popular, on-demand books processed within 2 working days of receipt: 97% (1991)

PROMPT PROCESSING OF PERIODICALS

Lubbock, TX
 99% of periodicals processed within 24 hours (1991)

PROMPT PROCESSING OF ACQUISITIONS

Houston, TX
 Ordering turnaround: 2 days; processing turnaround: 45 days; cataloging turnaround: 120 days (1992)

Oakland, CA
 90% available for public use within 1 month of receipt (1992)

Sunnyvale, CA
 80% available within 60 days (projected, 1994)

PROMPT MENDING/REBINDING

Oakland, CA
 Target: 75% repaired within 2 months (1993)

Sunnyvale, CA
 Target: 75% repaired within 30 days (1994)

RATE OF LOSSES FROM COLLECTION

Tacoma, WA
 Annual loss of 0.32% of circulation (1991)

ON-LINE CATALOG UPTIME

Houston, TX
 97% (1992)

Sunnyvale, CA
 95% (projected, 1994)

Process Targets and Miscellaneous Benchmarks

Efficient internal processes contribute to effective library services and patron satisfaction. The prompt reshelving of returned materials, prompt processing of newly received periodicals and other library acquisitions, and prompt restoration and return to service of damaged or worn volumes enhance the likelihood that patrons will secure the materials they are seeking. The performance targets and experience of selected municipalities on these and other service dimensions can provide a useful context for planning and assessing library performance (Table 13.22).

Note

1. Much of this section on the history of library standards is drawn, with few adaptations, from David N. Ammons (1995), "Overcoming the Inadequacies of Performance Measurement in Local Government: The Case of Libraries and Leisure Services," *Public Administration Review, 55* (January/February), pp. 37-47. Used with permission.

14

Parks and Recreation

In 1906, the fledgling Playground Association of America (PAA) called for playground space equal to 30 square feet per child. Although the formulators of this standard acknowledged the existence of "no inherent relation between space and children," they reported the establishment of such a requirement in London and thought it reasonable (cited in Gold, 1973, p. 144; Lancaster, 1983, p. 15). Thus began a long series of standards adopted by leading associations in the parks and recreation field that have established ratios between recommended park acreage or recreational facilities and population. These ratios are the best known, but not the *only* or least controversial, standards in parks and recreation.

History of Parks and Recreation Standards[1]

Following the adoption of the 30-square-feet-per-child standard for playground space by the PAA, a successor organization subsequently modified the playground standard to 1 acre per 2,000 schoolchildren (Krohe, 1990, p. 10). In time, rule-of-thumb ratios emerged for park acreage in general, with 10 acres of recreation space per 1,000 population eventually becoming the most widely accepted norm (Bannon, 1976, p. 209; Haley, 1985, p. 176; Kraus & Curtis, 1982, p. 117; Krohe, 1990, p. 11; Lancaster, 1983, p. 11). Other normative guides also have been cited as "traditional standards"—for example, the preservation of 10% of a jurisdiction's total area as parkland and the expenditure of $8 (in 1965 dollars) per capita for public recreation and parks (Haley, 1985, p. 176)—but have been less widely accepted.

Through the years, park standards were refined by the National Recreation and Park Association (NRPA) and now center on a recommendation "that a park system, at a minimum, be composed of a 'core' system of parklands, with a total of 6.25 to 10.5 acres of developed open space per 1,000 population" (Lancaster, 1983, p. 56; used with the permission of the National Recreation and Park Association). NRPA guidelines make further recommendations regarding an appropriate mix of park types (e.g., miniparks, neighborhood parks, and community parks), sizes, service areas, and acreages per 1,000 population (Table 14.1). In addition, the association offers standards regarding the availability of particular types of recreational facilities (Table 14.2). For example, the recommended availability of tennis courts is one court per 2,000 population, compared with one baseball diamond per 5,000, one football field per 20,000, and one swimming pool per 20,000 population (Lancaster, 1983, pp. 60-61).

Recent efforts of NRPA and the American Academy for Park and Recreation Administration have focused on the development of standards for an accreditation program for local public park and recreation agencies to be conducted through the National Commission on Accreditation for Local Park and Recreation Agencies (1992). The standards being developed in that effort, however, are directed less toward outputs, outcomes, and performance results than toward planning, organizational structure, and management processes.

Performance Measures

When Clarence Ridley and Herbert Simon (1943) offered advice on measuring municipal recreation activities in 1943, they suggested three possibilities: "units of cost; units of effort expended—facilities and personnel provided; and units of performance, including records of attendance" (p. 32; reprinted by permission).

Their advice has been at least partially followed in the hundreds of municipalities whose annual budgets dutifully report the number of ball fields they maintain, the number of full-time equivalent (FTE) employees on the payroll, and the number of participants in municipally operated basketball leagues, softball leagues, aerobics courses, and arts and crafts classes. Fewer cities by far report unit costs of recreation services.

The sets of performance measures recommended today for parks and recreation programs typically contain many elements found in earlier versions—unit costs, number and variety of program offerings, and participation levels. Some sets, such as the effectiveness measures recommended by the Urban Institute's Harry Hatry and colleagues, incorporate household or user surveys to tap satisfaction levels among service recipients and citizens in general (Table 14.3).

TABLE 14.1 Neighborhood and Community Park Acreage Standards

Component	Use	Service Area	Desirable Size	Acre per 1,000 Population	Desirable Site Characteristics
Minipark	Specialized facilities that serve a concentrated or limited population or specific group such as tots or senior citizens	Less than ¼-mile radius	1 acre or less	0.25 to 0.5 acre	Within neighborhoods and in close proximity to apartment complexes, town house development, or housing for the elderly
Neighborhood park/playground	Area for intense recreational activities, such as field games, court games, crafts, playground apparatus area, skating, picnicking, wading pools, etc.	¼- to ½-mile radius to serve a population up to 5,000 (a neighborhood)	15+ acres	1.0 to 2.0 acres	Suited for intense development; easily accessible to neighborhood population—geographically centered with safe walking and bike access; may be developed as a school-park facility
Community park[a]	Area of diverse environmental quality; may include areas suited for intense recreational facilities, such as athletic complexes, large swimming pools; may be an area of natural quality for outdoor recreation, such as walking, viewing, sitting, picnicking; may be any combination of the above, depending on site suitability and community need	Several neighborhoods; 1- to 2-mile radius	25+ acres	5.0 to 8.0 acres	May include natural features, such as water bodies, and areas suited for intense development; easily accessible to neighborhood served

Recommended combined acreage: 6.25 to 10.5 acres per 1,000 population

SOURCE: Excerpted from Roger A. Lancaster (Ed.), *Recreation, Park and Open Space Standards and Guidelines* (Alexandria, VA: National Recreation and Park Association, 1983), pp. 56-57. Used with the permission of the National Recreation and Park Association.

a. Community parks are distinguished not only from miniparks and neighborhood parks but also from regional parks or open space preserved as a conservancy.

A point of contention in the design of performance measures for parks and recreation and, ultimately, in the identification of benchmarks, centers on the question of whether recreational facilities and parks themselves are properly considered program inputs or outputs. As noted repeatedly in this volume, measures of program inputs are poor substitutes for measures of outputs or results as gauges of performance.

Consider an example or two drawn from other municipal operations. Police cars and jails are not the products of the police function; they are tools used by police department employees as they perform that function—they are *means* to the desired ends. Similarly, a wastewater treatment plant is not the product of the sewer operation; it, too, is a tool—in other words, a resource *input*—that is instrumental in achieving the desired product or output, clean water. The parks and recreation function, however, may constitute a special case that stands as an exception to the pattern represented by these examples.

Unlike police and sewer services, much of the parks and recreation function is designed as a self-service operation in which users simply avail themselves of municipally provided facilities. To a degree, therefore, parks and recreational facilities that are properly considered an input in the total recreation experience of community residents are nevertheless outputs as far as the municipality's role is concerned. To the degree that the public expects the city to provide facilities for "self-service" in the realm of recreation, the essential question becomes, "Are the appropriate facilities provided in sufficient quantity and quality?" Where recreation *services* are expected, the question becomes more complicated. In those cases, the facilities needed to provide the desired services are appropriately considered resource *inputs*.

Benchmarks

Benchmarks of potential usefulness to local government officials and citizens wishing to assess the adequacy of the local parks and recreation program may be found in the recommended standards of relevant associations and in the standardized targets and performance indicators of other municipalities. In this chapter, prospective benchmarks will be divided into five groups: parks and facility standards, maintenance standards for facilities and grounds, program operations standards, program offerings and administrative functions, and cost recovery.

PARKS AND FACILITY STANDARDS

Standards that prescribe park acreages or types and numbers of recreational facilities have collected their share of critics (for example, Gold, 1973; Shivers & Hjelte, 1971). Variations from one community to another in the age profile of residents, economic conditions, and culture—not to mention climate and population density—are likely to influence the array of facilities considered practical or even desirable. Critics contend that rigid standards ignore the need for flexibility in addressing objectives that may differ from one community to the next. Furthermore, they charge that because such standards often are developed without careful research, they may be meaningless or even harmful.

TABLE 14.2 Recreational Facility Development Standards

Activity/ Facility	Recommended Space Requirements	Recommended Size and Dimensions	Recommended Orientation	No. of Units per Population	Service Radius	Location Notes
Basketball						
1. Youth	2400-3036 sq. ft.	46'- 50' × 84'	Long axis north-south	1 per 5,000	¼-½ mile	Usually in school, recreation center, or church facility; safe walking or bike access; outdoor courts in neighborhood and community parks, plus active recreation areas in other park settings
2. High School	5040-7280 sq. ft.	50' × 84'				
3. Collegiate	5600-7980 sq. ft.	50' × 94' with 5' unobstructed space on all sides				
Baseball						
1. Official	3.0-3.85 acre minimum	Baselines—90'; pitching distance—60½'; foul lines—min. 320'; center field—400'+	Locate home plate so pitcher throwing across sun and batter not facing it. Line from home plate through pitcher's mound runs east-northeast.	1 per 5,000; lighted—1 per 30,000	¼-½ mile	Part of neighborhood complex; lighted fields part of community complex
2. Little League	1.2 acre minimum	Baselines—60'; pitching distance—46'; foul lines—200'; centerfield—200'-250'				
Football	Minimum 1.5 acres	160' × 360' with a minimum of 6' clearance on all sides	Fall season—long axis northwest to southeast; for longer periods, north to south	1 per 20,000	15-30 minutes travel time	Usually part of baseball, football, soccer complex in community park or adjacent to high school
Golf						
1. Par 3 (18-hole)	50-60 acres	Avg. length vary— 600-2700 yds.	Majority of holes on north-south axis	—	½ to 1 hour travel time	9-hole course can accommodate 350 people a day
2. 9-hole standard	Minimum 50 acres	Avg. length— 2250 yds.		1 per 25,000		18-hole course can accommodate 500-550 people a day
3. 18-hole standard	Minimum 110 acres	Avg. length— 6500 yds.		1 per 50,000		Course may be located in community or district park but should not be over 20 miles from population center

TABLE 14.2 *Continued*

Activity/ Facility	Recommended Space Requirements	Recommended Size and Dimensions	Recommended Orientation	No. of Units per Population	Service Radius	Location Notes
Golf— driving range	13.5 acres for minimum of 25 tees	900′ × 690′ wide; add 12′ width for each additional tee	Long axis southwest-northeast with golfer driving toward northeast	1 per 50,000	30 minutes travel time	Part of golf course complex; as a separate unit, may be privately operated
Ice Hockey	22,000 sq. ft. including support area	Rink 85′ × 200′ (minimum 85′ × 185′); additional 5,000 sq. ft. support area	Long axis north-south if outdoor	Indoor—1 per 100,000; outdoor—depends on climate	½ to 1 hour travel time	Climate important consideration affecting no. of units; best as part of multipurpose facility
Multiple recreation court (basketball, volleyball, tennis)	9,840 sq. ft.	120′ × 80′	Long axis of courts with *primary* use is north-south	1 per 10,000	1-2 miles	
¼-mile running track	4.3 acres	Overall width —276′; length — 600.02′; track width for 8 to 4 lanes is 32′	Long axis in sector from north to south to northwest-southeast with finish line at northerly end	1 per 20,000	15-30 minutes travel time	Usually part of high school, or in community park complex in combination with football, soccer, etc.
Soccer	1.7 to 2.1 acres	195′ to 225′ × 330′ to 360′ with a 10′ minimum clearance on all sides	Same as football	1 per 10,000	1-2 miles	Number of units depends on popularity; youth soccer on smaller fields adjacent to schools or neighborhood parks
Softball	1.5 to 2.0 acres	Baselines— 60′; pitching distance— 46′; min.; 40′ (women); fast pitch field radius from plate— 225′ between foul lines; slow pitch—275′ (men), 250′ (women)	Same as baseball	1 per 5,000 (if also used for youth baseball)	¼-½ mile	Slight difference in dimensions for 16″ slow pitch; may also be used for youth baseball

continued

Perhaps the standard with the highest profile of all in the parks and recreation field is NRPA's recommendation that combined acreage for neighborhood and community parks fall in the range of 6.25 to 10.5 acres of developed open space per 1,000 population (Lancaster, 1983, pp. 56-57)—more often expressed simply

TABLE 14.2 *Continued*

Activity/ Facility	Recommended Space Requirements	Recommended Size and Dimensions	Recommended Orientation	No. of Units per Population	Service Radius	Location Notes
Swimming pools	Varies on size of pool and amenities; usually ½- to 2-acre site	Teaching— minimum of 25 yds. × 45′ even depth of 3 to 4 ft.; competitive— minimum of 25 m × 16 m; minimum of 27 sq. ft. of water surface per swimmer; ratios of 2:1 deck vs. water	None— although care must be taken in siting of lifeguard stations in relation to afternoon sun	1 per 20,000 (pools should accommodate 3% to 5% of total population at a time)	15-30 minutes travel time	Pools for general community use should be planned for teaching, competitive, and recreational purposes with enough depth (3.4 m) to accommodate 1 m and 3 m diving boards; located in community park or school site.
Tennis	Minimum of 7,200 sq. ft. single court (2 acres for complex)	36′ × 78′; 12′ clearance on both sides; 21′ clearance on both ends	Long axis north-south	1 court per 2,000	¼-½ mile	Best in batteries of 2-4; located in neighborhood/ community park or adjacent to school site
Volleyball	Minimum of 4,000 sq. ft.	30′ × 60′; minimum 6′ clearance on all sides	Long axis north-south	1 court per 5,000	¼-½ mile	Same as other court activities (e.g., badminton, basketball, etc.)

SOURCE: Excerpted from Roger A. Lancaster (Ed.), *Recreation, Park and Open Space Standards and Guidelines* (Alexandria, VA: National Recreation and Park Association, 1983), pp. 60-61. Used with the permission of the National Recreation and Park Association.
NOTE: Other activities and facilities for which the National Recreation and Park Association has recommended standards include archery, badminton, beach areas, field hockey, handball, skeet and trap shooting, and trails.

as 10 acres per 1,000 population. More than one critic has pointed out that the park acreage needed in New York City for Manhattan to comply with accepted standards would exceed the entire acreage of that island (e.g., Clawson, 1963, p. 20).

Clearly, a one-size-fits-all standard will not be suitable for cases that deviate sharply from the norm around which it was designed. But even those who formulate standards often anticipate, and even recommend, adaptation where conditions warrant. In the NRPA publication that declares most park and recreational facility standards, Lancaster (1983) acknowledges the inappropriateness of blind application.

Ideally, the national standards should stand the test in communities of all sizes. However, the reality often makes it difficult or inadvisable to apply national standards without question in specific locales. The uniqueness of every community, due to differing geographical, cultural, climatic, and socio-economic characteristics, makes it imperative that every community develop its *own* standards for recreation, parks, and open space. (p. 37; used with the permission of the National Recreation and Park Association)

Even tailored standards may not endure the test of time, as conditions on which standards were based begin to change. If even the most carefully customized standards assume static demographic and socioeconomic status in a community, they, too, are likely to be imperfect. Parks remain, even as the neighborhoods around them change. A reasonable beginning point for parks and recreation planning, then, is the assumption that most communities will approximate the norm but that appropriate modifications to the standards should be made for significant community deviations. In extreme cases, a given standard may even be declared totally inapplicable.

The appropriately cautious application of park and recreational facility standards should take into consideration the nature and extent of a given community's deviation from the general characteristics of a typical U.S. community, focusing especially on those differences that are likely to affect leisure needs and tastes. Furthermore, communities attempting to apply national standards to their own parks and recreational facility ratios should be aware that *many* communities fall short of the prescribed marks. Therefore, a city that assumes that NRPA park acreage and facility standards are "minimum acceptable levels," rather than "targets of excellence," risks judging itself too harshly. The communities listed in Table 14.4 are widely respected for their municipal operations, yet a large number fall short of the popular 10 acre per 1,000 population standard or even the broader NRPA range of 6.25 to 10.5 acres per 1,000 population.

A degree of ambiguity afflicts many parks and recreation measures, making it advisable to consider alternate interpretations when weighing those measures. In reporting park acreage, for instance, some municipalities include natural preserves, land acquired for future parks, and all other open space in the possession of the city, whereas others report only developed parkland. The implications of that ambiguity, whispered via footnotes in Table 14.4, are shouted in Table 14.5—a tabulation compiled for a 1992 study performed by the city of Dallas, Texas, contrasting park acres and *developed* park acres per 1,000 residents in 10 cities. The ranking of Phoenix, for example, is either first or sixth, depending on whether undeveloped mountain acreage owned by that city is included in the tabulation. Four cities in the Dallas study easily exceed the standard of 10 acres per 1,000 population if undeveloped acres are included. Among the 10 cities, only Austin, Texas, complies if undeveloped acres are excluded.

Standards are sometimes presented as guidelines, sometimes as minimally acceptable levels of quality or performance, sometimes as norms, and sometimes as targets for ambitious municipalities to strive toward. Although parks and recreation specialists may imply a "minimally acceptable level" interpretation for park acreage standards, the class roll of solid but noncomplying municipalities suggests otherwise. Although some cities manage to achieve many of the parks and facility standards and thereby demonstrate their attainability, many others fall short of the mark.

In its parks and recreation master plan, the city of Tucson, Arizona, inventoried its parks and recreation facilities, compared those facilities with national standards,

TABLE 14.3 Effectiveness Measures for Park and Recreation Services

Overall objective: To provide for all citizens a variety of enjoyable leisure opportunities that are accessible, safe, uncrowded, physically attractive, and well maintained.

Objective	Quality Characteristic	Specific Measure[a]	Data Collection Procedure
Enjoyableness	Citizen satisfaction	• Percentage of households rating neighborhood park and recreation opportunities as satisfactory	Household survey
	User satisfaction	• Percentage of those households using community park or recreation facilities and programs who rate them as satisfactory	Household or user survey
	Use—participation rates	• Percentage of community households using (or not using) community park or recreation facilities at least once over a specific past period, such as three months	Household survey
	Use—attendance	• Number of visits at park and recreation sites	Attendance statistics
Fewer crowds	User satisfaction	• Percentage of user households rating community facilities as too crowded	Household or user survey
	Nonuser satisfaction	• Percentage of nonuser households giving crowded conditions as a reason for nonuse of facilities	Household survey
	Crowding factor	• Average peak-hour attendance divided by capacity	Attendance statistics and estimates of carrying capacity
Physical attractiveness	User satisfaction	• Percentage of user households rating physical attractiveness as satisfactory	Household or user survey
	Nonuser satisfaction	• Percentage of nonuser households giving lack of physical attractiveness as reason for nonuse	Household survey
	Facility cleanliness	• Percentage of user households rating cleanliness as satisfactory	Household or user survey
		• Percentage of facilities whose cleanliness is rated satisfactory by a trained observer	Trained observer ratings

TABLE 14.3 *Continued*

Objective	Quality Characteristic	Specific Measure[a]	Data Collection Procedure
Well-maintained facilities	Facility/equipment condition	• Percentage of user households rating the condition of facility/equipment as satisfactory	Household or user survey
		• Percentage of parks and other recreation facilities whose physical condition is rated as satisfactory (by a trained observer). Ratings should also be presented for specific features—lawns, playing areas, rest rooms, benches and picnic tables, and so forth.	Trained observer ratings
Safety	Hazardous facilities/equipment	• Number of serious injuries (say, per 10,000 visits)	Accident and attendance statistics
		• Percentage of facilities with one or more safety hazards	Trained observer ratings
	Criminal incidents	• Number of criminal incidents (say, per 10,000 visits)	Criminal incident statistics of some park and recreation agencies and most municipal police forces; attendance statistics
	User satisfaction	• Percentage of user households rating safety of facilities as satisfactory	Household or user survey
	Nonuser satisfaction	• Percentage of nonuser households giving lack of safety as a reason for nonuse of facilities	Household survey
Accessibility	Physical accessibility	• Percentage of citizens living within (or not within) 15 to 30 minutes of travel time of a community park or recreation facility, distinguished by type of facility and principal relevant mode of transportation	Counts from mapping latest census tract population figures against location of facilities, with appropriate travel-time radius drawn around each facility
	Physical accessibility— user satisfaction	• Percentage of user households rating physical accessibility as satisfactory	Household or user survey

continued

TABLE 14.3 *Continued*

Objective	Quality Characteristic	Specific Measure[a]	Data Collection Procedure
	Physical accessibility— nonuser satisfaction	• Percentage of nonuser households giving poor physical accessibility as a reason for nonuse	Household survey
	Hours/days of operation—user satisfaction	• Percentage of user households rating hours of operation as satisfactory	Household or user survey
	Hours/days of operation— nonuser satisfaction	• Percentage of nonuser households giving unsatisfactory operating hours as a reason for nonuse	Household survey
Variety of interesting activities	User satisfaction	• Percentage of user households rating the variety of program activities as satisfactory	Household or user survey
	Nonuser satisfaction	• Percentage of nonuser households giving lack of program variety as a reason for nonuse	Household survey
Helpfulness of staff	Staff helpfulness— user satisfaction	• Percentage of user households rating helpfulness or attitude of staff as satisfactory	Household or user survey
	Staff helpfulness—nonuser satisfaction	• Percentage of nonuser households giving poor staff attitude as a reason for nonuse	Household survey

SOURCE: Harry P. Hatry, Louis H. Blair, Donald M. Fisk, John M. Greiner, John R. Hall, Jr., and Philip S. Schaenman, *How Effective Are Your Community Services? Procedures for Measuring Their Quality,* 2nd edition (Washington, DC: Urban Institute and International City/County Management Association, 1992), pp. 36-37. Reprinted by permission.
a. Officials who wish to focus on amount of local dissatisfaction may substitute *unsatisfactory* for the term *satisfactory* in many of these measures.

and as recommended by NRPA standards promulgators themselves, established its own local standards for park acreage and accessibility (Table 14.6). In some cases, Tucson's holdings exceeded national standards, but in many cases they did not. The presentation of local facilities in the context of national standards affords community citizens and decision makers the opportunity to reconsider local priorities in light of professional standards and to decide which, if any, of those standards they believe applicable to their own community and worthy of pursuit.

National guidelines for types and numbers of recreational facilities provide a context for considering a community's array of facilities and the adequacy of individual components. Nevertheless, consumers of those standards should not assume too much. For example, the NRPA recommendation of one tennis court per 2,000 population—one of the lowest population-per-facility ratios—attests to the popularity of that sport and the substantial demand for courts in many commu-

TABLE 14.4 Park Acres per 1,000 Population in Selected Cities

	Reported Park Acres per 1,000 Population	
Little Rock, AR	32.11[a]	(1992)
Portland, OR	21.03[b]	(1992)
Carrollton, TX	20.43	(1993)
Orlando, FL	16.6	(1992)
Loveland, CO	15.14[c]	(1992)
Iowa City, IA	12.42	(1991)
Waukesha, WI	11.5	(1992)
Overland Park, KS	11.28	(1991)
Southfield, MI	10.33[d]	(1991)
Denton, TX	8.15	(1991)
Greenville, SC	7.76	(1991)
Burbank, CA	7.55	(1992)
Corpus Christi, TX	6.4	(1991)
Jacksonville, FL	5.06	(1992)
Longview, TX	4.9	(1992)
Boston, MA	4.35	(1992)
Portsmouth, VA	3.54	(1993)
Phoenix, AZ	1.96[e]	(1991)
Fort Smith, AR	1.53	(1992)
Averages reported in Haley study[f] of 152 cities:		
Overall average	10.7 (1980)	
Central cities	11.7 (1980)	
Suburbs	9.7 (1980)	
Percentage of 152 cities meeting standard of 10 acres per 1,000 pop.: 43%		

a. Including "park development."
b. Acres of "open space" per 1,000 population.
c. Including nearby mountain park (excluding the mountain park leaves 4.88 acres of "developed" park per 1,000 population in Loveland).
d. 6.15 acres per 1,000 population, if Southfield's nature preserves are excluded.
e. Acres of "developed" parks per 1,000 population (excludes mountain and other undeveloped property).
f. A. J. Haley, "Municipal Recreation and Park Standards in the United States: Central Cities and Suburbs, 1975-80," *Leisure Sciences, 7,* No. 2 (1985), pp. 175-188.

TABLE 14.5 Park Land in Selected Cities

	Park Acres per 1,000 Residents	*Developed Park Acres per 1,000 Residents*
Phoenix	30.82	1.99
Austin	26.50	13.96
San Diego	21.42	6.21
Dallas	17.47	5.10
Houston	7.48	4.60
San Antonio	5.75	5.08
Detroit	4.98	4.41
San Jose	4.09	1.57
El Paso	3.10	1.94
Philadelphia	1.58	1.51

SOURCE: City of Dallas (TX) Park and Recreation Department, *City Comparison Study,* April 20, 1992, pp. 8-9.

TABLE 14.6 Comparison of Actual Park and Recreation Facilities and Standards in Tucson, Arizona

Facility Type	1989 Total	Facility per 1989 Population	National Standards
All Parks	3,740 acres	9.4 acres per 1,000	N/A
Neighborhood	295 acres	0.7 acre per 1,000	2.5 acres per 1,000
District	1,493 acres	3.8 acres per 1,000	2.5 acres per 1,000
Regional	1,952 acres	4.9 acres per 1,000	20 acres per 1,000
Field Sports			
Baseball fields	25	1 per 15,892	1 per 5,000
Football fields	20	1 per 19,865	1 per 20,000
Little League	90	1 per 4,414	1 per 5,000
Rugby fields	4	1 per 99,325	N/A
Soccer fields	46	1 per 8,637	1 per 10,000
Softball fields	36	1 per 11,036	1 per 5,000
Court Sports			
Handball courts	18	1 per 22,072	1 per 20,000
Horseshoe pits	24	1 per 16,554	N/A
Multiple use courts	38	1 per 10,455	1 per 10,000
Shuffleboard courts	19	1 per 20,911	N/A
Tennis courts	147	1 per 2,703	1 per 2,000
Volleyball courts	21	1 per 18,919	1 per 5,000
Passive Recreation			
Nature trail	1	1 per 397,300	N/A
Picnic sites	311	1 per 1,277	N/A
Picnic *ramadas*	147	1 per 2,703	N/A
Active Recreation			
Bike paths	6	1 per 66,217	N/A
Exercise courses	3	1 per 132,433	N/A
Golf courses	5	1 per 79,460	1 per 50,000
Jogging paths	5	1 per 79,460	N/A
Playgrounds	76	1 per 5,228	N/A
Recreational centers	16	1 per 24,831	N/A
Running tracks	2	1 per 198,650	1 per 20,000
Swimming pools	22	1 per 18,059	1 per 20,000
Tucson Definitions/Standards:			
Park Size			
Neighborhood park		0-15 Acres	
District park		16-100 Acres	
Regional park		100+ Acres	
Service Area Size*			
Neighborhood park		0.5 mile radius from park	
District park		3 mile radius from park	
Regional park		3 mile radius from park	
General Plan Standards			
Neighborhood		2.5 acres per 1,000	
District		2.5 acres per 1,000	
Regional		5 acres per 1,000	

SOURCE: City of Tucson (AZ), *City of Tucson Parks and Recreation Master Plan 2000: Planning Our Recreational Future,* December 1989, pp. 44-45.
*Service areas for picnic *ramadas* and recreation centers are based on 20-minute travel time to the facility.

TABLE 14.7 Tennis Courts in Selected Cities

	Number of Tennis Courts per 100,000 Residents
Dallas	26.42
Austin	22.76
Detroit	22.57
El Paso	21.15
San Antonio	14.10
Philadelphia	14.06
Houston	13.49
San Diego	11.88
Phoenix	11.08
San Jose	9.08

SOURCE: City of Dallas (TX) Park and Recreation Department, *City Comparison Study,* April 20, 1992, p. 12.

nities. Nevertheless, a city aspiring to reach this standard should realize that many of its most admirable counterparts do not. Tucson, with one court per 2,703 residents, fell short. Similarly, all 10 of the cities in the Dallas study, needing a ratio of 50 tennis courts per 100,000 residents to meet the one court per 2,000 ratio, fell far short (Table 14.7). If local demand for tennis facilities warrants a community's pursuit of the national standard, achievement of that ratio would be commendable and noteworthy; nevertheless, the context provided by the ratios found in many other cities offers assurances that falling somewhere short is hardly a serious blemish.

In addition to standards that use population ratios to prescribe park acreages and recreational facility needs, other standards offer guidance on the quality of facilities. One such set of standards, offered by the American Public Health Association (APHA), provides guidance for swimming pool operation, as well as for facility design and equipment capacity (Table 14.8).

MAINTENANCE STANDARDS

A major responsibility of most parks and recreation departments is the maintenance of park properties, recreational facilities, and sometimes the grounds surrounding other municipal facilities. As noted by a city of Boston analyst following a disappointing search for clear "industry standards" addressing those duties, much of "the literature . . . on the performance evaluation of parks and recreation [is] of an extremely theoretical level (reminiscent of Veblen's *Theory of the Leisure Class*), and not pertinent to groundskeeping and Jiffy-John maintenance" (Terry, 1991, p. 14). Despite their elusiveness, however, standards, guidelines, and a few rules of thumb may be discovered in the recommendations of specialized professional associations and in the self-developed standards and performance reports of individual municipalities. For example, a report prepared by a management analysis team in Pasadena, California, concluded that a ratio of one park maintenance

TABLE 14.8 Selected Standards for Public Swimming Pools

Circulation system capacity:
Sufficient to clarify and disinfect entire volume of water at least 4 times per 24 hours

Filtration system capacity:
Sufficient to restore turbidity level of pool to 0.5 nephelometric turbidity units (NTUS) or less within 24 hours following peak bather load

Acceptable chemical and physical qualities of pool water:
- pH (hydrogen ion concentration) range of 7.2 to 8.0
- total alkalinity of 50 to 150 mg/L
- maximum chlorine level of 0.2 mg/L
- sufficient clarity to easily view a 6-inch black-and-white disc at the deepest point of the pool from poolside

Standard frequency of pool water testing for pH and residual disinfectant:
Within 1 hour of opening and at least 3 times per operating day from at least 2 locations (shallow and deep)

Standard ratio of attendant/lifeguards to pool area occupants: 1:75

Standard pool area load: The sum of (a) + (b) + (c), where
(a) = one person per 15 sq. ft. of water surface in the nonswimming areas, i.e., portions of the swimming pool with a water depth of 5 ft. or less
(b) = one person per 25 sq. ft. of water surface in the swimming areas, i.e., portions of the swimming pool with a water depth greater than 5 ft.
(c) = 10 persons per diving board

Minimum allowances/requirements for diving boards of various heights:

	Diving Board Height		
	2 ft. or less	2 ft. to 1 m	1 m to 3 m
Water depth	8.5 ft.	10 ft.	12 ft.
Length of diving well	12 ft.	12 ft.	13 ft.
Unobstructed height above each board	13 ft.	13 ft.	13 ft.
Distance from adjacent board/side wall	10 ft./10 ft.	10 ft./10 ft.	10 ft./12 ft.
Diving stand handrail	—	required	required

SOURCE: Adapted from American Public Health Association, *Public Swimming Pools: Recommended Regulations for Design and Construction, Operation and Maintenance.* Copyright © 1981 by the American Public Health Association. Adapted with permission.

employee for every 7 to 10 acres should produce "A-Level" service—in other words, "a high frequency maintenance service" (City of Pasadena Management Audit Team, 1986, p. 9.4). Standards of the maintenance-employee-per-park-acreage variety, however, are complicated by the question of developed versus undeveloped park acreage noted previously and are therefore relatively rare.

Tree Maintenance. Many parks and recreation departments are responsible for tree maintenance, including tree trimming and removal, on park property and sometimes on other municipal property throughout the community. A variety of standards adopted by the National Arborist Association (NAA) relevant to this function address quality of tree care (Table 14.9). A qualitative assessment of a city's tree services could be based on the selection by local authorities of particular NAA standards and recommended practices identified as especially relevant to

TABLE 14.9 Sample Standards/Recommended Practices for Tree Care

Treatment of cuts and wounds with wound dressing or paints has not been shown to be effective in preventing or reducing decay and is not generally recommended for that reason.

The use of climbing spurs (hooks, irons) is not an acceptable work practice for pruning operations on live trees.

Trees susceptible to serious infectious diseases should not be pruned at the time of year during which the pathogens causing the diseases or the insect vectors are most active. Similarly, if pruning wounds may attract harmful insects, pruning should be timed so as to avoid insect infestation.

To assist in new plant establishment use a high phosphorus fertilizer, such as those with nitrogen, phosphorus, and potash (N-P-K) ratios of 0-20-0, 0-46-0, 4-12-4, or 5-10-5.

Use fertilizers with N-P-K ratios of 3-1-2 or 3-1-1 for best response with established plantings.

For optimum plant growth, it is generally recommended to apply 3 lbs. of actual nitrogen per 1,000 sq. ft. of area under the branch spread of the tree per year, or 6 lbs. every two years. An alternative approach involves measuring the trunk diameter at $4\frac{1}{2}$ feet above grade ("diameter breast high" or DBH). Generally for optimum growth, apply $\frac{1}{4}$ lb. actual nitrogen per inch DBH to trees under 6 inches in diameter. The rate can be increased to $\frac{1}{2}$ lb. N per inch DBH for most trees over 6 inches DBH.

Autumn is an ideal time to fertilize, generally after the first hard freeze and until the moisture in the soil freezes and root activity ceases. In the southern areas of the country where the ground does not freeze, root growth in many cases will continue all winter long. Early spring, *before* budbreak is also an appropriate time. When leaves have fully expanded, fertilizing can continue until early July. However, treatments of readily available inorganic nitrogen between July and September could promote a late flush of growth that may not harden off before freezing temperatures in autumn, and injury could occur. Mid to late-summer fertilizer applications should be limited to correcting specific element deficiency problems.

SOURCES: National Arborist Association, *Pruning Standards for Shade Trees* (Amherst, NH: NAA, 1988), pp. A-2, A-3; National Arborist Association, *Standard for Fertilizing Shade and Ornamental Trees* (Amherst, NH: NAA, 1987), pp. C-1, C-2, C-3. Reprinted with permission.

municipal operations. The city would simply report the percentage of those selected standards and practices with which it complies.

Another organization concerned with tree care is the National Arbor Day Foundation, sponsor of the Tree City USA program. Unlike the standards and recommended practices of the NAA, which are directed more heavily toward technical guidance to tree care specialists, the standards promoted by the National Arbor Day Foundation are more directed toward local policy, governance, resource commitments, and ceremony (Table 14.10).

Reasonably precise guidance for assessing the quality of tree care is found in a checklist developed by the parks division of Sunnyvale, California (Table 14.11). The local standard adopted by Sunnyvale requires that at least 80% of that community's park trees be in compliance with the standards of safety, aesthetics, health, and utility specified in that checklist. Further guidance from individual cities may be found in local expectations regarding promptness of tree pruning and removal (Table 14.12). In emergency situations, Boston and Cincinnati expect tree work to be completed within 24 hours. Routine work is given much greater leeway.

Grass/Weeds/Turf Mowing. The acceptable height of grass in a given community is influenced by several factors, including local taste, resource availability, and

TABLE 14.10 Tree City USA Standards

STANDARD 1: A Tree Board or Department

A tree board is a group of concerned citizens, usually volunteer, charged by ordinance to develop and administer a comprehensive community tree management program for the care of trees on public property. Tree boards usually function with the aid of professional foresters. In communities with a population of more than 10,000, city forestry departments with salaried employees are often feasible. These departments may or may not be supported by advisory boards or administrative commissions.

STANDARD 2: A Community Tree Ordinance

The community tree ordinance needs to designate the tree board or department and give them the responsibility for writing and implementing the annual community forestry work plan. The ordinance should determine public tree care policies for planting, maintenance, and removals. Ideally, the city tree ordinance will make provisions for establishing and updating a list of recommended street tree species to be planted with spacing and location requirements. A sample tree ordinance may be obtained by writing the National Arbor Day Foundation.

STANDARD 3: A Community Forestry Program With an Annual Budget of at Least $2 per Capita

Many communities begin their program by taking an inventory of the trees growing on public property. The species, location, and condition of each tree are noted (i.e., healthy, needs pruning, should be removed, etc.) and the inventory data is summarized in a written report for presentation and approval by the city council. The report should be an objective analysis of the present state of the urban forest with recommendations for future management. The essential, ongoing activity for the care of trees along streets, in parks, and in other public places is the community forestry program. The annual work plan should address planting, watering and fertilizing, dead and hazardous tree removal, safety and fine pruning, and insect and disease control. To be named as a TREE CITY USA, a town or city must annually spend at least $2 per capita for its annual community forestry program. Consider all funds spent for tree care—budget for street tree department or board, park department's tree expenditures, dead tree removal, etc.

STANDARD 4: An Arbor Day Observance and Proclamation

An Arbor Day observance can be simple and brief or an all-day or all-week observance. A proclamation issued by the mayor must accompany the observance and declare the observance of Arbor Day in your community.

SOURCE: National Arbor Day Foundation, "Tree City USA Application" (Nebraska City, NE: National Arbor Day Foundation, n.d.). Reprinted by permission.

property use. Residents of an affluent community may expect their municipal soccer fields to be closely cropped, whereas slightly taller grass along roadways may be perfectly acceptable. Necessary mowing frequencies are influenced, of course, by desired grass height and also by local climate and rainfall. In establishing their own local standards, communities may find the turf height standards, mowing frequencies, and rule-of-thumb labor ratios of their counterparts helpful (Table 14.13). In addition, standards used for rating the quality of turf maintenance in a study of Los Angeles-area cities are shown in Table 14.14.

Park Structures, Fixtures, and Playgrounds. Sunnyvale, California, has established rigorous standards for the maintenance of park structures and fixtures (Table 14.15) and for playgrounds (Table 14.16). The parks division is expected to maintain at least 80% of its structures and fixtures in the conditions specified in the checklist. The playground maintenance standard requires repair or removal of unsafe equipment within 48 hours (Walker, 1989).

TABLE 14.11 Park Tree Quality Standards Checklist

Safety
- There should be no hanging/cracked limbs, or other hazards obvious from the ground.
- There should be no major structural damage evident.
- Limbs and/or foliage of semi-mature and mature trees should not exist below the height of ten feet over pedestrian rights-of-way or below the height of fourteen feet over vehicular rights-of-way.
- Tree stumps should not present trip hazards.

Aesthetics
- Trees should display at least some semblance of the form common to the species. Some allowance shall be made for natural "character," but grossly misshapen trees will not pass muster.
- There should be no stubs, dead "flags," or other unsightly distractions.
- Stumps should not be visible. They should either be removed to at least twelve inches below ground (e.g., in open turf) or hidden by existing vegetation (e.g., planter beds).

Health
- New trees should be planted in accordance with approved specifications.
- Trees should have stakes removed as soon as the trunk is self-supporting, or one year from the date of planting, whichever comes first. Approval to either maintain stakes for periods in excess of one year, or to remove trees which are not self-supporting, shall be secured from a program supervisor prior to acting.
- Obvious *and debilitating* pest and/or health problems (insects, disease, man-made problems) should be non-existent.

Utility
- Trees planted for a specific reason (e.g., to serve as visual screens, noise barriers, or to provide summer shade) should serve the intended purpose.
- Tree limbs should not interfere with buildings, utility wires, etc.
- Limbs and/or foliage should not obstruct park signs. Tree roots should not interfere with the utility of surrounding facilities or fixtures.

SOURCE: Robert A. Walker, *Parks Division Quality Standards Manual* (City of Sunnyvale, CA, 1989), pp. 9-10. Reprinted by permission.
NOTE: The City of Sunnyvale, California, has set as its target maintaining at least 80% of its park trees in compliance with the standards of safety, aesthetics, health, and utility specified in this checklist.

TABLE 14.12 Tree Pruning and Removal: Standards of Promptness From Selected Cities

Overland Park, KS
Target: Investigate within 36 hours and perform needed pruning within 24 hours to 30 days, depending on nature of problem

Orlando, FL
Target: Respond to citizen tree requests within 7 working days
Actual: 74% (1991)

Boston, MA
Percentage of emergency requests responded to within 24 hours: 100% (1992)

Cincinnati, OH
Target: Average of 60 days or less after request for routine pruning or removal
Actual: 34-day average (1991)
Target: By next working day following notification of tree damage from storms, accidents, or vandalism
Actual: 93% (1991)

TABLE 14.13 Grass/Weeds/Turf Mowing: Selected Cities

Turf Height	Mowing Frequencies	Labor Ratios
Cincinnati, OH Target: Maximum acceptable grass height along roadways, city property—4 inches **College Park, MD** Target: Maximum acceptable grass height along roadways, city property—3 inches **Milwaukee, WI** Target: Ensure weed growth does not exceed 12 inches, per state statute and city ordinance **Sunnyvale, CA** Target: 1.5 to 2 inches throughout parks	**Rock Hill, SC** Procedure: City-owned roadsides and plant strips—every 21 days; athletic fields—every 7 days; parks, neighborhood centers, utility sites—every 10 days; city hall lawn and selected park areas—every 7 days; selected properties for class C maintenance—once per month **Fort Collins, CO** Procedure: Mowing frequency in parks—30 times per year (1991) Target: Mow road shoulders at least 4 times per growing season **Reno, NV** Procedure: Mowing frequency in parks—once a week during growing season **Greenville, SC** Procedure: Mowing of city street right-of-way—40-day cycle (1991) **Tallahassee, FL** Procedure: Mow and trim street right-of-way 4 times annually	**Little Rock, AR** 2,500 to 3,500 square yards of undeveloped property cleared of brush and mowed per day by 2-person crew **Wichita, KS** Park and public area acres mowed per work-hour: 1.19 (1990); 1.32 (1991) Airfield acres mowed per work-hour: 3.83 (1990); 5.04 (1991) **Winston-Salem, NC** (Wastewater Collection Program) Work-hours per mile of right-of-way mowed: 14.57 (1991); 16.80 (1992) **Study of Los Angeles-area cities, 1984[a]** Work-hours (direct labor) per acre mowed: 5.27 (sample mean); range: 2.14 to 9.87

a. Barbara J. Stevens (Ed.), *Delivering Municipal Services Efficiently: A Comparison of Municipal and Private Service Delivery* (Prepared by Ecodata, Inc., for the U.S. Department of Housing and Urban Development, June 1984), p. 382.

Sports Facilities. Recreational facilities that serve as the setting for athletic contests usually require extraordinary maintenance. As in the case of other parks maintenance functions previously noted, a checklist developed by the Sunnyvale parks division may be used as a qualitative assessment tool for judging the maintenance of sports fields and other recreational facilities (Table 14.17). The assessment may be made on the basis of the average percentage of relevant standards with which each facility complies or by the percentage of facilities in total compliance with relevant standards.

Although every municipality may wish to design its own standards to reflect local preferences and conditions, it need not start from scratch. The maintenance procedures, standards, and rule-of-thumb labor ratios for the maintenance of various recreational facilities in other municipalities (Table 14.18), as well as

TABLE 14.14 Standards for Rating the Quality of Turf Maintenance

Ratings for Edging Work

1 = *Excellent:* Grass cut smoothly at sidewalk's edge; distinct straight line

2 = *Good:* Grass beginning to grow into edged area, although it has not grown more than $\frac{1}{2}$ inch

3 = *Fair:* Grass has grown 1 inch into edged area

4 = *Poor:* Grass has grown more than 1 inch into edged area

Ratings for Grass Color

1 = Medium green

2 = Light green

3 = Yellow green

4 = Brown

Ratings for Grass Height

1 = $\frac{1}{2}$ inch to $1\frac{1}{2}$ inches

2 = $1\frac{1}{2}$ inches to $2\frac{1}{2}$ inches

3 = $2\frac{1}{2}$ inches to 4 inches

4 = 4 inches to 6 inches

Ratings for Weeds

1 = *Excellent:* No weeds found in observation area

2 = *Good:* Weeds found in 10% or less of observation area

3 = *Fair:* Weeds found in 10% to one third of observation area

4 = *Poor:* Weeds found in more than one third of observation area

Ratings for Percentage of Grass Cover

1 = 100% to 75%

2 = 75% to 50%

3 = 50% to 25%

4 = 25% to 0%

SOURCE: Barbara J. Stevens (Ed.), *Delivering Municipal Services Efficiently: A Comparison of Municipal and Private Service Delivery* (Prepared by Ecodata, Inc., for the U.S. Department of Housing and Urban Development, June 1984), p. 379.

NOTE: Average quality rating in 1984 study of Los Angeles-area cities was 1.71 (range of 1.14 to 2.76).

guidelines devised by the NRPA (Table 14.19), may be useful to a community deciding on its own standards, procedures, and resource requirements.

PROGRAM OPERATIONS STANDARDS

Among the operating standards of relevance to public recreation programs are those established by APHA relating to public swimming pools. APHA standards address pool capacity, the proper ratio of attendants and lifeguards to pool area occupants, and chemical testing of the water (see Table 14.8). APHA also has established standards for lifesaving and first aid equipment at public swimming pools (Table 14.20).

Operating standards may also include proper training and credentials of operating personnel. A prime example is the special training required by most municipalities for swimming pool lifeguards. Many require American Red Cross certification or its equivalent (Table 14.21).

PROGRAM OFFERINGS AND ADMINISTRATIVE FUNCTIONS

The success of a parks and recreation department is influenced not only by the adequacy of its facilities and the productivity of its maintenance operations but

TABLE 14.15 Park Structures and Fixtures Quality Standards Checklist

BACKFLOW PREVENTION DEVICES should

- be locked to prevent tampering
- be freshly painted and graffiti-free
- be camouflaged attractively where possible
- be accessible to maintenance personnel

BACKSTOPS (portable) should

- be free of graffiti
- be securely anchored to the ground
- include framing, fabric, and hardware in good repair

BENCHES should

- offer a relatively smooth seating surface
- be secure and sturdy
- be free of unintended protrusions (e.g., nails, fasteners)
- be clean (especially seat and back supports)
- be properly sealed
- be graffiti-free

BIKE RACKS should

- not present a hazard to pedestrians
- be securely anchored
- be painted, where appropriate
- be clean and without graffiti
- be functional

BLEACHERS should

- offer smooth and sturdy seating surfaces

- offer railings for safety and support
- present skid-resistant foot surfaces
- be free of unintended protrusions
- be clean and free of graffiti, litter, and weeds
- be properly sealed and/or painted where applicable

BOLLARDS should

- be visible and obvious
- be tall enough to preclude pedestrian trip hazards
- be securely fastened and/or locked
- be free of sharp edges or unintended protrusions
- be clean and free of graffiti
- be properly sealed or painted where applicable
- be spaced to serve the intended purpose (deter traffic, prevent soil erosion, define separate user areas, etc.)

CIGARETTE BUTT CANS should

- be clean and free of litter and/or graffiti
- have sand replaced as needed

DRINKING FOUNTAINS should

- be securely anchored
- be handicap-accessible
- provide a steady flow of potable water when activated

- allow no water flow when not activated
- drain freely and completely
- be clean and free of debris
- be graffiti-free

DUMPSTERS AND THEIR ENCLOSURES

- should be properly signed
- should be free of graffiti
- dumpster should be enclosed
- dumpster should have a functional lid
- enclosure should include a functional gate
- enclosure should be clean and free of debris

FENCES AND GATES (chain link and wood)

- should be free of graffiti
- should be in good repair, with no boards, chain link, or hardware missing, bent, or broken
- should be sturdy and structurally sound
- should be properly sealed or painted (wood)
- gates should swing freely, with latches properly attached and functional

FLAGPOLES

- should be accessible and visible
- should include functional cables and hooks
- should be painted and free of graffiti
- flags should be in reasonable repair
- flags should be raised daily where appropriate

also by the efficiency and effectiveness with which its administrative functions are performed. Table 14.22 offers potential benchmarks ranging from selected administrative functions to program coverage, participant safety, and the space adequacy of swimming pools.

TABLE 14.15 *Continued*

LIGHT STANDARDS AND FIXTURES should	PLANTER BOXES, RAISED BEDS, AND CONTAINER PLANTS should be	TRELLISES, ARBORS, GAZEBO should
• be properly anchored or secured	• in good repair	• be free of graffiti
• be functional (bulbs and globes operational)	• free of graffiti	• be properly secured
• include shades in good condition	SIGNAGE should be	• include appropriate hardware properly attached
• be clean and fee of graffiti, cobwebs, birds' nests, etc.	• free of graffiti	• be structurally sound
• be numbered for maintenance repairs	• visible and legible	• be painted/sealed
• present maintenance personnel a secure and accessible hand hole	• painted or sealed where appropriate	UTILITY BOXES should be
PAY PHONES should be	• properly secured	• clean and free of graffiti
• secure and operational	• located so as not to present a hazard to park visitors	• painted the standard color
• clean and free of graffiti	STATUES, SCULPTURES, AND ARTWORK should	• accessible and identifiable
	• be clean and free of graffiti	• locked and securely attached to ground or structure
	• be properly secured	WASTE CONTAINERS should
	• have no loose or missing parts	• be clean and free of graffiti
		• be locked so as not to impede traffic or present hazards
		• be free of major cracks, holes, etc.
		• include functional lids and liners

SOURCE: Robert A. Walker, *Parks Division Quality Standards Manual* (City of Sunnyvale, CA, 1989), pp. 17-20. Reprinted by permission.
NOTE: The City of Sunnyvale, California, has set as its target maintaining at least 80% of its park structures and fixtures in the condition specified in this checklist.

COST RECOVERY

Especially during periods of fiscal constraint, the question of how much of a recreation program's costs the community should recoup from fees can be a high-profile concern. Awareness of the cost recovery targets and results in other communities can provide a helpful context for local deliberations (Table 14.23).

Note

1. Much of this section on the history of parks and recreation standards is drawn, with few adaptations, from David N. Ammons (1995b), "Overcoming the Inadequacies of Performance Measurement in Local Government: The Case of Libraries and Leisure Services," *Public Administration Review, 55* (January/February), pp. 37-47. Used with permission.

TABLE 14.16 Playground Quality Standards Safety Checklist

GENERAL

- All playgrounds should contain an approved surface material [see *Play for All Guidelines* (c) 1987] spread over the entire playing area to a minimum depth of ten inches (preferred 12 inches) and to within six inches of the curbing top.
- Surface materials should be free of hazardous materials (e.g., broken glass, animal feces).
- All playgrounds should be free of trip hazards (e.g., roots, obstacles).
- All concrete footings should be covered by surface material—none should be exposed.

ALL PLAYGROUND APPARATUS should be free of

- loose footings or loose structural supports
- sharp edges
- worn parts
- missing parts

METAL PARTS should be free of

- enlarged bolt holes or loose or protruding nuts and bolts
- bent, broken, or severely worn pipe
- exposed ends of tubing intended to be plugged or capped

WOODEN PARTS should be free of

- splinters
- structural cracks (as opposed to cosmetic cracks or natural "checking")
- dry rot or insect (e.g., termite) damage that impacts the structural integrity of the apparatus

SYNTHETIC PARTS (e.g., FIBERGLASS, PLASTIC, RUBBER, etc.) should be free of

- cracks, rot, or wear

SWINGS

- Seats, straps, and rivets should be sturdy and show no signs of deterioration.
- Chains, S-hooks, swing hangers, and clevis assemblies should be in good working order and show limited signs of wear. Chain links and S-hooks should be replaced before wearing through $\frac{1}{4}$ of the metal's diameter.
- S-hooks should be completely closed.
- Swing hangers and other moving parts should move freely.

SLIDES

- Apparatus should be free of structural stress or damage.
- Ladders should be sturdy and complete (no missing rungs or rails).
- Bedway exits should be parallel to the ground.
- Bedways, bedrails, and handrails should be free of foreign objects, holes, and rough edges.

WHIRLS

- Surface should not be worn slick or have holes.
- Bearings should move freely.
- Handrails should be sturdy and free of holes or rough edges.

ROCKING AND BOUNCING EQUIPMENT

- All moveable parts should perform their intended function.
- Spring castings should not be loose.
- Seats should be structurally sound.
- Axis of seesaws should be free of pinch or crush points.

CLIMBERS

- Should have tight fittings; bars and pipes should be snug.
- All parts should be present and intact and serve the intended purpose.

SOURCE: Robert A. Walker, *Parks Division Quality Standards Manual* (City of Sunnyvale, CA, 1989), pp. 28-30. Reprinted by permission.

TABLE 14.17 Recreational Facilities Quality Standards Checklist

ATHLETIC FIELDS: Ball Diamonds

- Entire area should be free of glass and litter.
- Bleachers should be clean, graffiti-free, and in safe condition.
- Backstops and chain link fences must be in safe condition—i.e., properly attached, no protrusions, snags, or holes.
- Adequate lighting should be provided.
- Permanent scoreboards should be in good condition—graffiti-free and functional.
- Playing field should be free of washboard effects, holes, and depressions.
- A minimum of 1" of approved surface material should be maintained along base paths (all fields) and throughout infields where appropriate (additional amounts to be determined by park staff).
- Turf infield should be maintained at 1.5" height.
- Field markings and boundaries should be accurate and clearly defined (i.e., batters box, 1st base line, foul lines—see schematic for appropriate type of field).
- All lines should be of uniform width.
- All plates should be firmly planted and level with surrounding surface material.
- Concrete dugouts should be free of debris, soil, and graffiti.
- Litter receptacle should be safe, functional, and attractive.
- Distinction between infield and outfield should be accurately measured and clearly defined.

ATHLETIC FIELDS: Soccer Fields, Multipurpose Fields, Outfields

- Playing field should be reasonably level.
- There should be no holes or large dips.
- The area should be free of litter.
- During the soccer season, fields should be clearly lined according to the appropriate schematic.
- Soccer goals should be functional, secure, and in good repair. Poles should be secure and free of metal burrs.

BASKETBALL COURTS

- Fence and gates should be in good repair. Appropriate signs should be visible and securely attached.
- Nets should be in good repair.

- Hoops and backboards should be in good repair and securely fastened.
- Court surfaces should be in playable condition: smooth, without large cracks, clean, and with clearly visible and well-defined lines.
- Court lamps should be present and operational.
- Timers should be set properly and in working condition.

BOWLING GREEN

NOTE: The City's intent is to establish quality standards that provide user groups with facilities that meet *their* needs. Because the needs of lawn bowlers *cannot* be met by normal lawn surfaces, these quality standards differ greatly from those related to general park turf. It is important that staff not judge one by standards intended for the other.

- Turf should be maintained at the lowest uniform height possible.
- Turf should be level with the height of perimeter base boards (if not level, it should be kept above rather than below board height).
- Playing surface should be firm and level.
- Playing surface should be as dry as continued maintenance of the turf will allow.
- Turf *color* is of little consequence: Staff should attempt to maintain a green lawn but should keep in mind that there is more than one way to accomplish this (e.g., use of dyes) and that wherever a conflict exists between standards of moisture and standards of color, highest priority shall be awarded to maintaining a surface "as dry as continued maintenance of the turf will allow."
- Turf should be free of weeds, insects, [and] disease.
- Turf should be free of litter or other debris.
- Sand gutter should be maintained at a uniform depth that allows a minimum of 3" clearance between sand and top of perimeter base boards—kept weed free; boards kept in good repair.
- Surrounding hardscape and fencing should be kept clean, graffiti-free, and functional.
- Area should be secured.

HANDBALL

- Surface should be in playable condition: smooth, clean, with any lines clearly marked.

continued

TABLE 14.17 *Continued*

HORSESHOES

- The pit should be of regulation size.
- Material and sand mixture around stake should be close to level with the surrounding soil.
- The stake must be secure.
- The stake should be free of sharp edges.
- The header boards should be secure and in good condition.
- Backboards should be secure and free from major damage. The support post should be secure in the ground.
- The ground should be level and free from potholes, glass, debris, litter, and weeds.

PICNIC SITES

- Pedestrian surface should be level and free of trip hazards, litter, and weeds.
- Picnic areas should drain properly; there should be no standing water.
- Picnic tables and benches should be clean—free of staples, graffiti, deep carving, food residues, etc.
- Picnic tables and benches should be properly finished, dependent on material of table.
- Picnic tables and benches should be stable and secured to surface. There should be no loose hardware or sharp surfaces.
- BBQ pits should be clean and in usable condition (ashes kept to a minimum, hardware of grill in good repair, moving parts operating properly).
- All electrical outlets should be functional and covered.
- Sinks should be clean and drain properly. All plumbing fixtures should be present and functional.
- Drinking fountains should be clean and functional.

- Trash cans should be clean, functional, and safe.
- All picnic structures and fixtures should be clean and graffiti-free.

SHUFFLEBOARD

- Area should be free of litter, graffiti, and grit.
- All lines should be clearly visible.
- Playing surface should be smooth and free of large cracks.
- Area should be free of trip hazards.

TENNIS

- Fence and gates should be in good repair. Appropriate signs should be visible and securely attached.
- Nets should be in good repair and at proper tension. They should measure 36" from ground to top of net at center strap, or be capable of adjustment to that height.
- Poles should be properly aligned.
- Court surfaces in playable condition: smooth, without large cracks, clean, with clearly visible and well-defined lines.
- Court reservation clock should be operating properly.
- Court lamps should be present and operating. Timer should be set properly and in working condition.

VOLLEYBALL

- Area should be free of glass, litter, and animal droppings.
- Support poles should be secure and free of burrs.
- Support poles' hardware should be in working order.
- Sand should be spread evenly within court; there should be no piling up of sand near the edges.

SOURCE: Robert A. Walker, *Parks Division Quality Standards Manual* (City of Sunnyvale, CA, 1989), pp. 22-26. Reprinted by permission.

TABLE 14.18 Maintenance Standards and Labor Ratios: Selected Facilities

GOLF COURSE

Turf height (Sunnyvale, CA)
Target: Greens, $\frac{3}{16}$ inch, tees and fairways, 1 inch; roughs, 1.5 inch

Sand trap depth (Sunnyvale, CA)
Target: 4 inches

Course maintenance (Reno, NV)
Procedures and performance: Mow fairway twice per week (May-Oct.), greens three times per week (May-Oct.); aerate and top-dress greens twice per month (May-Oct.); fertilize greens once per month (May-Oct.), non-greens areas once in April and once in September.

Percentage of irrigation repairs made within 1 hour of notification: 90% (1990)

Percentage of equipment repairs made within 1 day: 85% (1990)

Labor ratio (Wichita, KS)
Rounds played per labor-hour of golf course maintenance: 3.32 (1990); 3.90 (1991)

RECREATION FIELD

Labor Ratio (Wichita, KS)
Maintenance labor-hours per sports field: 221.5 (1990); 187.1 (1991)

TABLE 14.19 Labor Ratios for Selected Parks and Recreation Maintenance Activities

Task	Labor-Hours	Task	Labor-Hours
Mowing 1 Acre,		**Flower Bed Preparation**	
Flat Medium Terrain		Cultivating combined	
at Medium Speed		shrubbery and flower bed	0.9 per 100 sq. ft.
20" walking	2.8 per acre	Spring bed preparation	3.3 per 1,000 sq. ft.
24" walking	2.2 per acre	**Planting**	
30" riding	2.0 per acre	Annuals from a flat	0.10 per 1,000 sq. ft.
72" (6-foot) riding	0.35 per acre	Weed, no mulch	1.0 per 1,000 sq. ft.
Bush hog	1.25 per acre	Mulch	0.83 per 1,000 sq. ft.
Trim		Fall bed cleanup	
Gas powered (weedeater)	1.0 per 1,000 lin. ft.	and preparation	6.6 per 1,000 sq. ft.
Planting Grass		**Ball Fields**	
Cut and plant sod by		Mowing—riding	
hand (1½" strips)	1.0 per 1,000 sq. ft.	with E-10 mower	0.5 per ball field
Cut and plant sprigs		Mowing/trimming with	
by hand (not watered)	10.9 per 1,000 lin. ft.	push-behind power mower	1.0 per ball field
Seed, by hand	0.5 per 1,000 sq. ft.	Drag infield	0.75 per ball field
Overseeding,		Clean fields, fans area,	
reconditioning	0.8 per acre	and players areas	2.0 per ball field
Fertilize Turf		Drag infield, line field	
24": sifter spreader	0.16 per 1,000 sq. ft.	plus rake	2.0 per ball field
Hand push spreader 36"	2.96 per acre	Regrade, repair, and	
Tractor towed		reconstruct	8.0 per ball field
spreader 12"	0.43 per acre	Football, soccer field,	
Weed Control		lining, general	
Spraying herbicide		maintenance	2.5 per field
w/fence line truck,		**Tennis Courts**	
tank sprayer 2 ft.		Tennis nets, check repairs;	
wide (1" either		repair includes windscreen	1.0 per court
side of fence)	0.45 per 1,000 sq. ft.	Recoat color surface	32.0 per court
Leaf Removal		**Swimming Pool Preparation**	
Hand rake leaves	0.42 per 1,000 sq. ft.	Paint pool and deck	
Vacuum 30"	0.08 per 1,000 sq. ft.	(epoxy paint rolled on,	
Planting Trees		including scraping,	
Plant tree 5-6 ft. ht.	0.45 per tree	priming, painting)	5.0 per 1,000 sq. ft.
Plant tree 2-2½" dia.	1.0 per tree	Caulking—pool and deck	120.0 each
Tree Removal		**Picnic Facilities**	
Street tree removal	13.0 per tree	Picnic grill—	
Street tree stump removal	3.5 per tree	check and clean	0.8 each
Park tree removal	5.0 per tree	Picnic grill—repair/	
Park tree stump removal	2.0 per tree	replace	4.0 each
Shrub Maintenance		**Buildings**	
Prune shrubs		6-stool comfort	
(deciduous) mature	0.50 per shrub	station—cleaning	
Prune shrubs		and maintenance	
(evergreen) mature	1.0 per shrub	(2 laborers)	1.6 per building
Hedge, trimming		Large double	
by hand, includes cleanup	2.85 per 100 linear ft.	latrine/cleaning	1.0 per building
Hedge trimming, electric;			
includes cleanup	1.50 per 100 linear ft.		

SOURCE: Excerpted from Ron Donahue (Ed.), *Park Maintenance Standards* (Alexandria, VA: National Recreation and Park Association, 1986), pp. 18-27. Used with the permission of the National Recreation and Park Association.

TABLE 14.20 Standards for Lifesaving and First Aid Equipment at Public Swimming Pools

Standard Lifesaving Equipment

(a) One ring buoy not more than 15 in. (38 cm) in diameter or similar flotation device to which shall be attached a 60-ft. (18 m) length of $\frac{3}{16}$ in. (5 mm) rope or two pineapples (tightly rolled balls of rope) composed of $\frac{1}{4}$-in. (6 mm) rope the length of which is at least $1\frac{1}{2}$ times the maximum width of the swimming pool; and

(b) One life pole or shepherd's crook type pole having blunted ends and a minimum length of 12 ft. (3.7 m).

At least one unit of lifesaving equipment, consisting of at least (a) and (b) above, shall be provided for each 2,000 sq. ft. (186 sq. m) or fraction thereof of water surface area of the swimming pool.

Standard First Aid Items

2 units—1-inch adhesive compress	1 unit—eye dressing packet
2 units—2-inch bandage compress	4 units—plain absorbent gauze, $\frac{1}{2}$ sq. yard
2 units—3-inch bandage compress	3 units—plain absorbent gauze, 24-inch × 72-inch
2 units—4-inch bandage compress	4 units—triangular bandages, 40-inch
1 unit—3-inch × 3-inch plain gauze pad	1 unit—bandage scissors, tweezers
2 units—gauze roller bandage	

SOURCE: Adapted from American Public Health Association, *Public Swimming Pools: Recommended Regulations for Design and Construction, Operation and Maintenance.* Copyright © 1981 by the American Public Health Association. Adapted with permission.

TABLE 14.21 Rescue Skills for Lifeguard Certification

Nonswimming Assist (Rescue Tube)

- Extension Assist From Deck

Entries

- Stride Jump
- Compact Jump

Approaching the Victim

- Modified Crawl
- Breaststroke
- Rescue Kicks
  Scissors Kick
  Inverted Scissors Kick
  Elementary Backstroke Kick
  Rotary Lick

Victims at or Near Surface

- Swimming Extension Rescue
- Active Victim Rear Rescue
- Passive Victim Rear Rescue

Submerged Victims

- Feet-First Surface Dive
- Submerged Victim Rescue

Multiple-Victim Rescue

Removal From Water

- Two-lifeguard Lift
- One-lifeguard Lift

Providing Emergency Care

SOURCE: Excerpted from American Red Cross, *Lifeguarding Today* (St. Louis: Mosby Lifeline, 1995), pp. 102-125. Reprinted courtesy of the American Red Cross. All rights reserved in all countries.

TABLE 14.22 Odds and Ends: Miscellaneous Parks and Recreation Benchmarks From Selected Cities

PROMPT DEPOSITING OF REVENUE

Denton, TX
90% within 24 hours of receipt (1991)

PROMPT RESPONSE TO CITIZEN REQUESTS/COMPLAINTS

Corvallis, OR
Target: Within 24 hours

Reno, NV
85% within 1 working day; 100% within 3 days (1990)

Denton, TX
80% within 48 hours

PROMPT PROCESSING OF SPECIAL PERMITS

Reno, NV
99% of park, facility, field, alcohol, and special event permits processed within 48 hours (1990)

YOUTH PROGRAM COVERAGE

Savannah, GA
9.8% of local youth participating in recreation department athletic program; 32.6% in recreation program overall (1991)

PARTICIPANT SAFETY

Winston-Salem, NC
0.02 serious injuries (requiring medical attention) per 1,000 participants at recreation centers (1992)

SPACE ADEQUACY OF PUBLIC SWIMMING POOLS

San Luis Obispo, CA
283 square feet of pool surface per 1,000 population (1992)

TABLE 14.23 Recreation Program "Cost Recovery" Benchmarks

Percentage of Overall Recreation Program Expenses Recovered Through Fees	Percentage of Swimming Pool Expenses Recovered Through Fees	Percentage of Athletic/ Sports Program Expenses Recovered Through Fees	Percentage of Golf Course Expenditures Recovered Through Revenues	Percentage of Other Program/Facility Expenses Recovered Through Fees
Fairfield, CA At least 33% (target) **San Luis Obispo, CA** 40% (1990) **Denton, TX** 40% (1991) **Cincinnati, OH** 45% (1991) **Loveland, CO** 74% (1991)	**Raleigh, NC** 10% to 79% (1991) **Tallahassee, FL** 19.9% (1991) **Oak Ridge, TN** 25% to 40% (1992) **Winston-Salem, NC** 36% (1992) **San Luis Obispo, CA** 40% (1990) **Lubbock, TX** 42% (1991) **Overland Park, KS** 50% (target) **Duncanville, TX** 69.5%	**College Station, TX** 40%-70% for adult sports (target); 0%-30% for youth sports (target) **Savannah, GA** 49% for adult sports (1991) **Duncanville, TX** 64% (excluding staff salaries) **Tallahassee, FL** 65.1% for major adult sports, 36.0% for major youth sports (1991) **San Luis Obispo, CA** 80% (1990) **Lubbock, TX** Direct costs recovered: 129% for softball leagues; 122% for volleyball leagues; 101% for basketball leagues (1991)	**Tallahassee, FL** 67.1% (1991) **San Luis Obispo, CA** 80% (1990) **Winston-Salem, NC** 85% (1991) **Orlando, FL** 100% (target); 87.6% (1991); targeted profit margin of 25% or greater on pro shop merchandise sales **Corpus Christi, TX** 108% (1991) **Overland Park, KS** 112% (1992)	**Savannah, GA** At least 30% of tennis program costs (target); 27% (actual, 1991) **Tallahassee, FL** 40.7% for park center room rental, 35.6% for tennis center, 23.8% for gymnastics program, 17.7% for arts and crafts, and 8.4% for multipurpose centers (1991) **Alexandria, VA** 82% of recreation center costs (est., 1992) **Winston-Salem, NC** 84% of stadium operating costs (1992); 149% of coliseum operating costs (1992) **Vancouver, WA** At least 90% of operating and debt service expenses of tennis and racquetball center (target) **Waukesha, WI** 100% of operating costs (excluding administration) for youth programs, 100% (including administration) for adult programs (target) **Overland Park, KS** 120% of adult programs; 40% of youth programs; 200% of tennis and aerobics costs (targets)

SOURCE: Expanded from David N. Ammons (1995). "Overcoming the Inadequacies of Performance Measurement in Local Government: The Case of Libraries and Leisure Services," *Public Administration Review*, 55 (January/February 1995). p. 46. Used with permission.

15

Personnel Administration

Professionalism in local government personnel administration implies, at minimum, competence, diligence, and integrity. More formally, professional personnel administration is associated with five fundamentals:

1. the use of merit systems, or merit principles, in employment decisions and personnel rules;
2. the administration of the personnel function by a personnel department;
3. the introduction of performance appraisals and provision for merit pay;
4. formal disciplinary and grievance procedures; and
5. an official affirmative action policy and a formal policy against sexual harassment. (Fox, 1993, p. 7)

One basis for judging personnel administration in a given municipality, then, is the extent to which it complies with these five elements of professionalism. A more highly professional personnel department may be presumed to be superior to one that is less professional.

For the most part, the standards of professional personnel administration prescribe structures, policies, and processes, rather than outputs or outcomes. As such, they are *means* to desired ends, rather than ends in themselves. Officials wishing to judge the adequacy of local operations may find far more useful gauges for that purpose among the performance statistics and targets of other municipal personnel departments.

Objective Indicators of Employee Morale

As with financial administration, in which performance is sometimes assessed by examining the overall financial condition of a community rather than by judging the performance of various financial activities, personnel administration is sometimes assessed by examining the general health of an organization from a human resources perspective. Although assessments based on broad indicators offer the advantage of relative simplicity, they can be misleading and, therefore, should be used with care.

Two general indicators of employee morale are turnover and absenteeism. If the personnel department is involved in the recruitment of job candidates; the selection, placement, training, and nurturing of new employees; and the adoption of good supervisory practices throughout the organization, then rates of employee turnover and absenteeism may be reasonable indicators of its performance. Good performance should lead to the hiring of employees who are well suited to and enjoy their work. Such employees are unlikely to seek new job opportunities elsewhere, to be involuntarily terminated from employment, or to experience high rates of absenteeism. Careful selection, proper training, suitably challenging work, and good supervision all contribute to individual job satisfaction that *ceteris paribus* should be reflected in long tenure and low rates of turnover and absenteeism.

Average length of employment at the city of Fort Collins, Colorado, was 8.5 years in 1991. Although few cities join Fort Collins in reporting tenure statistics, several provide information on turnover and sick leave use. On the basis of such reports, annual turnover rates in the 6% to 10% range do not appear to be unusual (Table 15.1).

TABLE 15.1 Employee Turnover Rates in Selected Cities

City	Employee Turnover Rate
Irving, TX	5.02% (1991)
Fort Collins, CO	5.9% (1991)
Fayetteville, AR (rate of voluntary turnover)	6.7% (1991); 6.3% (1992)
Tucson, AZ	6.51% (1989); 7.97% (1990); 7.9% (1991)
Raleigh, NC (turnover rate for new hires[a])	8% (1991)
Fort Worth, TX	8.7% (1991) (target: 7% or less)
Longview, TX	9% (1991); 10% (1992)
Shreveport, LA	10% (1991)
Denton, TX	11% (1991)
Waukesha, WI	17.4% (1991); 12.4% (1992)
Duncanville, TX	16% (1990)

a. Defined by the city of Raleigh as separation within the first 6 months of employment.

TABLE 15.2 Rates of Sick Leave Use: Selected Cities

City	Target	Annual Sick Leave Use Rates	
		Employees in General	*Specific Departments*
Boston, MA		10.0 days per employee (1992)	Parks & recreation: 10.81 days; public works: 13.78 days; personnel dept.: 4.64 days (1992)
Cincinnati, OH	(a) less than 5 days per data processing employee; (b) 5% of total hours or less for sanitation employees		Data processing: 3.4 days; sanitation: 3.7% (1991)
Fort Collins, CO		40 hours per employee (1991)	
New York City			Water & sewer: 3.36%; parks & recreation: 3.75%; sanitation: 3.85%; finance: 3.63%; personnel: 3.80%; police: 2.26%; transportation: 3.96% (1992)
Orlando, FL	(a) not more than 2.5% for public works and parks & recreation employees; (b) not more than 2% for human resources dept. employees; (c) not more than 2% for city employees as a whole		Public works: 1.74% ; parks & recreation: 2.24%; human resources dept: 2% (1991)
Savannah, GA	(a) 6 days or less per facilities maintenance employee; (b) 5.8 days or less per public development employee; (c) 5.03 days or less per police employee; (d) 7.4 days or less per fire department employee; (e) 4.5 days or less per leisure services employee; (f) 5.1 days or less per sanitation employee		Facilities maintenance: 6.10 days; public development: 5.8 days; police: 5.03 days; fire: 7.8 days; leisure services: 4.5 days; sanitation: 5.4 days (1991)
Tucson, AZ		6.76 days per employee (1990)	

Municipalities reporting rates of sick leave use typically do so either as days of sick leave used per employee annually or as the percentage of total hours consumed by sick leave (Table 15.2). Cities that reported sick leave use rates in the form of "days used" generally reported 5 to 8 days per employee—sometimes higher for workers exposed to inclement weather and other adverse field conditions and lower for office employees. For cities reporting use rates as a percentage of total time for the average employee, rates generally fell within the 2% to 4% range.

Judgments based on turnover and absenteeism should be rendered cautiously. As in the case of judgments on financial management based on overall measures of financial health, important factors other than the quality of personnel administration may be influencing those numbers. Low rates of turnover and absenteeism often accompany good supervision and high morale, but low rates could also reflect a tight local economy and few job options. Furthermore, it is important to recognize that some degree of turnover—preferability a modest rate—is healthy. An organization that hires good employees should expect to lose some of them to opportunities outside the organization. A moderate rate of turnover opens opportunities for promotion, without which the organization's rising stars probably could not be retained.

Hiring New Employees

In some cities, duties associated with the recruitment, selection, and hiring of new employees are decentralized and handled almost entirely by operating departments. In other cases, a central personnel department plays an important role and appropriately may be judged on the municipality's overall proficiency in handling those duties.

In the absence of nationally accepted standards for the performance of employment activities, municipal officials are left to establish their own local standards. In its effort to generate strong candidate pools, for example, the city of Orlando, Florida, strives to secure five or more qualified applicants for at least 90% of its vacancies. In labor markets having fewer job opportunities, an even higher standard may be appropriate.

Some personnel departments establish local standards for promptness in conducting qualifying examinations. Dayton, Ohio, for example, tries to schedule at least 80% of all exams within 120 days from the time they were requested. Oakland, California, considers 135 days to be a reasonable lead time for conducting an exam for sworn positions and 90 days to be reasonable for an exam for nonsworn positions.

Outstanding personnel departments waste little time in referring qualified applicants to hiring departments. Otherwise, delays in the process may lead to the loss of top candidates, who decide not to wait and instead accept offers from more expedient employers. Municipalities reporting the time it takes the personnel department to refer job candidates generally fall into two categories: those requiring no eligibility list prior to forwarding applications, thereby permitting a quick turnaround, and those requiring the personnel department to carefully review candidate qualifications and establish an eligibility list (Table 15.3). When an eligibility list is not required, it is reasonable to expect applications to be forwarded

TABLE 15.3 Prompt Referral of Job Candidates

When Eligibility List Is Not Required	*When Eligibility List Is Prepared*
Greenville, SC Target: Referral of applications to departments within 1 day following cutoff date Actual: 100% (1990) **Oklahoma City, OK** Target: Referral of qualified applicants to departments within 3 days of vacancy closing Actual: 2.6-day average (1991); 3-day average (1992) **Oak Ridge, TN** Target: Forward applications to hiring manager within 3 days following application deadline **Raleigh, NC** Target: Referral of applications within 5 workdays of application closing date Actual: 91% (1991) **Alexandria, VA** Average number of working days between application closing date and referral of qualified applicants to departments: 5 workdays (1990); 6 workdays (1991); 9 workdays (1992); 10 workdays (1993)	**Savannah, GA** Target: Certify qualified applicants for civil service positions within 5 weeks of requisition Actual: 100% (1988); 90% (1989); 96% (1990); 90% (1991) Target: Certify qualified applicants for professional/administrative positions within 6 weeks of requisition Actual: 75% (1988); 90% (1989); 94% (1990); 90% (1991) **Anaheim, CA** Target: Recruit and provide lists of eligible applicants within 50 days after recruitment begins **Phoenix, AZ** Target: Establish eligibility lists within an average of 50 workdays from the recruitment order Actual: Average of 42 workdays (estimate, 1993) **Portland, OR** Target: 2 months to establish eligibility lists Actual: 2.25 months (estimate, 1992) **Cincinnati, OH** Targets: (a) 120 days to issue eligibility lists from structured exams; (b) 60 days from unstructured exams Actual: (a) 75-day average with structured exam; (b) 50-day average with unstructured exam (1991) **San Luis Obispo, CA** Target: 60 days to establish eligibility list Actual: 90-day average (1990)

to the hiring department within a few days. When an eligibility list is required, periods ranging from 5 weeks to 3 months are not uncommon due primarily to the testing process.

Some cities record the time consumed from beginning to end in the recruitment and hiring process. Others report only a major component—for example, the time from the application deadline to the hiring of the selected candidate. Decatur, Illinois, holds the speed record among the performance documents examined, with an average turnaround time of less than 38 days for filling vacancies (Table 15.4). Some cities establish different timetables depending on the type of recruitment. For example, Oak Ridge, Tennessee, expects to proceed from application deadline to selection in 3 months for managerial positions, in 2 months for public safety employees, and in 1 month for all others.

TABLE 15.4 Prompt Filling of Job Vacancies

Decatur, IL
  Average turnaround time for vacancies: 37.8 days (1991)

Reno, NV
  Percentage of recruitments completed within 60 days: 65%; within 120 days: 100% (1990)

Corvallis, OR
  Percentage of positions filled within 90 days of vacancy: 79% (1992)

Classification and Compensation Analysis

Centralized personnel departments typically have responsibility for conducting studies or audits to determine the appropriateness of a given position classification and associated compensation for a particular employee or set of employees. Often, such studies are triggered by requests from supervisors or department heads who suspect that a classification or pay level is out of line. How quickly should the personnel department be expected to respond? Several cities attempt to complete position audits within 30 days of the request (Table 15.5).

Employee Grievances

The city of Raleigh, North Carolina, reported 1.8 grievances per 100 employees during a recent year. Although such a figure could serve as a useful benchmark of employee relations, several other measures permit a direct assessment of personnel administration performance in dealing with employee grievances.

TABLE 15.5 Prompt Position Audits for Reclassification and Compensation Requests

Oklahoma City, OK
  Target: Completed within 30 days
  Actual: 100% (1992)

St. Petersburg, FL
  Target: Completed within 30 days

Portland, OR
  Target: 90% completed within 30 days

Oakland, CA
  Percentage completed within 30 days: 80% (1991); 74% (1992)

Boston, MA
  100% completed within 48 days of request by department head or union official (1992)

Reno, NV
  70% of classification studies completed within 90 days; 80% within 120 days; 85% within 210 days (1990)

TABLE 15.6 Prompt Hearing and Resolution of Employee Grievances: Selected Cities

City	Time Frame for Hearing Appeals	Time Frame for Resolution of Grievances
Boston, MA	93.1% of grievance hearings scheduled within 3 weeks of receipt (1992)	71% of decisions issued within 2 weeks of hearing (1992)
Raleigh, NC		Targets: Grievances involving dismissals, demotions, or suspensions to be resolved within 33 workdays; all others within 75 workdays
		Actual: 67% of first category within 33 workdays; 87% of all others within 75 (1991)
Dayton, OH	Target: Schedule hearings within 45 days of receipt of request	
Oakland, CA	Target: Administrative hearing and opinion within 30 days	Target: Process appeals within 60-day maximum
Cincinnati, OH	Target: Schedule disciplinary appeals within 2 months of request	

Some cities schedule hearings and resolve grievances more promptly than others. Boston, for example, schedules most grievance hearings within 3 weeks and usually has decisions within 2 weeks of the hearing (Table 15.6). Raleigh attempts to resolve grievances involving dismissals, demotions, or suspensions within 33 workdays and all others within 75 workdays.

Although promptness in dealing with grievances is important, even more important is the successful resolution of grievances, appeals, and complaints. Reno, Nevada, for example, reported that 40% of its arbitration decisions during one recent year were favorable to the city. Oakland, California, reported that 75% of its employee grievances were resolved at the third step of its grievance procedures (resolution by the employee relations officer). Savannah, Georgia, strives each year to resolve at least 95% of Civil Service Board appeals, employee grievances, and Equal Employment Opportunity Commission (EEOC) complaints without adverse ruling by the Civil Service Board, EEOC, the Department of Labor, or the courts.

Employee Training

Many other duties are performed by the typical personnel department, but one of particular interest here is employee training. Although many municipal employees receive the bulk of their training in formal college or trade school programs, in training offered by consultants, or in training opportunities offered in conjunction with professional conferences, personnel departments often provide oversight and frequently arrange on-site training opportunities.

Of particular relevance to the training function was a 1993 recommendation of the National Commission on the State and Local Public Service (1993), popularly known as the Winter Commission. That commission encouraged states and localities to appropriate a "learning budget" equal to at least 3% of total personnel costs (p. 42). Although the recommended learning budget prescribes resource inputs and, as such, is not truly a performance benchmark, some municipalities may nevertheless wish to compare their own employee development appropriations with the Winter Commission's 3% standard.

16

Police

Despite an abundance of potential benchmarks for police services, their use as measuring rods for judging or comparing municipal police departments has been controversial. When headlines in the local newspaper, for instance, proclaim that the community's crime rate is below the state or national average, the police chief and other local officials are understandably proud. A low rate of crime is touted as another example of the community's high quality of life and a testimony to its competent police department. On the other hand, when local news media bemoan a community's "crime problem," charge that police efforts are ineffective, and cite higher-than-average crime rates to substantiate their contentions, several predictable arguments against the use of crime rates invariably are raised by police officials and supporters of the local police force.

Nationally collected crime statistics, formally known as Uniform Crime Reports (UCR), are not above challenge. Critics point out that crime statistics can only be as accurate as the data provided to the Federal Bureau of Investigation (FBI) by individual police departments. If individual departments are sloppy in their data collection or reporting, or if they intentionally misrepresent their performance statistics, the quality of the UCR is thereby diminished. Furthermore, the UCR includes statistics only on *reported* crimes. Victimization studies indicate that many crimes go unreported and, therefore, are excluded. Finally, critics of the use of UCR crime rates for evaluating police departments correctly contend that a host

184

of community factors other than police performance also contribute to a community's rate of crime.

So, where does this leave a local government official searching for appropriate police benchmarks? Should the FBI's nationally collected UCR crime rates be used or not? The simple answer is, "yes, they should"—but users should be clear about the limitations of UCR data, and in most cases, the national average may not be the best figure to use. Often, more precise community benchmarks can be found.

FBI Crime Statistics

More than 16,000 law enforcement agencies serving 95% of the U.S. population participate in the uniform crime reporting program of the FBI (U.S. Department of Justice, 1994, p. 1). Their collective input generates an extensive set of data providing insights into four categories of violent crime (murder and non-negligent manslaughter, forcible rape, robbery, and aggravated assault) and four property crimes (burglary, larceny-theft, motor vehicle theft, and arson). These crimes (sometimes excluding arson) constitute what are known as "Part I" crimes, with more than 14 million incidents reported annually.

The FBI itself (U.S. Department of Justice, 1994) warns against "simplistic and incomplete analyses" (p. iv) that merely compare crime index figures for one community with those of another. Unless cities are carefully matched or unless composite statistics take region and community size into consideration, too many other factors may explain the crime rate difference—even assuming equal diligence in reporting. Police officials who contend that their department's efforts are only one component in the battle against crime are correct. Nevertheless, those officials must remember that their department's reason for existence is to reduce the incidence of crime. It is reasonable for the public and public officials to desire a local crime rate near, or perhaps below, the norm for similar communities. Most police officials probably expect that themselves.

Recommended Performance
Measures and Standards

Appropriately, most sets of recommended performance measures for police departments have not focused exclusively on crime rates and the incidence of particular types of crime but have also incorporated indicators of other dimensions of police performance. A set of indicators published recently by the Governmental Accounting Standards Board (GASB) as part of its series on service efforts and accomplishments (SEA) reporting includes indicators of input, output, outcomes, and efficiency (Table 16.1). Measures focusing more extensively on crime control

TABLE 16.1 Recommended Service Efforts and Accomplishments Indicators: Police[a]

Inputs	Percentage of crimes cleared
Budget expenditures	Response time
Equipment, facilities, vehicles	Citizen satisfaction
Number of personnel; hours expended	**Efficiency**
Outputs	Cost per case assigned; cost per crime cleared
Hours of patrol	Personnel hours per crime cleared
Responses to calls for service	**Explanatory Variables**
Crimes investigated	Population by age group
Number of arrests	Unemployment rate
Persons participating in crime prevention	Number of households; number of business firms
activities	Percentage of population below poverty level
Outcomes	Land area
Deaths and bodily injury resulting from crime	Dollar value of property within jurisdiction
Value of property lost due to crime	Demand
Crimes committed per 100,000 population	Calls for service
	Cases assigned

SOURCE: Excerpted from Allan Drebin and Marguerite Brannon, *Police Department Programs: Service Efforts and Accomplishments Reporting—Its Time Has Come* (Norwalk, CT: Governmental Accounting Standards Board, 1992), pp. 8-9. Reprinted by permission.
a. The recommended indicators presented in this exhibit are illustrative. They are intended to serve as a starting point for use in the development of a comprehensive set of SEA indicators for external reporting of an entity's results of operation. This exhibit does not provide illustrations of indicator disaggregation or of comparison data such as trends, targets, or other comparable entities. Both disaggregation and comparison data are important aspects of SEA reporting.

service effectiveness have been developed by Harry Hatry and his colleagues at the Urban Institute (Table 16.2).

As demonstrated repeatedly in this volume, performance measurement lies at the foundation of benchmarking. Examining the performance records of a collection of municipalities on important aspects of police work could lead to the identification of best-in-class performers for those work activities. Using such a process to develop performance standards, however, is slow and imperfect, because it depends both on the collection and conscientious reporting of performance measures by individual municipalities and on the tedious compilation of measures from those jurisdictions.

A direct attempt to establish standards for police services has been undertaken in an effort to encourage accreditation among law enforcement agencies. The Commission on Accreditation for Law Enforcement Agencies (CALEA) had accredited more than 265 law enforcement agencies by 1993 on the basis of compliance with approximately 900 standards. The availability of so many standards might seem to be a boon to benchmarking efforts, but most of the CALEA standards address processes rather than performance outputs or outcomes. Often, they call for the preparation of a plan or a written directive addressing various aspects of police operations.

Although having written policies or directives governing everything from the use of deadly force to the conducting of staff meetings is a good idea, policies and directives do not guarantee a high-achieving department. For the most part, process standards make poor benchmarks unless they focus on the outputs or outcomes of those processes.

Staffing Norms—If You Insist!

Like the adoption of recommended procedures or processes, the employment of police officers and support personnel is a matter of strategy—a choice made by local decision makers who anticipate that the employment of those personnel will be an effective *means* of achieving important program objectives. It is not an end in itself. Accordingly, the employment of law enforcement personnel should not be granted status as an output objective or even as a measure of service quality. Police department staffing is a form of input. It reflects police performance only indirectly, if at all, and should not be regarded as a primary benchmark. The major benchmarking value of staffing information is as a diagnostic device in exploring possible reasons for performance differentials.

Despite potential hazards in relying too much on staffing norms, staffing information is reported here for three reasons. First, staffing information does have diagnostic value. If a police department consistently ranks below its counterparts on various performance indicators, the possibility of understaffing is one potential factor worth exploring. Second, public officials simply are interested in staffing levels and frequently are buffeted at budget time by claims of inadequate staffing. And third, officials who use national staffing data—or who have it used *on them*—should be acquainted with existing information and be aware of how an appropriate norm for their community may differ from the national average.

The Bureau of Justice Statistics of the U.S. Department of Justice (1992) reports that in 1990, municipal police departments with 100 employees or more had an average of 21 full-time sworn officers per 10,000 residents (p. x). Individual cities, however, vary from the mean on the basis of economics, perceived crime problem, and community values. The Washington, D.C., Metropolitan Police, for example, had 74 sworn officers per 10,000 in 1990—three-and-one-half times the national average. Other cities had fewer than the national average. Stockton, California, for example, had only 11 sworn officers per 10,000 population.

FBI information differs slightly from Bureau of Justice figures—and reports municipal police staffing levels per 1,000 rather than per 10,000 population. According to FBI statistics for 1993 (U.S. Department of Justice, 1994), municipal police departments averaged 2.2 officers per 1,000 population (22 per 10,000; Table 16.3). Higher numbers were found among the largest cities (i.e., 250,000 or greater) and the smallest cities (i.e., less than 10,000 population). Lower staffing levels were found among cities of intermediate size. Furthermore, higher staffing levels were reported by northeastern and southern cities, and lower staffing levels were reported by midwestern and western cities. Total police department staffing—sworn officers and civilians—averaged 2.8 per 1,000 population.

TABLE 16.2 Recommended Measures of Effectiveness: Crime Control Services

Overall Objective: To promote the safety of the community and a feeling of security among the citizens, primarily through the deterrence/prevention of crime and the apprehension of offenders, providing service in a fair, honest, prompt, cooperative, helpful, sensitive, and courteous manner, to the satisfaction of the citizens.

Objective	Quality Characteristic	Specific Measure	Data Collection Procedure
Prevention of crime	Reported crime rates	Number of reported crimes per 1,000 population, total and by type of crime	Incident reports
	Victimization rates	Number of reported plus unreported crimes per 1,000 households (or residents or businesses), by type of crime	Household survey
	Different households and businesses victimized	Percentage of (a) households, (b) businesses victimized	Household survey, business survey
	Physical casualties	Number and rate of persons (a) physically injured, (b) killed in course of crimes or nontraffic, crime-related police work—including victims and police	Incident reports
	Peacekeeping in domestic quarrels	Percentage of domestic quarrels and other disturbance calls with no arrest and no second call within x hours	Dispatch records, incident reports
		Number of injuries to (a) citizens (after police arrival), (b) police per 100 domestic quarrel calls	Incident reports
Apprehension of offenders	Crimes "solved" at least in part	Percentage of reported crimes cleared, by type of crime and whether cleared by arrest or by "exception"	Incident reports
	Completeness of apprehension	Percentage of "person-crimes" cleared, by type of crime[a]	Incident reports, arrest reports
	Quality/effectiveness of arrest	Percentage of adult arrests that survive preliminary court hearing (or state attorney's investigation) and percentage dropped for police-related reasons, by type of crime	Arrest and court records
		Percentage of adult arrests resulting in conviction or treatment (a) on at least one charge, (b) on highest initial charge, by type of crime	Arrest and court records
	Stolen property recovery	Percentage of stolen property that is subsequently recovered: (a) vehicles, (b) other property	Incident reports, arrest or special property records
Responsiveness of police	Response time	Percentage of emergency or high-priority calls responded to within x minutes and percentage of nonemergency calls responded to within y minutes	Dispatch records
	Perceived responsiveness	Percentage of (a) citizens, (b) businesses that feel police respond fast enough when called	Household survey, business survey, complainant survey

TABLE 16.2 *Continued*

Objective	Quality Characteristic	Specific Measure	Data Collection Procedure
Feeling of security	Perceived safety	Percentage of (a) citizens, (b) businesspersons who feel safe (or unsafe) walking in their neighborhoods at night	Household survey, business survey
Fairness, courtesy, helpfulness/ cooperativeness, honesty[b]	Fairness	Percentage of (a) citizens, (b) businesses that feel police are generally fair in dealing with them	Household survey, business survey, complainant survey
	Courtesy	Percentage of (a) citizens, (b) businesses that feel police are generally courteous in dealing with them	Household survey, business survey, complainant survey
	Helpfulness/ cooperativeness	Percentage of (a) citizens, (b) businesses that feel police are generally helpful, cooperative, and sensitive to their concerns	Household survey, complainant survey
	Honesty	Number of reported incidents or complaints of police misbehavior, and the number resulting in judgment against the local government or employee—by type of complaint (civil charge, criminal charge, other service complaints)—per 100 police	Police and mayor's office records
		Percentage of citizens who feel police are in general honest and can be trusted[b]	Household survey, complainant survey
	Citizen satisfaction with police handling of miscellaneous incidents	Percentage of persons requesting assistance for other than serious crimes who are satisfied (or dissatisfied) with police handling of their problems, categorized by reason for dissatisfaction and by type of call	Complainant survey
	Citizen satisfaction with overall performance	Percentage of (a) citizens, (b) businesses that rate police performance as excellent or good (or fair or poor), by reason for satisfaction (or dissatisfaction)	Household survey, business survey, complainant survey
	Police safety	Number of injuries to police officers (a) per 100 officers, (b) per 100 calls	Police injury reports

SOURCE: Excerpted from Harry P. Hatry, Louis H. Blair, Donald M. Fisk, John M. Greiner, John R. Hall, Jr., and Philip S. Schaenman, *How Effective Are Your Community Services? Procedures for Measuring Their Quality,* 2nd edition (Washington, DC: Urban Institute and International City/County Management Association, 1992), pp. 72-73. Reprinted by permission.
a. One person committing four crimes or four persons committing one crime would be four "person crimes." When the number of offenders involved in a crime is unknown, as may frequently happen with such crimes as burglary, "one" criminal can be assumed for this statistic (or the historical average number of offenders for that type of crime could be used).
b. A satisfactory approach to measuring the degree of corruption, malfeasance, or negligence is lacking. Data on the number of complaints received by the city on these problems should be examined, particularly when their number increases substantially.

TABLE 16.3 Law Enforcement Staffing Levels in U.S. Cities, 1993

	Full-Time Law Enforcement Employees and Law Enforcement Officers per 1,000 Inhabitants, by Population Cluster												
	All Cities		250,000 or greater		100,000-249,999		50,000-99,999		25,000-49,999		10,000-24,999		Less than 10,000	
Region	Employees	Officers	Employees	Officers	Employees	Officers	Employees	Officers	Employees	Officers	Employees	Officers	Employees	Officers
All cities	**2.8**	**2.2**	**3.6**	**2.8**	**2.4**	**1.9**	**2.2**	**1.7**	**2.2**	**1.7**	**2.3**	**1.8**	**3.5**	**2.7**
Northeast	**3.0**	**2.5**	**5.0**	**3.9**	**3.2**	**2.7**	**2.4**	**2.0**	**2.2**	**1.9**	**2.1**	**1.8**	**2.3**	**1.9**
New England	2.5	2.0	4.7	3.5	3.3	2.7	2.3	2.0	2.1	1.8	2.1	1.8	2.6	2.0
Middle Atlantic	3.2	2.6	5.1	3.9	3.1	2.6	2.4	2.0	2.3	2.0	2.1	1.8	2.2	1.9
Midwest	**2.6**	**2.1**	**4.0**	**3.3**	**2.2**	**1.8**	**1.9**	**1.5**	**1.9**	**1.5**	**2.1**	**1.6**	**2.8**	**2.2**
East North Central	2.7	2.2	4.3	3.6	2.3	1.9	2.1	1.6	2.0	1.5	2.1	1.7	2.8	2.2
West North Central	2.3	1.8	3.3	2.4	2.1	1.6	1.6	1.3	1.7	1.4	2.0	1.6	2.7	2.2
South	**3.3**	**2.5**	**3.4**	**2.6**	**2.7**	**2.1**	**2.8**	**2.1**	**2.7**	**2.0**	**2.7**	**2.1**	**5.0**	**3.7**
South Atlantic	3.9	3.0	4.4	3.4	3.0	2.2	3.2	2.4	3.0	2.3	3.2	2.5	6.1	4.6
East South Central	3.1	2.4	3.0	2.3	2.8	2.0	2.7	2.1	2.6	2.0	2.7	2.1	4.0	3.2
West South Central	2.8	2.1	2.9	2.2	2.3	1.8	2.3	1.7	2.2	1.7	2.3	1.8	4.1	2.9
West	**2.4**	**1.7**	**2.6**	**1.9**	**2.0**	**1.4**	**1.9**	**1.3**	**2.0**	**1.4**	**2.1**	**1.6**	**4.3**	**3.0**
Mountain	2.5	1.8	2.6	2.0	2.2	1.6	1.8	1.3	2.1	1.6	2.3	1.7	3.8	2.8
Pacific	2.3	1.7	2.6	1.9	1.9	1.3	1.9	1.3	2.0	1.4	2.1	1.5	4.7	3.3

SOURCE: U.S. Department of Justice, Federal Bureau of Investigation, *Crime in the United States 1993: Uniform Crime Reports* (Washington, DC: Government Printing Office, 1994), pp. 289-290.

TABLE 16.4 Patrol Availability: Selected Cities

City	Patrol Availability Factor[a]
Savannah, GA	40% (target)
Peoria, AZ	37.7% (1991)
Sacramento, CA	34% (1990)
San Antonio, TX	32.7% (1991)
Phoenix, AZ	Available patrol time per unit: 19.8% (1989); 16.2% (1991)

a. Percentage of shift that an officer is available for proactive patrol (excludes assigned responses, administrative time, miscellaneous duties, etc.). For example, Savannah tries to devote 40% of patrol officers' time to preventive patrol, directed patrol, and community-oriented policing (City of Savannah, *1992 Program of Work*, p. 69).

Cities often report police staffing levels not only as a sign of their commitment to public safety but also as an indicator of officer availability in the community. Higher levels of staffing are presumed to mean more officers "on the street." In reality, however, the number of officers employed is only one of two important factors in that equation. A second and often overlooked ingredient is police management proficiency in actually getting available officers out of the station and onto the street—something called the "patrol availability factor" (Table 16.4). How much of a patrol officer's time should be spent on patrol rather than on assigned responses or various administrative duties? The LCC (1994) suggests that officers in high-service-level departments are able to devote at least 45% of their time to patrolling the field uncommitted; officers in medium-service departments have 30% to 45% for uncommitted patrol; and those in low-service departments have less than 30% (p. 34). The evidence from reporting cities, however, indicates that departments able to commit one third of the typical patrol officer's time to actual patrol are doing rather well.

Officer-to-population ratios imply that resident head counts can serve as good indicators of demand for law enforcement services. Although a relationship between population size and the need for police officers clearly exists, population provides only a general clue to likely demand. More precise indicators are available through direct measures of demand. Accordingly, ratios of calls for service per officer or arrests per officer provide demand and workload information that in many ways may be more revealing than staffing ratios based simply on population (Table 16.5).

Crime Rates

Variations in crime rates are influenced not only by police efforts and economic factors but also by whether a community is rural or urban, its population, and the state or region in which it is located. Crime rates tend to be lower in cities lying outside metropolitan statistical areas and lower still in rural counties (Table 16.6).

TABLE 16.5 Officer Demand: Indicators From Selected Cities

CALLS FOR SERVICE

Tallahassee, FL
  1,712 per uniformed officer (1991)

San Jose, CA
  1,546 per sworn officer (1989)

Rochester, NY
  734 per uniformed officer (1991)

Corpus Christi, TX
  0.68 per hour per police department employee (1991)

ARREST RATIO

Rochester, NY
  47.2 arrests per uniformed officer (1991)

Phoenix, AZ
  Arrests per sworn officer: 37.8 (1989); 38.3 (1990); 35.7 (1991)
  Part I arrests per sworn officer: 11.5 (1989); 11.9 (1990); 10.3 (1991)

Crime rates tend to be higher in big cities than in small towns (Table 16.7). And crime rates tend to be greater in the South and West than in other regions (Table 16.8).

With variables such as population and region possessing such obvious significance for crime rate tendencies, should not those factors be taken into consideration as a city chooses—or establishes—an appropriate benchmark? One way to do so is by calculating a "projected norm" using the crime rate for a given community's

TABLE 16.6 Crime Rates for Metropolitan, Suburban, and Rural Areas, 1993

			Rates per 100,000 Inhabitants							
			Violent Crimes					*Property Crimes*		
	Crime Index	*Sub-total*	*Murder & Non-Neg. Manslaughter*	*Forcible Rape*	*Robbery*	*Aggravated Assault*	*Subtotal*	*Burglary*	*Larceny -Theft*	*Motor Vehicle Theft*
United States, total	5,482.9	746.1	9.5	40.6	255.8	440.1	4,736.9	1,099.2	3,032.4	605.3
Metropolitan statistical areas	6,045.1	852.2	10.6	43.2	312.0	486.4	5,192.9	1,182.3	3,289.3	721.4
Cities outside metropolitan areas	5,303.4	504.4	5.3	39.1	71.3	388.8	4,799.0	992.8	3,581.7	224.5
Rural counties	1,971.1	222.4	5.4	24.9	16.4	175.7	1,749.0	633.4	1,005.7	109.9

SOURCE: U.S. Department of Justice, Federal Bureau of Investigation, *Crime in the United States 1993: Uniform Crime Reports* (Washington, DC: Government Printing Office, 1994), p. 59.

TABLE 16.7 Crime Rates in U.S. Cities by Population Cluster, 1993

		Violent Crimes				Property Crimes			
			Murder &						*Motor*
	Crime	*Sub-*	*Non-Neg.*	*Forcible*		*Aggravated*		*Larceny*	*Vehicle*
	Index	*total*	*Manslaughter*	*Rape*	*Robbery*	*Assault*	*Subtotal*	*Burglary*	*-Theft*	*Theft*
All cities	6,888.5	975.0	12.1	47.6	367.6	547.7	5,913.5	1,293.2	3,800.5	819.8
1 million or greater	8,505.6	1,920.0	27.7	50.9	954.5	887.0	6,585.6	1,470.5	3,465.0	1,650.2
500,000-999,999	9,119.7	1,376.5	21.6	74.6	632.0	648.4	7,743.2	1,647.3	4,832.1	1,263.9
250,000-499,999	10,215.4	1,705.2	23.6	80.9	686.6	914.0	8,510.2	2,003.3	4,986.8	1,520.2
100,000-249,999	8,034.0	1,093.1	13.2	58.1	379.8	642.0	6,940.9	1,631.2	4,409.2	900.5
50,000-99,999	6,390.5	773.4	7.5	45.1	246.1	474.7	5,617.1	1,236.2	3,695.7	685.2
25,000-49,999	5,590.4	582.3	5.2	36.0	158.7	382.7	5,007.8	1,071.7	3,458.8	477.2
10,000-24,999	4,788.7	448.8	4.2	32.1	97.8	314.7	4,339.9	878.0	3,129.4	332.5
Less than 10,000	4,798.2	429.0	3.9	26.7	72.3	326.2	4,369.2	829.8	3,299.6	239.8

SOURCE: U.S. Department of Justice, Federal Bureau of Investigation, *Crime in the United States 1993: Uniform Crime Reports* (Washington, DC: Government Printing Office, 1994), pp. 196-197.

population cluster and adjusting the rate on the basis of that community's state or region (Table 16.9). By doing so, a city can gauge how its actual crime rate compares with the rate that could be expected on the basis of its size and locale.

Consider, for example, the plight of city officials in Toledo, Ohio, as they might have attempted to assess their crime statistics for 1993. Toledo, a city of 331,416 inhabitants, reported 3,191 violent crimes that year for a rate of 962.8 per 100,000 population (U.S. Department of Justice, 1994, p. 104). They could have compared their local mark with the national average of 746.1 or with the state average of 504.1 violent crimes per 100,000 population and have been discouraged; or they could have stacked the local figure up against the national average of 1,705.2 violent crimes per 100,000 population among cities of 250,000 to 499,999 population and have been rather pleased. Which average should they have used? Recognizing that cities of Toledo's size face a more formidable crime challenge than smaller communities, they might have started with the average crime rate for their population cluster but then could have adjusted that rate because Toledo is located in a state with a lower-than-average rate of reported crimes. Through such an adjustment, a more appropriate projected norm of 1,152.1 violent crimes per 100,000 population could have been calculated (see Table 16.9). Because Toledo's actual rate of 962.8 fell comfortably below the projected norm of 1,152.1, community leaders might have congratulated police officials on their success and then

TABLE 16.8 Crime Rates by Region and State, 1993

		Rates per 100,000 Inhabitants								
		Violent Crimes					Property Crimes			
Area	Crime Index Total	Subtotal	Murder and Non-Negligent Manslaughter	Forcible Rape	Robbery	Aggravated Assault	Subtotal	Burglary	Larceny-Theft	Motor Vehicle Theft
United States	**5,482.9**	**746.1**	**9.5**	**40.6**	**255.8**	**440.1**	**4,736.9**	**1,099.2**	**3,032.4**	**605.3**
Northeast	**4,612.6**	**713.0**	**8.2**	**28.4**	**322.6**	**353.9**	**3,899.6**	**878.2**	**2,357.6**	**663.8**
New England	4,431.2	537.6	4.1	31.4	140.9	361.2	3,893.6	925.3	2,366.2	602.1
Connecticut	4,650.4	456.2	6.3	24.4	196.7	228.7	4,194.2	978.1	2,620.6	595.5
Maine	3,153.9	125.7	1.6	26.6	21.3	76.3	3,028.2	719.0	2,174.7	134.4
Massachusetts	4,893.9	804.9	3.9	33.4	175.7	592.0	4,089.0	1,001.7	2,271.3	816.1
New Hampshire	2,905.0	137.8	2.0	44.4	27.3	64.1	2,767.2	515.1	2,058.0	194.0
Rhode Island	4,499.0	401.7	3.9	28.6	101.1	268.1	4,097.3	1,040.9	2,410.1	646.3
Vermont	3,972.4	114.2	3.6	39.8	9.0	61.8	3,858.2	874.3	2,851.2	132.6
Middle Atlantic	**4,675.6**	**773.9**	**9.6**	**27.3**	**385.6**	**351.4**	**3,901.7**	**861.8**	**2,354.7**	**685.2**
New Jersey	4,800.8	626.9	5.3	28.1	296.0	297.5	4,174.0	974.0	2,486.1	714.0
New York	5,551.3	1,073.5	13.3	27.5	561.2	471.5	4,477.8	998.6	2,644.2	835.0
Pennsylvania	3,271.4	417.5	6.8	26.5	179.0	205.1	2,853.9	582.0	1,831.7	440.2
Midwest	**4,805.8**	**601.6**	**7.6**	**42.3**	**204.0**	**347.7**	**4,204.3**	**899.6**	**2,850.4**	**454.3**
East North Central	4,953.2	661.3	8.3	45.9	235.1	372.0	4,291.9	909.9	2,882.8	499.2
Illinois	5,617.9	959.7	11.4	34.6	381.2	532.6	4,658.2	1,015.5	3,084.0	558.7
Indiana	4,465.1	489.1	7.5	39.1	119.8	322.6	3,976.0	852.0	2,695.9	428.1
Michigan	5,452.5	791.5	9.8	71.1	238.5	472.1	4,661.0	982.7	3,063.2	615.0
Ohio	4,485.3	504.1	6.0	49.1	192.7	256.3	3,981.2	878.1	2,667.7	435.3
Wisconsin	4,054.1	264.4	4.4	25.2	113.4	121.4	3,789.7	663.0	2,762.0	364.7
West North Central	4,454.4	459.3	5.9	33.8	129.8	289.8	3,995.1	875.1	2,772.8	347.2
Iowa	3,846.4	325.5	2.3	24.4	53.9	244.8	3,521.0	730.7	2,599.4	190.8

(Continued)

Kansas	4,975.3	496.4	6.4	40.1	123.6	326.3	4,478.8	1,132.2	3,024.0	322.7
Minnesota	4,386.2	327.2	3.4	35.2	112.7	175.8	4,059.0	844.5	2,872.0	342.6
Missouri	5,095.4	744.4	11.3	36.2	241.8	455.2	4,351.0	1,025.5	2,777.8	547.7
Nebraska	4,117.1	339.1	3.9	27.8	55.4	252.0	3,778.0	663.5	2,912.9	201.6
North Dakota	2,820.3	82.2	1.7	23.5	8.3	48.7	2,738.1	373.2	2,216.2	148.7
South Dakota	2,958.2	208.4	3.4	44.5	15.0	145.6	2,749.8	549.2	2,086.0	114.5
South	**5,983.3**	**802.0**	**11.3**	**45.2**	**235.8**	**509.7**	**5,181.3**	**1,285.6**	**3,339.2**	**556.5**
South Atlantic	6,333.5	871.6	10.8	42.8	276.0	541.9	5,461.9	1,361.5	3,528.0	572.5
Delaware	4,872.1	685.9	5.0	77.0	186.7	417.1	4,186.3	892.0	2,979.0	315.3
Dist. of Columbia	11,761.1	2,921.8	78.5	56.1	1,229.6	1,557.6	8,839.3	1,995.5	5,449.0	1,394.8
Florida	8,351.0	1,206.0	8.9	53.8	357.6	785.7	7,145.0	1,835.4	4,413.9	895.7
Georgia	6,193.0	723.1	11.4	35.4	248.0	428.3	5,469.8	1,307.3	3,568.7	593.8
Maryland	6,106.5	997.8	12.7	44.0	434.7	506.4	5,108.7	1,132.8	3,292.5	683.4
North Carolina	5,652.3	679.3	11.3	34.3	192.4	441.3	4,973.0	1,515.8	3,168.8	288.5
South Carolina	5,903.4	1,023.4	10.3	52.3	187.3	773.4	4,880.0	1,309.2	3,226.8	344.0
Virginia	4,115.5	372.2	8.3	32.1	142.0	189.8	3,743.3	667.7	2,790.1	285.5
West Virginia	2,532.6	208.4	6.9	20.1	43.0	138.5	2,324.2	599.1	1,563.5	161.5
East South Central	4,528.4	640.8	10.3	41.0	159.1	430.5	3,887.5	1,068.3	2,429.0	390.2
Alabama	4,878.8	780.4	11.6	35.1	159.5	574.3	4,098.4	1,088.6	2,672.0	337.8
Kentucky	3,259.7	462.7	6.6	34.3	90.4	331.4	2,797.0	740.1	1,840.7	216.2
Mississippi	4,418.3	433.9	13.5	42.6	139.3	238.4	3,984.4	1,285.8	2,363.5	335.1
Tennessee	5,239.5	765.8	10.2	49.9	220.1	485.5	4,473.8	1,182.6	2,700.2	591.0
West South Central	6,227.9	778.7	12.7	51.3	213.0	501.7	5,449.2	1,283.7	3,541.7	623.8
Arkansas	4,810.7	593.3	10.2	42.4	124.9	415.8	4,217.5	1,099.3	2,795.7	322.5
Louisiana	6,846.6	1,061.7	20.3	42.3	283.6	715.4	5,784.9	1,368.3	3,802.9	613.7
Oklahoma	5,294.3	634.8	8.4	49.3	121.8	455.3	4,659.4	1,235.0	2,943.7	480.7
Texas	6,439.1	762.1	11.9	55.0	224.4	470.8	5,677.0	1,297.3	3,687.3	692.3

(Continued)

195

TABLE 16.8 Continued

		Rates per 100,000 Inhabitants								
		Violent Crimes					Property Crimes			
Area	Crime Index Total	Subtotal	Murder and Non-Negligent Manslaughter	Forcible Rape	Robbery	Aggravated Assault	Subtotal	Burglary	Larceny-Theft	Motor Vehicle Theft
West	6,220.0	844.6	9.9	42.9	283.1	508.7	5,375.3	1,221.5	3,359.8	794.1
Mountain	5,929.4	589.3	6.4	43.5	129.7	409.8	5,340.0	1,116.9	3,707.7	515.5
Arizona	7,431.7	715.0	8.6	37.8	162.9	505.7	6,716.7	1,465.5	4,387.4	863.8
Colorado	5,526.8	567.3	5.8	45.8	116.7	399.0	4,959.5	1,009.8	3,499.4	450.3
Idaho	3,845.1	281.8	2.9	35.3	16.9	226.7	3,563.3	668.8	2,711.1	183.4
Montana	4,790.0	177.5	3.0	27.9	32.4	114.2	4,612.5	714.2	3,652.1	246.2
Nevada	6,180.1	875.2	10.4	60.9	340.1	463.9	5,304.9	1,245.0	3,321.6	738.3
New Mexico	6,266.1	929.7	8.0	52.1	138.4	731.1	5,336.4	1,421.2	3,510.1	405.1
Utah	5,237.4	301.0	3.1	44.6	58.6	194.7	4,936.3	790.8	3,903.4	242.2
Wyoming	4,163.0	286.2	3.4	34.3	17.2	231.3	3,876.8	643.2	3,078.7	154.9
Pacific	6,323.8	936.0	11.2	42.6	338.1	544.2	5,387.8	1,258.9	3,235.1	893.8
Alaska	5,567.9	760.8	9.0	83.8	122.4	545.6	4,807.2	816.9	3,539.4	450.9
California	6,456.9	1,077.8	13.1	37.7	405.1	621.8	5,379.1	1,327.0	3,029.1	1,023.0
Hawaii	6,277.0	261.2	3.8	33.6	103.6	120.1	6,015.8	1,135.7	4,429.4	450.8
Oregon	5,765.6	503.1	4.6	51.3	129.6	317.6	5,262.5	1,024.8	3,656.9	580.7
Washington	5,952.3	514.6	5.2	64.4	137.1	307.9	5,437.7	1,067.2	3,914.4	456.1

SOURCE: U.S. Department of Justice, Federal Bureau of Investigation, *Crime in the United States 1993: Uniform Crime Reports* (Washington, DC: Government Printing Office, 1994), pp. 60-67.

TABLE 16.9 Establishing Community Benchmarks from UCR Information

Uniform Crime Report (UCR) statistics often reveal broad differences in key variables (e.g., staffing levels, crime rates, and clearance rates) among regions, states, and population clusters. Using the "national average" as a community benchmark may be a poor choice.

A better baseline for crime and clearance rates, for example, may be calculated as a "projected norm," as follows:

$$\text{Projected Norm} = \text{Statistic for population cluster} \times \frac{\text{Statistic for state or region}}{\text{national average}}$$

Then, a community benchmark may be set in relation to that baseline.

For example, a Pennsylvania community of 40,000 that wishes to set a suitable benchmark for reported robberies would first calculate the projected norm as a baseline. On the basis of 1993 statistics, the following calculation would be made:

$$\text{Projected Norm} = (158.7 \text{ robberies per } 100,000 \text{ population}) \times \frac{179.0}{255.8} = 111.1 \text{ robberies per } 100,000 \text{ population}$$

If the community's rate of reported robberies exceeds 111.1 per 100,000 inhabitants, then the projected norm might serve as a suitable benchmark. If the community's rate is less than the projected norm, a more useful benchmark might be set at 90% or 80% of the projected norm.

A similar technique may be used for establishing community benchmarks for local clearance rates or other performance statistics using UCR information.

challenged them to achieve a benchmark set at, say, 80% of the projected norm in the future.

In contrast, Tallahassee's rate of 2,034 violent crimes per 100,000 population in 1993 (U.S. Department of Justice, 1994, p. 104) was far greater than the national average of 746.1, the Florida average of 1,206, or the population cluster average of 1,093.1 violent crimes per 100,000 population. Applying the prescribed calculations, however, yields a projected norm of 1,766.9 violent crimes per 100,000 population, a figure that recognizes that Tallahassee police officials are situated in a state and region that have higher rates of reported crimes than the national average. The projected norm appears to be a much more reasonable—and potentially obtainable—benchmark for Tallahassee than any of the three more usual choices.

Response Time

Municipalities often judge their police department in part on the promptness with which they respond to emergencies. High-service-level departments, according to the LCC (1994), should respond to emergencies within 5 minutes (p. 34). Coincidentally, among the 34 municipalities providing response time figures for this chapter, the median response time was 5 minutes (Table 16.10). The quickest response times were reported by Duncanville, Texas, with an average of 2.3 minutes, and Corvallis, Oregon, in which 79% of emergency responses were within 3 minutes.

TABLE 16.10 Average Response Time to Police Emergencies: Selected Cities

Duncanville, TX
2.3 minutes (1990)

Blacksburg, VA
2.77 minutes (1994)

Sanford, NC
3 minutes (1991)

Corvallis, OR
76% within 3 minutes (1991); 79% within 3 minutes (1992)

Lubbock, TX
Average dispatch-to-arrival time: 3.2 minutes (1991)

Corpus Christi, TX
Estimate: 3 minutes 27 seconds (1992)

Fort Collins, CO
4 minutes (1991)

Overland Park, KS
Target: 4 minutes

Phoenix, AZ
Target: 4.2 minutes
Actual: Emergency responses within 5 minutes: 72.4% (1989); 69.4% (1990); 64.1% (1991)
Emergency responses within 10 minutes: 97.1% (1989); 96.8% (1990); 94.3% (1991)
Average response time: 3 minutes (1992); 4 minutes (1993)

Sunnyvale, CA
Targets: 4-minute average; 90% within 4 minutes from time of dispatch; 90% within 7 minutes from initial call

Vancouver, WA
3-5 minute average

Denton, TX
4.88 minutes (1991)

Long Beach, CA
72.5% in 5 minutes or less (1992)

Reno, NV
Target: Within 5 minutes

Tucson, AZ
Target: 5 minutes or less

Gresham, OR
5 minutes (1992)

Houston, TX
5.0 minutes (12/92)

Newnan, GA
5 minutes (1992)

Santa Fe, NM
5 minutes (1992)

San Antonio, TX
5.19 minutes (1991)

Peoria, AZ
5.34 minutes (1991)

Portland, OR
5.35 minute average receipt-to-arrival time (1991)[a]

Cincinnati, OH
5.5 minutes, including dispatch time (1991)

Savannah, GA
Target: Every response in 6 minutes or less

Jacksonville, FL
Target: 95% in 6 minutes or less

St. Petersburg, FL
Target: 95% in 6 minutes or less

Dayton, OH
6 minutes

Chandler, AZ
Target: 5-minute average or less
Actual: 6.1 minutes (1991)

Boston, MA
6.3 minutes (1992)

Grand Prairie, TX
6.6 minutes (YTD, 12/31/92)

Charlotte, NC
Target: 90% within 7 minutes
Actual: 89% within 7 minutes (1987)

Sacramento, CA
7 minutes (median, 1991)

Shreveport, LA
7.38 minutes (1991)

Chesapeake, VA
7.7 minutes (1991)

New York City
9.9 minutes (1992)

a. Portland, Oregon, reports average emergency "travel times" (dispatch to arrival) of 5.25 minutes (1988), 5.20 minutes (1989), 4.85 minutes (1990), and 4.75 minutes (1991).

Cities often differentiate between calls requiring emergency response and those of a less urgent nature. When nonemergency calls are included in response statistics, the average response times are often in excess of 10 minutes (Table 16.11).

TABLE 16.11 Average Response Time to All Calls—Emergency and Nonemergency: Selected Cities

City	Average Response Time
Blacksburg, VA	2.77 minutes to emergencies; 4.93 minutes to routine calls (1994)
Vancouver, WA	3-5 minutes for emergencies; 5-7 minutes for nonemergencies
Corpus Christi, TX	Estimate: 3 minutes 27 seconds to emergencies; 8 minutes 51 seconds to routine calls (1992)
Grand Prairie, TX	8.9 minutes (YTD, 12/31/92)
Sacramento, CA	10 minutes (median, 1991)
Chandler, AZ	Targets: Average of 5 minutes or less to Priority I calls; 15 minutes or less to nonemergency Priority II calls; no more than 30 minutes to Priority III calls Actual: 6.1-minute average to Priority I calls; 11.0 minutes to Priority II; 18.3 minutes to Priority III (1991)
Charlotte, NC	Targets: (a) 90% of emergency calls within 7 minutes; (b) 90% of "immediate" calls within 10 minutes Actual: (a) 89%; (b) 89% (1987)
Fayetteville, AR	11 minutes (1991); 11 minutes (1992)
Peoria, AZ	5.34 minutes for Priority I calls; 12.04 minutes for Priority II (1991)
Denton, TX	13.41 minutes (1991)
Houston, TX	14.4 minutes (weighted average, Dec. 1992)
Fort Collins, CO	Average of 4 minutes to emergency calls, 7 minutes to urgent calls, 17 minutes to routine calls; 28% of calls delayed (stacked) more than 5 minutes (1991)
Dayton, OH	18 minutes
San Antonio, TX	18 minutes (1991)
Overland Park, KS	Targets: Median response time of 4 minutes for Priority I calls, 5 minutes for Priority II, and 20 minutes for Priority III calls
Reno, NV	Targets: Within 5 minutes to emergency calls; within 25 minutes to nonemergency calls
Phoenix, AZ	Targets: Average of 4.2 minutes for Priority I emergencies; 11.5 minutes for Priority II nonemergencies; 38.5 minutes for Priority III; 38 minutes for telephone "call-backs"
Savannah, GA	Targets: Response time no greater than 6 minutes for emergency calls; 9 minutes for "immediate" calls; 2 hours for "delayed" calls

Clearance Rates

Cases are said to be "cleared by arrest" when primary suspects have been arrested for that offense and turned over to the court or "cleared by exception" when the victim refuses to cooperate, extradition is denied, the offender is deceased, or other extraordinary circumstances preclude the placing of charges. The police discontinue their pursuit of suspects in such cases. Obviously, a high clearance rate is desired. Clearance rates tend to be higher in small cities (Table 16.12) and to vary

TABLE 16.12 Clearance Rates in U.S. Cities by Population Cluster, 1993

		Percentage of Offenses Cleared by Arrest or by Exception										
		Violent Crimes						Property Crimes				
City Population	Crime Index[a]	Subtotal	Murder and Non-Negligent Manslaughter	Forcible Rape	Robbery	Aggravated Assault	Subtotal[a]	Burglary	Larceny-Theft	Motor Vehicle Theft	Arson
All cities	21.2	42.5	64.7	52.0	23.2	54.5	17.6	12.6	20.4	12.8	14.4
1 million or greater	18.5	36.6	60.8	51.2	20.2	52.2	13.1	9.8	16.5	8.5	6.0
500,000-999,999	17.9	37.6	65.1	55.4	21.0	50.9	14.3	12.1	15.7	12.3	13.1
250,000-499,999	19.0	39.6	62.5	51.0	23.0	50.8	14.9	10.8	17.7	11.2	13.5
100,000-249,999	21.1	45.0	64.0	52.9	26.3	55.0	17.3	12.2	19.9	13.7	16.5
50,000-99,999	22.0	44.9	68.3	49.4	25.9	54.1	18.9	13.0	22.1	12.4	16.8
25,000-49,999	23.3	48.4	71.2	49.9	28.5	56.4	20.4	13.9	23.0	16.5	18.0
10,000-24,999	25.6	53.3	73.4	52.5	30.7	60.3	22.8	15.4	25.0	21.4	23.2
Less than 10,000	24.0	57.3	76.2	55.0	29.9	63.4	20.8	15.8	21.5	28.5	23.6

SOURCE: U.S. Department of Justice, Federal Bureau of Investigation, *Crime in the United States 1993: Uniform Crime Reports* (Washington, DC: Government Printing Office, 1994), pp. 208-209.
a. Excludes arson figures (reported by fewer agencies than other categories of crime).

TABLE 16.13 Clearance Rates by Region, 1993

| | | Percentage of Offenses Cleared by Arrest or by Exception | | | | | | | | | |
| | | Violent Crimes | | | | | Property Crimes | | | | |
Region	Crime Index[a]	Subtotal	Murder and Non-Negligent Manslaughter	Forcible Rape	Robbery	Aggravated Assault	Subtotal[a]	Burglary	Larceny-Theft	Motor Vehicle Theft	Arson
Total, all agencies	**21.1**	**44.2**	**65.6**	**52.8**	**23.5**	**55.5**	**17.4**	**13.1**	**19.8**	**13.6**	**15.4**
Northeast	**20.0**	**40.0**	**66.2**	**53.1**	**22.7**	**54.6**	**16.2**	**12.8**	**19.2**	**10.2**	**12.6**
New England	21.7	51.4	63.9	53.0	24.7	62.2	17.5	13.3	19.7	15.5	16.9
Middle Atlantic	19.5	37.5	66.5	53.2	22.5	52.2	15.8	12.6	19.1	8.8	11.6
Midwest	**20.2**	**39.1**	**61.4**	**50.2**	**19.2**	**50.4**	**17.5**	**11.0**	**19.9**	**15.1**	**12.2**
East North Central	19.7	37.7	58.5	49.3	18.3	49.4	16.8	10.6	19.3	14.4	11.7
West North Central	21.9	44.9	72.7	52.9	24.0	54.0	19.3	12.3	21.6	18.1	13.7
South	**21.9**	**47.7**	**71.0**	**55.9**	**26.5**	**56.7**	**17.9**	**14.6**	**19.3**	**17.0**	**19.9**
South Atlantic	21.8	47.3	71.6	57.6	25.6	57.0	17.7	15.0	18.7	18.0	20.9
East South Central	23.1	47.8	74.6	54.2	24.8	56.9	19.0	15.5	21.1	15.9	16.8
West South Central	21.8	48.5	69.4	54.0	29.0	56.0	18.1	13.8	20.0	15.8	19.6
West	**21.4**	**45.7**	**59.2**	**49.1**	**23.5**	**57.7**	**17.5**	**12.3**	**20.8**	**11.7**	**14.6**
Mountain	22.8	46.9	68.8	42.9	23.1	55.0	20.1	11.9	23.2	16.0	20.2
Pacific	21.0	45.4	57.3	51.2	23.6	58.4	16.7	12.5	19.9	10.9	13.6

SOURCE: U.S. Department of Justice, Federal Bureau of Investigation, *Crime in the United States 1993: Uniform Crime Reports* (Washington, DC: Government Printing Office, 1994), pp. 210-211.
a. Excludes arson figures (reported by fewer agencies than other categories of crime).

TABLE 16.14 Parking Regulations Enforcement: Performance Targets and Actual Experience of Selected Cities

COMPLIANCE WITH PARKING REGULATIONS

Savannah, GA
Target: No more than 10% of vehicles parked in violation of regulations in controlled parking district
Actual: 11% in violation (1988); 10% in violation (1989); 7% in violation (1990); 8% in violation (1991)

PARKING ATTENDANT/ OFFICER OUTPUT

Corpus Christi, TX
751 parking citations monthly per attendant (1991)

PARKING CITATION ERROR RATE

Corpus Christi, TX
0.29% of all citations (1991)

RATE OF PARKING CITATIONS CONTESTED

Fort Collins, CO
0.4% (1991)

COLLECTION RATE FOR PARKING FINES

Raleigh, NC
84% (1991)

San Luis Obispo, CA
83% (1990)

Boston, MA
74% for tickets issued 6 months ago; 78% for tickets issued 12 months ago; 83% for tickets issued 24 months ago (1992)

Auburn, AL
Target: 80% (1993)

Iowa City, IA
79.9% (1989); 86.9% (1990); 79.8% (1991)

Long Beach, CA
73.0% (1992)

Reno, NV
73% (1990)

Fort Lauderdale, FL
Target: 75%
Actual: 69% (1991)

Shreveport, LA
47.5% (1991); 54.5% (estimate, 1992)

APPREHENSION OF SCOFFLAWS

Corpus Christi, TX
9.2% of vehicles on scofflaw list immobilized (1991)

slightly by region (Table 16.13). Projected norms for clearance rates may be calculated in the same fashion as norms for crime rates.

Potential Benchmarks for Selected Police Activities

Among the performance targets and performance records reported by various municipalities are potential performance benchmarks for a variety of police functions. For example, effectiveness in the enforcement of parking regulations was reported in the form of error rates for parking citations and apprehension rates for scofflaws (Table 16.14).

Police effectiveness in pursuing thefts may be judged in part by the recovery of stolen items. Recovery rates of 25% to 40% are not uncommon (Table 16.15).

TABLE 16.15 Recovery of Stolen Property: Selected Cities

City	Recovery Rate of Stolen Property
League City, TX	20.7% (1990); 83.8% (1991)
Chandler, AZ	34.9% of value (1991)
Shreveport, LA	30% (1991)
Dunedin, FL	28% (1992)
Avon, CT	27% (1986); 51% (1987); 26% (1988); 60% (1989); 26% (1990)
Greenville, SC	34.3% (1989); 25% (1990)
Overland Park, KS	25% (target)
Winston-Salem, NC	12.9% (1991); 8.7% (1992)

Police officers deal with many situations involving discretion and requiring good judgment. Occasional complaints are inevitable. How promptly are serious matters handled? And what are the results? Internal affairs investigations against officers generally were completed in 1 to 3 months among the cities reporting such statistics (Table 16.16) and typically involved disciplinary actions in one case out of five.

TABLE 16.16 Internal Affairs Investigations: Performance Targets and Actual Experience in Selected Cities

PROMPT INVESTIGATION OF COMPLAINTS AGAINST OFFICERS

Oakland, CA
95% of investigations by Citizens' Complaints Board completed within 30 days (1992)

Tucson, AZ
Targets: 50% of investigations completed within 5 workdays; 80% within 10; 90% within 15; 95% of all formal citizen complaints within 30 days

Fort Collins, CO
90% completed within 30 days (1991)

Long Beach, CA
88% of internal affairs cases completed within 30 days (1992)

Alexandria, VA
Internal affairs cases completed within 30 days: 37% (1989); 72% (1990); 73% (1991)

Boston, MA
88.3% of requests or complaints resolved within 90 days (1992)

New York City
Civilian Complaint Review Board investigations resolved within 90 days: 81% (1991); 83% (1992)

Sunnyvale, CA
Targets: At least 45% of investigations completed within 2 months of complaint; 60% within 3 months; 75% within 4 months

INVESTIGATIONS RESULTING IN DISCIPLINARY/CORRECTIVE ACTION

Fayetteville, AR
82% (9 of 11 investigations, 1991)

Fort Collins, CO
28% (7 of 25 investigations, 1991)

Cincinnati, OH
26% (of 554 investigations, 1991)

Wichita, KS
18% (23 of 128 investigations, 1990); 23% (31 of 133 investigations, 1991)

Greenville, SC
Sustained complaints against officers: 23% (10 of 44 investigations, 1989); 20% (12 of 60 investigations, 1990)

TABLE 16.17 Timely Analysis of Crime Evidence: Charlotte, North Carolina

Type of Evidence	Average Turnaround Time	
	Target	Actual (FY87)
Fingerprints	4 days	5 days
Photographic evidence	3 days	5 days
Trace evidence (hair, blood, etc.)	20 days	25 days
Firearms, tool, and tread marks	12 days	28 days
Questioned documents	12 days	17 days
Drug evidence	10 days	16 days

SOURCE: City of Charlotte (North Carolina), *FY88-FY89 Objectives*, p. 20.

How quickly should analysis of crime evidence be completed? The expectations and experience of Charlotte, North Carolina, are shown in Table 16.17. Potential benchmarks for the police records function are provided in Table 16.18.

TABLE 16.18 Police Records: Performance Targets and Actual Experience of Selected Cities

PROMPT PROCESSING OF POLICE REPORTS INTO RECORDS SYSTEM

Oklahoma City, OK
Percentage of reports entered within 4 hours of receipt: 25% (1991); 50% (1992)

Lubbock, TX
95% of Priority I case reports processed within 6 hours; 95% of Priority II case reports processed within 8 hours; 60% of Priority III case reports processed within 12 hours (1991)

Overland Park, KS
Target: Within 24 hours of receipt

Oakland, CA
Target: Within 64 hours
Actual: 100% within 64 hours (1989-1992)

Corvallis, OR
Target: Data entry within 5 days

Savannah, GA
Target: Process all police reports, including data entry, by the 7th day of the following month
Actual: 100% (1988-1993)

PROMPT RESPONSE TO REQUESTS FOR POLICE REPORTS

Overland Park, KS
Target: Fulfill in-person requests within an average of 10 minutes

Tucson, AZ
Targets: Process 90% of phone and counter requests within 15 minutes; 90% of written requests within 10 workdays

Largo, FL
Targets: Fulfill internal requests within 24 hours; external requests within 36 hours

Raleigh, NC
Target: Process mail requests within 3 workdays

PROMPT RECORD CHECKS

Savannah, GA
Targets: Complete 90% of record checks within 24 hours; 100% within 48 hours
Actual: Completed within 24 hours: 95% (1991); 95% (1992); 90% (1993)
Actual: Completed within 48 hours: 100% (1988-1993)

Reno, NV
Target: Respond to at least 75% of public requests for preemployment information on criminal history and insurance reports within 5 workdays
Actual: 75% (1990)
Target: Complete records review of business license applicants within 3 workdays
Actual: 100% (1990)

17

Property Appraisal

Property taxes constitute a major source of revenue for most cities. The rationale for that tax, although imperfect, is simple: Those having the greatest stake in the community and those having the greatest ability to pay should bear a commensurate share of the cost of municipal government and its services. The value of property is used as the gauge for both criteria.

Although critics have challenged the fairness of the property tax and the validity of property value as a proxy for ability to pay, state restrictions have left municipal governments with few viable options for replacing such a major source of revenue, even if local officials wished to do so. The objective in most communities is to administer the property tax fairly and efficiently, extracting as little of the revenues for tax administration as possible.

Property Appraisal and
Tax Administration Benchmarks

Fairness of the property tax depends on fairness of property appraisals. Persons owning properties of greater value should pay higher taxes than those owning properties of lesser value; persons owning properties of equal value should pay equal property taxes.

TABLE 17.1 Selected Professional Standards for Property Tax Administration

Accurate Appraisal of Market Value
- 100% of current market value

Frequent Ratio Studies
- Monitor the level and uniformity of appraised values through ratio studies at least annually

Frequent Reinspection
- All properties should be physically inspected at least every 6 years

Prompt Updating of Records
- (1) Update ownership and legal description information in tax office records within 30 days of recorded sale; (2) Parcel splits and combinations should be noted on cadastral maps within one month of a new deed's recording

Accessible Property Records
- Accessible (i.e., indexed) by parcel identifier, address, and owner

Sufficient Resources and Staffing
- Rules of thumb: Approximately 1.5% of property tax collections or at least $10 (1991 dollars) per parcel
- Staffing level that provides an appraiser-to-parcel ratio as follows:

    1 appraiser per 1,000-1,500 parcels in small jurisdictions (fewer than 10,000 parcels)

    1:2,500—cities with 10,000 to 20,000 parcels

    1:3,000 to 1:3,500—cities with more than 20,000 parcels

    Higher in the largest jurisdictions

SOURCE: Richard R. Almy, Robert J. Gloudemans, and Garth E. Thimgan, *Assessment Practices: Self-Evaluation Guide* (Chicago: International Association of Assessing Officers, 1991). Reprinted by permission.

How, then, may the performance of a municipal property tax office be judged? In several ways, according to the International Association of Assessing Officers (IAAO; Almy, Gloudemans, & Thimgan, 1991). The IAAO has established a set of professional standards for property tax administration, including, for example, the maintaining of municipal appraisals at 100% of current market value (Table 17.1). The extent to which appraisers are able to achieve that high level of accuracy may be tested by the periodic calculation of "sales ratios" that compare the prices from actual sales transactions with the appraised values of those properties. No municipality should expect absolute accuracy, but some have demonstrated an ability to come reasonably close (Table 17.2).

The IAAO recommends that ratio studies be conducted frequently, that property reinspection occur at least every 6 years, that property records be updated and accessible, and that sufficient resources—including personnel—be committed to the tax administration function. Offered as rules of thumb for adequate budget and staffing are resource commitments equal to 1.5% of property tax collection and staffing levels ranging from one appraiser per 1,000 parcels in small jurisdictions to one appraiser per 3,500 (or more) parcels in the largest cities. The reasonableness of these rules of thumb, however, may be drawn into question by municipalities that have managed to perform well despite more modest resource commitments.

TABLE 17.2 Accurate Appraisal of Market Value

HOW CLOSE TO MARKET VALUE?

International Association of Assessing Officers[a]
 Recommended target: 100% of current market value

Boston, MA
 Accuracy rate of data quality and assessment: 97% (1992)[b]

Milwaukee, WI
 Assessment levels as a percentage of market value: 96.39% (1993)

Alexandria, VA
 Assessment-to-sales ratio: 85.7% (1989); 82.7% (1990); 93.3% (1991); 95.2% (1992); 97.0% (1993)

a. Richard R. Almy, Robert J. Gloudemans, and Garth E. Thimgan, *Assessment Practices: Self-Evaluation Guide* (Chicago: International Association of Assessing Officers, 1991), p. 3.
b. Based on 5% random sample of reviewed abatement filings.

For example, Alexandria, Virginia, has reported an appraiser-to-assessable-parcels ratio ranging from 1:4,881 in 1991 to 1:6,615 in 1989.

The work of a property appraiser varies in difficulty, depending on the type of property appraised and the type of appraisal activity conducted. The IAAO has offered guidance to municipal officials attempting to determine a fair day's work for appraisal operations by declaring typical production rates according to appraisal activity and property type (Table 17.3). According to this guideline, for example, it is reasonable to expect an appraiser doing "new work" on newly constructed homes to appraise 8 to 10 homes during one full day. Similarly, a "field canvass" of 5 to 10 commercial properties should be expected to consume a full day's work for a typical appraiser.

In addition to the market test of appraisal accuracy provided by periodic ratio studies, yet another gauge of quality is the extent to which the public deems property appraisals to be acceptable. Some communities boast extremely low rates

TABLE 17.3 Typical Production Rates for Property Assessor (Parcels per Day)

Property Type	Type of Appraisal Activity			
	New Work	Field Canvass	Field Review	Model Review
Homes	8-10	16-20	50-100	200-250
Apartments	2-4	5-10	15-20	100
Commercial	2-4	5-10	15-20	100
Agricultural	2-4	4-6	NA	50
Vacant land	30-50	50-75	100-150	400

SOURCE: Richard R. Almy, Robert J. Gloudemans, and Garth E. Thimgan, *Assessment Practices: Self-Evaluation Guide* (Chicago: International Association of Assessing Officers, 1991), p. 20. Reprinted by permission.
NOTES: *New work* is fieldwork necessary to describe and appraise new construction.
Field canvass is an orderly parcel-by-parcel detailed inspection program.
Field review is a drive-by inspection.
Model review is an in-office review of computer-assisted mass appraisal (CAMA) model value estimates, including the flagging of selected parcels for field review or canvass.

TABLE 17.4 Appeals of Property Appraisal

MINIMAL APPEALS AS AN INDICATION OF ACCEPTABILITY

Milwaukee, WI

  Assessment objections as percentage of real and personal property assessments: 1.76% (1993)

  Formal assessment appeals as percentage of real and personal property assessments: 0.20% (1993)

Alexandria, VA

  Appeals to appraisal department, as percentage of total parcels: 1.4% (1989); 2.4% (1990); 2.7% (1991)

  Appeals to Board of Equalization, as percentage of total parcels: 0.5% (1989); 1.2% (1990); 2.4% (1991)

DEFENDABILITY OF APPRAISALS ON APPEAL

Alexandria, VA

  Percentage upheld/adjusted at departmental appeal: 59%/41% (1989); 70%/30% (1990); 65%/35% (1991); 62%/38% (1992); 60%/40% (1993)

  Percentage upheld/adjusted at Board of Equalization appeal: 50%/50% (1989); 49%/51% (1990); 53%/47% (1991); 47%/53% (1992); 45%/55% (1993)

---

of assessment objections and formal appeals (Table 17.4). Perhaps even more significant is a favorable record for having municipal appraisals upheld when appeals are made.

Office Space Standards

Although much of an appraiser's work is conducted in the field, suitable office space is also needed. The IAAO recommends minimally 50 to 100 square feet of office space per appraiser, with greater allocations for supervisors, other administrators, and the chief assessor (Table 17.5). The question of how much space to provide is, of course, a resource input issue that may influence performance or results, but office space standards should not be confused with *performance* standards.

TABLE 17.5 Space Standards for Property Assessors' Offices

Position	Space Requirement
Assessor	200 square feet of private space (approximately 14 ft. by 14 ft.)
Chief deputy or manager	170 square feet (approximately 12 ft. by 14 ft.)
Supervisor	150 square feet (approximately 12 ft. by 12 ft.)
Appraiser	50 to 100 square feet of space

SOURCE: Richard R. Almy, Robert J. Gloudemans, and Garth E. Thimgan, *Assessment Practices: Self-Evaluation Guide* (Chicago: International Association of Assessing Officers, 1991), p. 16. Reprinted by permission.

18

Public Health

Although public health services are a function of county governments in many parts of the country, they are provided by city governments in some regions. Services typically span a broad spectrum but often include health clinics and the inspection of food establishments.

Recommended Performance Measures

Public health was among the local government services examined recently in the SEA reporting project of the GASB. On the basis of that study, assorted measures of input, output, outcome, and efficiency were recommended for several public health activities, including programs addressing maternal and child health care, sexually transmitted diseases, AIDS, chronic diseases, and the control of stress and violent behavior (Carpenter, Ruchala, & Waller, 1991).

Cities attempting to establish local standards for public health and wishing to have the benefit of external benchmarks as they proceed should anticipate encountering some difficulty in assembling information on the performance targets and actual experience of municipalities in public health operations. Many municipalities simply do not engage in that type of service, leaving it instead to county governments, where more abundant performance information may reside.

TABLE 18.1 National Public Health Targets

Public Health Focus	National Target
Reduction of death rate from coronary heart disease	No more than 100 per 100,000 people (baseline: 135 per 100,000 in 1987)
Reduced incidence of overweight	No more than 20% among people 20 years of age and older; 15% among adolescents (baseline: 26% for 20- to-74-year-olds in 1976-1980)
Reduced cancer death rate	No more than 130 per 100,000 people (baseline: 133 per 100,000 in 1987)
Food choices in restaurants	At least 90% offering identifiable low-fat, low-calorie food choices (baseline: 70% in 1989)
School and child care breakfasts and lunches	At least 90% with menus consistent with *Dietary Guidelines for Americans*
Reduced lung cancer death rate	No more than 42 per 100,000 people (baseline 37.9 per 100,000 in 1987)
Reduction in cigarette smoking	No more than 15% of people aged 20 or older (baseline: 29% in 1987)
Reduction in adolescent pregnancy	No more than 50 per 1,000 girls aged 17 or younger (baseline: 71.1 per 1,000 girls 15 through 17 years old in 1985)
Reduction of suicides	No more than 10.5 per 100,000 people (baseline: 11.7 per 100,000 in 1987)
Reduction in maltreatment of children	Less than 25.2 incidents of maltreatment per 1,000 children younger than age 18 (baseline: 25.2 per 1,000 in 1986)
Reduced incidence of foodborne infections	Incidence rates from key foodborne pathogens:

	Target	1987 Baseline
Salmonella species	16 per 100,000	18
Campylobacter jejuni	25 per 100,000	50
Escherichia coli 0157:H7	4 per 100,000	8
Listeria monocytogenes	0.5 per 100,000	0.7

Public Health Focus	National Target
Dental care for children	Untreated caries in permanent or primary teeth among no more than 20% of children aged 6 through 8; no more than 15% among 15-year-olds (baseline: 27% of children, 6 through 8, and 23% of 15-year-olds in 1986-1987)
Optimal water fluoride levels	Between 0.7 and 1.2 parts fluoride per 1 million parts water, depending on daily air temperature (baseline: 62% in 1989)
Use of topical or systemic fluorides	At least 85% of persons who do not receive optimally fluoridated public water (baseline: 50% in 1989)
Reduced infant mortality	No more than 7 per 1,000 live births (baseline: 10.1 per 1,000 in 1987)
Prenatal care	At least 90% of all pregnant women receiving prenatal care in first trimester (baseline: 76% of live births in 1987)

SOURCE: Adapted from American Public Health Association, *Healthy Communities 2000: Model Standards,* 3rd edition. Copyright © 1991 by the American Public Health Association. Adapted with permission.

TABLE 18.2 Immunization of Children in the Community: Selected Cities

Cincinnati, OH
Percentage of school children meeting immunization requirements: 99% (1991)

Boston, MA
Percentage of children immunized by the time they start school: More than 97% (1992)

Alexandria, VA
Percentage of children with immunizations that are current (based on sample of records): 83% (1989); 88% (1990); 94% (1991); 95% (1992); 95% (1993)

Houston, TX
Target: At least 90% of children under age 2 having had complete immunization series

New York City
Percentage of new students completely immunized: 88% (1991); 88% (1992)

Jacksonville, FL
Percentage of 2-year-olds completely immunized: 63% (estimate, 1992)

A readily available external peg for local conditions, however, has been provided through the development of national public health targets, many of which are output or outcome related. A sample of national public health targets is provided in Table 18.1. A community wishing to establish its own local standards may simply adopt the national target or, if the target seems overly ambitious given local conditions and current public health record, a point between the community's current mark and the national target.

Information from other municipalities on immunizations of community children (Table 18.2) and food service inspection programs (Tables 18.3 and 18.4) also may be useful. Once again, however, local conditions may influence community leaders to establish standards that differ from those of their counterparts.

TABLE 18.3 Frequency of Food Service Inspections: Selected Cities

Alexandria, VA
Inspection visits per food service establishment (annual): 15 (1989); 13 (1990); 15 (1991); 14 (1992); 15 (1993)

Overland Park, KS
Target: At least 3 routine inspections of all food service establishments annually

Wichita, KS
Inspections per establishment: 1.8 (1990); 1.5 (1991)

TABLE 18.4 Food Service Inspections Program Statistics: Garland, Texas, 1990-1991

Permitted establishments per full-time equivalent employee:	278.10
Inspections per citation:	11.85
Inspections per closure:	64.44
Inspections per permitted establishment:	1.77

SOURCE: Pat Fowler, "The Establishment of Health Indicators in Municipal Health Departments," *Texas Town & City,* 79 (February 1992), pp. 20-24. Reprinted by permission.

19

Public Transit

Many cities operate some form of public transit service. Medium-sized and large cities often provide bus services with regular routes on major thoroughfares and principal arteries. Many of the largest cities also offer heavy rail, commuter rail, and light rail transit service. Alternatively, relatively small cities sometimes offer transportation services on demand—occasionally on a contractual basis with local taxi companies or other private entities—rather than by means of regular transit routes. The focus of this chapter is primarily on bus services, a staple among cities that offer regular public transit services.

Performance Measures and Standards

Research undertaken on behalf of the GASB revealed a fairly wide assortment of productivity indicators espoused by transit authorities or currently in use (Wallace, 1991; Table 19.1). If collected widely and tabulated centrally, many of these indicators would have considerable potential as performance benchmarks.

Relevant standards for bus systems recommended in a U.S. Department of Transportation report (Attanucci, Jaeger, & Becker, 1979) address major facets of transit operations (Table 19.2). The authors of that report contended, for example, that passenger fares should constitute at least 20% of operating expenditures, that

TABLE 19.1 Productivity Indicators for Mass Transit

Passengers per vehicle
Passengers per vehicle hour
Passengers per revenue hour
Revenue passengers per vehicle hour
Passengers per trip
Passengers per revenue mile
Passengers per mile
Passengers per service mile
Revenue passengers per mile
Operating revenue per operating expense
Revenue vehicle hours per operating expense
Revenue:cost by route
Revenue per vehicle hour
Passenger miles (and trips) per directional miles
Passenger miles (and trips) per vehicle operated in maximum service
Passenger miles (and trips) per revenue vehicle hour
Passenger trips per employee
Passenger trips per nonrevenue mile
Vehicle hours per employee
Vehicle miles per peak vehicles required
Vehicle miles per maintenance employee
Annual vehicle miles per annual maintenance expense
Annual vehicle miles per gallon of fuel (mpg) or per kilowatt hour of power
Maintenance work done a second time
Miles between work orders (bus repairs through inspection)
Miles per mechanic equivalent

SOURCE: Wanda A. Wallace, *Mass Transit: Service Efforts and Accomplishments Reporting—Its Time Has Come* (Norwalk, CT: Governmental Accounting Standards Board, 1991), p. 11. Reprinted by permission.

buses should neither depart from a stop ahead of schedule nor be more than 5 minutes late, and that buses should average at least 1.5 passengers per vehicle mile.

In some cases, more extensive sets of operating standards have been established by local transit agencies themselves. A set developed by the Williamsport

TABLE 19.2 Recommended Standards for Bus Systems

Fares as a percentage of operating expenditures:	20 to 50%
Adherence to route schedule:	Zero minutes early to 5 minutes late
Bus stop spacing:	660 to 2,000 feet apart
Passengers per vehicle mile:	At least 1.5
Service frequency:	At least every 30 minutes during peak periods; at least hourly during off-peak periods

SOURCE: John P. Attanucci, Leora Jaeger, and Jeff Becker, *Bus Service Evaluation Procedures: A Review—Short Range Transit Planning,* Special Studies in Transportation Planning (Washington, DC: U.S. Department of Transportation, April 1979), pp. 10-13.

TABLE 19.3 Selected Public Transit Performance Standards of the Williamsport Bureau of Transportation[a]

Criterion	WBT Standard	WBT Actual (FY 1989)	Pennsylvania Average (1989)
Percent employee attendance	≥ 97%	98.8%	NA
Vehicle-miles per employee	≥ 15,000	16,631	15,219
Vehicle-miles per maintenance employee	≥ 80,000	85,631	91,233
Vehicle-hours per operator	≥ 1,700	1,757	1,669
Vehicle-miles per vehicle	≥ 28,000	31,512	26,008
Expense per vehicle-mile	≤ $2.75	$2.43	$2.75
Expense per vehicle-hour	≤ $40.61	$34.57	$36.92
Vehicle miles between road calls	≥ 3,500	5,596	5,024
Collision accidents per 100,000 vehicle-miles	≤ 3.0	0.84	2.42
Percent on-time performance[b]			
Peak periods	≥ 90%	93.3%	NA
Non-peak	≥ 95%	100.0%	NA
Percent transfers	≤ 10%	6.5%	6.2%
Passenger trips per vehicle-hour	≥ 28	27.5	23.2
Net cost per passenger	≤ $0.90	$0.77	$0.78
Operating ratio (revenue/expense)	≥ 35%	43%	51%

SOURCE: Reprinted from Theodore H. Poister, "Productivity Monitoring: Systems, Indicators, and Analysis." In Marc Holzer (Ed.), *Public Productivity Handbook* (New York: Marcel Dekker, 1992), p. 208. Reprinted by courtesy of Marcel Dekker, Inc.

a. Williamsport Bureau of Transportation (WBT) of Williamsport, Pennsylvania.

b. Percentage of trips departing from a bus stop within 5 minutes of scheduled time.

(Pennsylvania) Bureau of Transportation (WBT) is displayed in Table 19.3. Also shown are the WBT performance marks relative to those standards, as well as the statewide averages of other Pennsylvania transit agencies.

National Norms and Benchmarks

National data compiled by the U.S. Department of Transportation, Federal Transit Administration (FTA), provide operating statistics aggregated for large and small agencies. Those statistics, for example, indicate that transit agencies serving populations of less than 200,000 in 1993 derived an average of 21% of their operating funds from passenger fares, whereas transit agencies serving larger populations averaged 37%. Overall, the average transit bus nationwide was between 8 and 9 years old. Furthermore, transit agencies averaged 1.74 to 3.04 unlinked passenger trips per vehicle revenue mile and 24.2 to 38.9 unlinked passenger trips per vehicle revenue hour for bus services, depending on the population of the urban area served (Table 19.4).[1] Although summary figures for

TABLE 19.4 Selected Operating Statistics of Public Transit Agencies, 1993

Service Population	N	Revenue From Fares as a Percentage of Operating Funds Expended	Average Fleet Age in Years				Unlinked Passenger Trips per Vehicle Revenue Mile				Unlinked Passenger Trips per Vehicle Revenue Hour			
			Motor Bus	Heavy Rail	Commuter Rail	Light Rail	Motor Bus	Heavy Rail	Commuter Rail	Light Rail	Motor Bus	Heavy Rail	Commuter Rail	Light Rail
Transit agencies serving populations greater than 200,000	291	37%	8.3	17.8	18.8	14.4	3.04	4.05	1.58	6.96	38.90	82.78	53.44	96.90
Transit agencies serving populations of less than 200,000	232	21%	9.3	NA	NA	5.0	1.74	NA	NA	5.50	24.20	NA	NA	25.54
30 largest transit agencies	30	40%	8.5	17.7	17.1	16.9	3.65	4.07	1.59	7.78	43.85	82.90	53.77	102.61

SOURCES: U.S. Department of Transportation, Federal Transit Administration (1994a), *Transit Profiles: Agencies in Urbanized Areas Exceeding 200,000 Population,* December 1994, pp. 287-291; U.S. Department of Transportation, Federal Transit Administration (1994b), *Transit Profiles: Agencies in Urbanized Areas With a Population of Less Than 200,000,* December 1994, pp. 213-216; U.S. Department of Transportation, Federal Transit Administration (1994c), *Transit Profiles: The Thirty Largest Agencies,* December 1994, pp. 63-65.

TABLE 19.5 Cost Recovery Targets and Experience Among Selected Municipal Transit Systems

City	Cost Recovery via Operating Revenue
Alexandria, VA	50.6% (1990 estimate); 41.8% (1991 estimate); 42.2% (1992 estimate); 42.2% (1993 estimate)
Winston-Salem, NC	43% (1991); 39.1% (1992)
Lubbock, TX	39.4% (1991)
Phoenix, AZ	Target: Fare box revenues equal to 30% of costs
Iowa City, IA	34% (1988); 32% (1989); 28% (1990); 30% (1991)
Tallahassee, FL	26.2% (1991)
Wichita, KS	25.0% (1990); 22.6% (1991)

transit agencies serving populations in excess of 200,000 showed an average cost recovery of 37% from fares, statistics for the 30 largest agencies revealed a slightly higher 40% cost recovery. Detailed FTA information on individual agencies suggests that cost recovery of more than half of operating expenses is relatively rare among those that rely heavily on bus operations. In fact, only 11 transit agencies that relied heavily on bus operations acquired at least half their operating funds through passenger fares in 1993 (U.S. Department of Transportation, 1994a, 1994b).[2] More than half those agencies operated in the nation's densely populated northeastern region.

TABLE 19.6 Passengers per Service Hour and Revenue Mile: Selected Municipalities

PASSENGERS PER SERVICE HOUR

Charlotte, NC
Passenger boardings per revenue hour: 39.8 (5 months, 1994)

San Luis Obispo, CA
Passengers per service hour: 39 (1990)

Lubbock, TX
Average city route passengers per hour: 18.6 (1991)

Wichita, KS
Passenger trips per revenue hour: 18.9 (1990); 18.2 (1991)

Waukesha, WI
Passengers per vehicle hour: 13.8 (1992)

PASSENGER TRIPS PER REVENUE MILE

Iowa City, IA
Passengers per revenue mile: 2.75 (1988); 2.68 (1989); 2.75 (1990); 2.95 (1991)

Winston-Salem, NC
Passenger trips per revenue mile: 2.50 (1991); 2.43 (1992)

Fort Collins, CO
Passengers per revenue mile: 1.63 (1991)

Decatur, IL
Passengers per mile: 1.52 (1991)

Waukesha, WI
Passengers per vehicle mile: 1.44 (1992)

Lubbock, TX
Passenger trips per revenue mile: 1.29 (1991)

Wichita, KS
Passenger trips per revenue mile: 1.24 (1990); 1.23 (1991)

Sioux Falls, SD
Passenger trips per revenue mile: 1.22 (1991)

TABLE 19.7 Routes on Schedule: Selected Cities

Blacksburg, VA 100% on time (1994)	**Charlotte, NC** Target: 85% of peak-hour trips and 95% of nonpeak trips on time Actual: 94.3% of peak-hour trips and 99.1% of nonpeak trips on time (1987)
Oklahoma City, OK 99.34% on schedule (1991); 98.59% on schedule (1992)	
Lubbock, TX 95.2% on schedule (1991)	**Houston, TX** 91% on time (1983)[a]
Wichita, KS 95% on schedule (1990); 94.1% on schedule (1991)	**Boise, ID** 90.4% within 5 minutes of schedule (1991)
	Raleigh, NC 79.43% on schedule (1991)
Phoenix, AZ Target: 95% on schedule	

a. Theodore H. Poister and Robert P. McGowan, "Municipal Management Capacity: Productivity Improvement and Strategies for Handling Fiscal Stress." In *The Municipal Year Book 1984* (Washington, DC: International City Management Association, 1984), p. 209.

Perhaps more representative of municipal transit operations—and consistent with the Department of Transportation report's prescription that 20% to 50% of expenditures be recovered through fares—are the statistics reported for seven cities in Table 19.5. Among these cities, only Alexandria, Virginia, approaches the 50% cost recovery mark.

Selected municipal transit operations report 13.8 to 39.8 passengers per service hour and 1.22 to 2.75 passengers per revenue mile (Table 19.6). At least one factor likely to influence such numbers is the degree to which a transit operation maintains its routes on schedule. Almost all responding municipalities that reported on-time

TABLE 19.8 Avoiding Bus Accidents: Selected Cities

City	Accidents per 100,000 Bus miles
Blacksburg, VA	1.3 (1994)
Knoxville, TN	1.65 (1991)
Phoenix, AZ	2.34 (1991)
Wichita, KS	4.07 (1990); 2.5 (1991)
Lubbock, TX	2.8 (1991)
Houston, TX	2.9 (1983)[a]
Winston-Salem, NC	2.3 (1991); 2.99 (1992)
Raleigh, NC	3.6 (1991)
Charlotte, NC	Target: Fewer than 3.5 Actual: 3.63 (8 months, 1987)
Decatur, IL	4.77 (1991)

a. Theodore H. Poister and Robert P. McGowan, "Municipal Management Capacity: Productivity Improvement and Strategies for Handling Fiscal Stress." In *The Municipal Year Book 1984* (Washington, DC: International City Management Association, 1984), p. 209.

statistics indicated an on-schedule rate in excess of 90%, with some reporting performance targets or actual on-time results of 95% or greater (Table 19.7).

Safety rate—that is, an agency's ability to avoid bus accidents—is yet another potential performance benchmark for transit operations. The Phoenix (Arizona) Transit System was awarded first place in the National Safety Council's Fleet Safety Contest for 1991 among systems serving populations of 1 million or greater for its record of only 23.4 accidents per million miles of operation (City of Phoenix, *The Phoenix Budget 1993-94,* p. 16). Among the responding municipalities that reported transit system accident rates, only Blacksburg, Virginia, and Knoxville, Tennessee, operating in smaller population categories, reported lower marks (Table 19.8).

Notes

1. Each time a passenger climbs aboard a bus, it is considered an "unlinked trip." If traveling across the city requires a passenger to ride a total of three buses, the journey counts as three unlinked trips.

2. Omitted from the list were the Chicago Commuter Rail Operation, the New Jersey-Port Authority Transit, the New York Metro North Commuter Railroad, the Pierce County (Washington) Ferry Operations, the San Diego Trolley, the Seattle Monorail, and several other transit operations that had high cost recovery rates through passenger fares but that relied little, if at all, on bus services as a component of their operations.

20

Public Utilities

Electricity, natural gas, water, and sewer services often are provided by municipalities. The extent of direct municipal involvement, however, often differs from one utility to another and from one community to another. Relatively few cities generate electricity. Most municipalities that provide electrical services purchase electricity and resell it to their customers. Natural gas services typically follow a similar pattern. On the other hand, many municipalities assume a more comprehensive role in the treatment and distribution of potable water and the collection and treatment of wastewater.

Federal and state regulations pertaining to the four common utilities form a set of basic standards governing operations and product quality. Some of these regulations will be noted in this chapter.[1] In addition, the operating norms of the utility industry, as well as the performance targets and actual experience of utility counterparts, can serve as the basis for benchmarks in the various service specialties.

Electricity and Natural Gas

Compared with the number of municipalities that operate water and sewer systems, relatively few own electric or gas utility systems. For those that do, relevant benchmarks in the form of industry norms are fairly abundant.

TABLE 20.1 Publicly Owned Electric Utilities Operating Statistics, by Customer Size: 1991

	Total	Median Values, by Customer Size Categories					
		2,000-5,000 Customers	5,000-10,000 Customers	10,000-20,000 Customers	20,000-50,000 Customers	50,000-100,000 Customers	More Than 100,000 Customers
Debt to total assets	0.280 (392)	0.280 (26)	0.231 (127)	0.277 (123)	0.304 (79)	0.339 (19)	0.589 (18)
Operating ratio	0.844 (395)	0.902 (27)	0.870 (129)	0.847 (123)	0.827 (79)	0.768 (19)	0.731 (18)
Net income per revenue dollar	$0.042 (395)	$0.037 (27)	$0.044 (129)	$0.042 (123)	$0.047 (79)	$0.053 (19)	$0.033 (18)
Uncollectible accounts per revenue dollar	$0.0023 (214)	$0.0008 (11)	$0.0019 (52)	$0.0020 (65)	$0.0022 (52)	$0.0043 (16)	$0.0038 (18)
Retail customers per non-power-generation employee	285 (230)	262 (12)	285 (59)	288 (67)	291 (58)	270 (16)	238 (18)
Retail customers per meter reader	3,919 (220)	2,383 (12)	3,194 (56)	3,719 (62)	4,305 (56)	5,409 (16)	5,727 (18)
OSHA incidence rate (per 100 employees)	4.9 (207)	4.3 (10)	4.5 (50)	7.0 (62)	5.0 (54)	3.7 (14)	3.4 (17)

SOURCE: Adapted from American Public Power Association, *Selected Financial and Operating Ratios of Public Power Systems, 1991* (Washington, DC: APPA, March 1993), pp. 9-10. Used by permission.
NOTE: The number of responses for each category is shown in parentheses.

TABLE 20.2 Publicly Owned Electric Utilities Operating Statistics, by Region: 1991

| | Median Values by Region | | | |
	Northeast	*Southeast*	North Central/ *Plains*	*Southwest/West*
Debt to total assets	0.281 (28)	0.255 (153)	0.262 (93)	0.344 (118)
Operating ratio	0.889 (29)	0.899 (153)	0.833 (94)	0.788 (119)
Current ratio	2.31 (28)	1.62 (153)	3.02 (93)	2.73 (117)
Times interest earned	1.93 (25)	4.24 (100)	2.46 (65)	2.30 (88)
Net income per revenue dollar	$0.038 (29)	$0.032 (153)	$0.050 (94)	$0.057 (119)
Uncollectible accounts per revenue dollar	$0.0026 (17)	$0.0026 (73)	$0.0014 (50)	$0.0026 (74)
Retail customers per non-power-generation employee	287 (18)	272 (79)	295 (53)	288 (80)
Retail customers per meter reader	3,793 (16)	3,619 (77)	3,780 (50)	4,545 (77)
OSHA incidence rate (per 100 employees)	6.6 (16)	5.3 (68)	4.2 (48)	4.7 (75)

SOURCE: Adapted from American Public Power Association, *Selected Financial and Operating Ratios of Public Power Systems, 1991* (Washington, DC: APPA, March 1993), p. 11. Used by permission.
NOTE: The number of responses for each category is shown in parentheses.

Financial and operating statistics reveal, for example, that the average publicly owned electric utility has a debt equal to approximately one fourth of its total assets, that it nets four cents of every revenue dollar, and that it has almost 300 retail customers for every non-power-generation employee and almost 4,000 retail customers per meter reader (Table 20.1). Median values for these and other operating statistics differ by utility size and region (Table 20.2).

A recent survey of more than 100 electric and gas utilities revealed performance norms for a variety of important indicators (Mercer Management Consulting, 1992; used by permission). Lead time from an initial customer request for new electric or gas service until completion of installation averaged 15 days for residential customers, 22 days for commercial customers, and 43 days for industrial customers. The average customer of the responding utilities experienced approximately 1.5 interruptions in electric service in 1990, with the average interruption lasting 92 minutes. Altogether, the average customer was without power for approximately 123 minutes during the year. When power outages occurred, the typical utility had a troubleshooter or electric crew on the scene in approximately 35 minutes. Most utilities reported average response times between 10 minutes and 1 hour. Similarly, the average response time to gas leak calls was 30 minutes, with top-performing utilities responding in less than 20 minutes.

All public utilities are concerned about *line loss.* In the case of gas and electrical distribution systems, line loss is the difference between the amount of natural gas or electricity metered at the head of the system and the amount actually received by (i.e., sold to) customers. Assuming accurate metering, the tightest gas systems

TABLE 20.3 Gas and Electric Line Loss: Selected Cities

Gas Line Loss	Electric Distribution Line Loss
Corpus Christi, TX 0.6% (1991)	**Denton, TX** 4.03% (9/88); 4.96% (9/89); 5.71% (9/90); 4.57% (9/91); 4.84% (5/92)
Tallahassee, FL Target: Less than 2% unaccounted Actual: 0% (1991); 2% (estimated, 1992)	**Oak Ridge, TN** Difference between KWH purchased and KWH sold: 5.8% (1991); 5.96% (1992)
	Lubbock, TX Line loss: 6.65% (1991)

and most efficient electric distribution systems will have the lowest rates of line loss. The gas utility in Tallahassee, Florida, for example, strives to keep its losses at less than 2% (Table 20.3). The typical line loss for electrical distribution systems tends to be higher.

Water Treatment and Distribution

The residents of a community typically know little about the process of purifying water or maintaining distribution lines and may willingly leave the details of formulating standards to local government officials, but that does not mean their expectations are ambiguous. As one authority notes,

> For the public, the goals of water supply and distribution are simple. The supply should be ample and dependable. The water should be tasteless and odorless, without ingredients that are harmful to health. It should be reasonably soft and cool. Its cost should not be excessive.
>
> For the local government, the goals are more detailed. The volume of water must be sufficient not only to satisfy domestic, commercial, and industrial requirements but also for fire protection. The pressure should be maintained at not less than 30 to 40 pounds per square inch (107 to 276 kilopascals) throughout the system. Fire hydrants must be reliable, as must be the valves in the system. Water mains must be free from leaks so that as much as is practicable of the water that has been purified and pumped at rather substantial cost actually will be used by the customers. Water rates must be adequate to make the system economically viable. (William S. Foster, *Handbook of Municipal Administration and Engineering,* Copyright © 1978 by McGraw-Hill, p. 16.20. Reprinted with permission of McGraw-Hill, Inc.)

Water system performance targets differ from place to place. Charlotte, North Carolina, attempts to maintain a daily treated water capacity exceeding average daily demand by a factor of at least 1.5. Portland, Oregon, strives to maintain in storage the equivalent of three times average daily demand. Fort Collins, Colorado, which reported a water supply-to-demand ratio of 2.15 in 1991, attempts to provide sufficient water supply to meet the demand of a once-in-50-year drought.

TABLE 20.4 Summary Statement on Water Quality

AWWA [American Water Works Association] advocates the principle that water of the highest quality should be delivered to all consumers and encourages each public water supplier to provide water of a quality that is as close to the AWWA goals as is feasible. In addition,

- Water delivered to the consumer for domestic purposes should contain no pathogenic organisms and be free from biological forms that may be aesthetically objectionable. It should be clear and colorless and should have no objectionable taste or odor. It should not contain concentrations of chemicals that may be physiologically harmful, aesthetically objectionable, or economically damaging.
- The water should neither be corrosive nor leave excessive or undesirable deposits on water-conveying structures, including pipes, tanks, and plumbing fixtures.
- The water should be obtained from the best sources available and should be treated as necessary to ensure safety and consistent quality.
- Existing and potential sources of community water supply should be protected from contamination to the highest degree feasible.
- A water utility has the responsibility to be aware of the nature of its source water. Therefore, water utilities should adequately monitor their source waters to determine their quality.
- Care should be taken to avoid and minimize the introduction of unnecessary constituents during the treatment and delivery process.
- In the distribution process, every reasonable effort, including backflow prevention, should be made to protect water from degradation.
- Water utilities should undertake adequate monitoring and periodic sanitary surveys to ensure that water quality objectives are met on a continuous basis.

SOURCE: Reprinted from *American Water Works Association 1992-1993 Officers and Committee Directory,* by permission. Copyright © 1992, American Water Works Association.

Not every system subscribes to the water pressure prescriptions offered above by Foster (1978). Tallahassee, Florida, for example, attempts to maintain system pressure of at least 20 pounds per square inch (psi). Portland, Oregon, strives for pressure between 20 and 110 psi. On the other hand, Sunnyvale, California, has a system target of 40 to 105 psi, and Houston, Texas, strives for 50 psi.

Perhaps not surprisingly, cities differ in their expectations regarding fire hydrant reliability and the speed at which deficiencies should be corrected. New York City reported finding 1.32% of its hydrants broken or otherwise inoperative in 1991 and 1.09% in 1992. Savannah, Georgia, discovered not a single unreported dry hydrant at a fire scene between 1988 and 1991. When fire hydrants needed to be repaired, the work was done within 14 days in Savannah. Portland, Oregon, repaired fire hydrants and returned them to service within 5 days. Lubbock, Texas, reported that in 1991, it completed such work within 24 hours in every case.

In contrast to the typical citizen's imprecisely expressed desire for an ample and dependable supply of tasteless and odorless water, the prescriptions of the American Water Works Association (AWWA) embodied in its statement of objectives are more precise and more narrowly focused on water quality (Table 20.4). The AWWA summary statement, addressing matters ranging from water supply contamination to the corrosion of plumbing fixtures, is supplemented by more precise standards of water quality (Table 20.5).

TABLE 20.5 Recommended Standards of Water Quality

Element/Characteristic	Recommended Standard
Chloride	Operating level of less than 250 mg/L in finished water; goal of less than 100 mg/L
Chlorine residual	Maintain a detectable free chlorine residual or 1.0 mg/L of combined chlorine throughout the distribution system; goal of free chlorine residual of at least 0.5 mg/L or combined chlorine of at least 2.0 mg/L
Color	Less than 5 cobalt platinum units (cpu) true color in finished water; goal of less than 3 cpu
Foaming agents	Sufficiently low to prevent foaming and other aesthetic problems (Environmental Protection Agency prescribes less than 0.5 mg/L)
Manganese	Less than 0.05 mg/L in finished water
Nitrate	Less than 10 mg/L of nitrate as nitrogen in finished water
Phenols	Less than 0.002 mg/L of phenols at the point of chlorination
Taste and odor	Less than 3 threshold odor number (TON) in finished water; taste should be nonoffensive
Turbidity	No more than 0.5 nephelometric turbidity units (ntu) of turbidity in filter plant effluent; goal of no more than 0.2 ntu
Zinc	Less than 5 mg/L in finished water; goal of less than 2 mg/L

SOURCE: Reprinted from *American Water Works Association 1992-1993 Officers and Committee Directory,* by permission. Copyright © 1992, American Water Works Association.

Although the significance of many of the AWWA water quality standards may be more fully understood by chemists and lab technicians than by city managers, elected officials, or general citizens, awareness of their importance even among nonspecialists may be increasing. A few key characteristics are being reported by a growing number of cities. Turbidity, for example, refers to the cloudiness of water. The AWWA recommends no more than 0.5 nephelometric turbidity units (ntu) as a standard. Although Fort Collins pursues an even more demanding standard of 0.1 ntu, Portland and Savannah have set a less ambitious local performance target of 1 ntu, and Oklahoma City attempts to control turbidity to levels of 3 ntu or less.

When water systems fail to attain local objectives for water quality, officials may hear the news as quickly from angry citizens as from lab technicians. Foul-tasting water, brown water, or no water at all is sure to prompt several calls. An effective water treatment and distribution system, however, should hear few such complaints—the rate in Milwaukee was 0.039 water quality complaints per million gallons purified and pumped in 1993. Despite best efforts, a few complaints are virtually inevitable. When they do occur, top water departments respond quickly (Table 20.6).

Water departments deal not only with a steady stream of miscellaneous service requests but also with intermittent emergencies, often in the form of broken water mains, and with the perpetual quest to minimize line loss. A survey conducted by the Urban Institute in the early 1980s provides national data on both these major

TABLE 20.6 Responsiveness to Water Treatment and Distribution Complaints and Requests: Selected Cities

City	Response to Complaint or Miscellaneous Service Request
Duncanville, TX	Percentage of service orders worked within 24 hours: 96% (1990)
Savannah, GA	Target: Respond to complaints within 1 workday Actual: 100% (1988-1991)
Tallahassee, FL	Target 1: Respond to water quality complaints within 24 hours Actual: 100% Target 2: Respond within 48 hours to requests to identify water line locations Actual: 100% (1991)
Charlotte, NC	Target: Correct 85% of minor water leaks within 24 hours Actual: 88% within 24 hours (1987)
Phoenix, AZ	Target: Respond to at least 98% of contractor-requested utility locations, shutdowns, and taps within 2 days of notification
Shreveport, LA	Percentage of water and sewer complaints handled in 3 days or less: 90% (1991)
Fort Worth, TX	Target: Within 3 days Actual: Within 5 days (1991)
New York City	Average time to resolve leak complaints requiring excavation (days): 24.1 (1991); 25.0 (1992)

concerns (Peterson, Miller, Godwin, & Shapiro, 1984). The average number of water main failures among 34 major U.S. cities was 229 per 1,000 miles of pipe. A similar study conducted by the U.S. General Accounting Office (1980) reported a nearly identical average of 231 breaks per 1,000 miles of pipe. Local officials responding to the Urban Institute survey cited severe weather and ground movement as major causes of breaks, followed by corrosive soils, damages caused by construction, and the age of pipe (Peterson et al., 1984, p. 38). Cities experiencing substantial population growth were especially likely to sustain a large number of water main failures (Table 20.7).

Line loss is water that is unaccounted—the difference between the amount of treated water entering the distribution system and the amount metered (and usually billed) for use by individual customers or other authorized users. Steady leakage and intermittent main breaks are major culprits, but unaccounted water may also result from inaccurate meters, theft, and unmetered water used for fire fighting, hydrant flushing, street cleaning, and other legitimate municipal purposes. Urban Institute researchers found the average of unaccounted-for water among 49 major cities to be 15% in 1980 (Peterson et al., 1984, p. 36). Highest rates of line loss were reported in northeastern cities, cities experiencing high levels of distress, and cities with older housing stock (Table 20.7). Urban Institute researchers contended that unaccounted-for water exceeding 15% to 20% is generally considered to be a problem. AWWA rules of thumb place the acceptable rate of unaccounted-for water between 5% and 25% (U.S. General Accounting Office, 1980, p. 28).

TABLE 20.7 Water Main Failures and Unaccounted Water: 1980

City Groupings	Water Main Failures: Average Number of Failures per 1,000 Miles	N	Average Unaccounted-for Water: Percentage of Water Unaccounted for	N
Region				
Northeast	168	8	26.5	11
North Central	228	13	12.3	13
South	275	11	14.6	13
West	229	13	8.4	13
Size				
Large	207	19	15.6	31
Small	259	26	13.8	29
Population change: 1970-1980				
Growing	282	17	11.2	20
Declining	197	28	16.2	30
Size of system				
Less than 500 miles	244	15	NA	—
Between 500 and 1,500 miles	298	10	NA	—
Greater than 1,500 miles	234	20	NA	—
Age of housing				
Young	287	12	NA	—
Middle-aged	222	18	NA	—
Old	190	15	NA	—
Distress				
High	NA	—	23.1	14
Moderate	NA	—	13.0	17
Not distressed	NA	—	9.4	19
Age of housing				
Old	NA	—	21.0	15
Middle-aged	NA	—	12.1	20
Young	NA	—	10.1	15

SOURCE: Urban Institute survey, cited in George E. Peterson, Mary John Miller, Stephen R. Godwin, and Carol Shapiro, *Guide to Benchmarks of Urban Capital Condition* (Washington, DC: Urban Institute Press, 1984), pp. 36, 38. Reprinted by permission.

Among municipal documents examined for this chapter, several reported performance targets and experience with main breaks and unaccounted water losses. Cincinnati, Ohio, and Corvallis, Oregon, for example, each reported the number of water main breaks experienced per 100 miles of water main. Corvallis had 18 main failures per 100 miles of water main in 1991—virtually the same as Cincinnati's 18.3 breaks per 100 miles the same year—but only 11 breaks per 100 miles in 1992. Several cities emphasize quick response and prompt repair of main

TABLE 20.8 Water Main Failures and Unaccounted Water: Performance Targets and Experience of Selected Cities

Contending With Main Breaks	*Unaccounted Water Losses*
QUICK RESPONSE	**Wichita, KS**
Denton, TX	3.1% (1990); 1.7% (1991)
Average response time: 15 minutes (1991)	**Addison, TX**
Oklahoma City, OK	Target: 5% or less
Target: Average response time of 20 minutes to emergency main breaks	Actual: 3.8% to 6% (1988-91)
Portland, OR	**Corvallis, OR**
Target: Respond to system emergencies within 30 minutes	Target: 5% or less (water treated but not metered)
Actual: 100% within 30 minutes (1991)	Actual: 8.88% (1992)
Sunnyvale, CA	**Irving, TX**
Target: Respond to at least 95% of emergency calls within 30 minutes (1994)	6.04% (1992)
St. Charles, IL	**Fort Collins, CO**
Target: Respond to all service interruptions within 1 hour	Target: 7% or less
St. Petersburg, FL	Actual: 7% (1991)
Target: Respond to 75% of reported damage to the system within 3 hours	**Grants Pass, OR**
Actual: 76.4% (1990)	7.73% (YTD, 12/92)
Decatur, IL	**Chandler, AZ**
Target: Average response time of 4 hours for water main breaks	8.7% (1991)
Actual: 4.42 hours (1991)	**Corpus Christi, TX**
Peoria, AZ	9.8% (1991)
Targets: Respond to emergency leaks within 1 day; respond to nonemergency leaks within 5 days	**Tallahassee, FL**
PROMPT RESTORATION	Target: 10% or less
Dayton, OH	Actual: 9.9% (1991)
Target: Hold 80% of emergency mainline shutdowns to a 2-hour maximum	**St. Petersburg, FL**
Oklahoma City, OK	10.1% (1990)
Target: Repair main breaks in an average of 5.5 hours or less	**Cincinnati, OH**
Actual: 4.2 hours (1991); 4.5 hours (1992)	17% (1987); 16% (1988); 15% (1989); 14% (1990); 15% (1991)
Tucson, AZ	
Target: Complete 90% of emergency repairs within 5 hours; no repair exceeding 8 hours if water service interrupted	**Duncanville, TX**
Hurst, TX	18% (1990)
Target: Average repair time of 5 hours or less for main breaks	**Oak Ridge, TN**
Actual: 6-hour average (1990)	28.8% (1992)
Rock Island, IL	
Target: Restore water service within an average of 12 hours after a main break	
Actual: 7.2-hour average (1990); 7.2-hour average (1991)	
Portland, OR	
Target: Complete all system emergency repairs within 24 hours of call	
Actual: 100% compliance (1991)	
Milwaukee, WI	
Percentage of breaks back in service within 24 hours:	
Mains less than 16 inches: 94% (1991); 98% (1992); 93% (1993)	
Mains 16 inches or greater: 71% (1991); 92% (1992); 93% (1993)	
Columbus, OH	
Target: Repair 95% of breaks or leaks within 48 hours of notification	

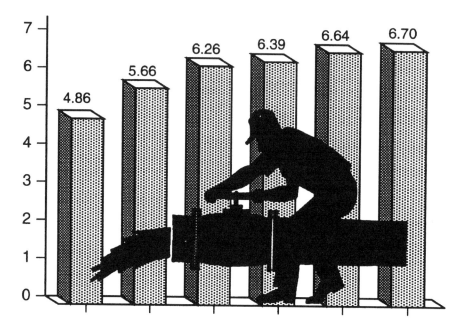

Figure 20.1. Pipeline Miles per Water and Sewer Employee in Charlotte, North Carolina
SOURCE: City of Charlotte (North Carolina), *FY95 Operating Plan,* p. ES-21.

breaks—some striving to maintain an average response time of less than 1 hour and others targeting completion times within several hours (Table 20.8). Line losses—water sent into the distribution system but not subsequently metered or billed—are reported by many cities, with most experiencing losses of 5% to 10%.

In addition to the work associated with water treatment and maintenance of the water distribution system, another key aspect of water services—especially from the standpoint of system revenues—is the maintenance and reading of water meters. To ensure the system's financial integrity, meters must operate at a high degree of accuracy. To achieve and sustain such accuracy, many municipalities establish regular programs of testing and replacement. Fayetteville, Arkansas, for example, tests at least 6% of its water meters annually. Sunnyvale, California, attempts to test its large meters—2-inch meters or larger—on a 2-year cycle. Decatur, Illinois, replaces any meter exceeding 10 years of age. In Sunnyvale, ⅝-inch to 1-inch disc meters are on a 15-year replacement schedule and 1½- to 2-inch disc meters are on a 10-year replacement schedule.

Even if meters are registering water flow accurately, meter reading errors can negate that precision. Therefore, many cities take steps to ensure low error rates by meter readers. The following error rates, for example, were reported: .05 of 1% (0.05%) in Charlotte, North Carolina; 0.06% in Chandler, Arizona; 0.16% in College Station, Texas; 0.22% in Fort Collins; 0.3% in Lee's Summit, Missouri; 0.346% in Wichita, Kansas; 0.5% in Houston and Lubbock, Texas; and 0.847% in Winston-Salem, North Carolina.

TABLE 20.9 Production Ratios for Water Services: Selected Cities

WATER TESTING	SERVICE INSTALLATION
Duncanville, TX Average worker-hours per sample collected: 0.2	**Wichita, KS** Average worker-hours per service installation: 15.57 (1991)
Wichita, KS Number of lab tests per worker-hour: 5.64 (1991)	**Winston-Salem, NC** Worker-hours per ¾-inch water connection installed: 15.29 (1991); 16.85 (1992)
WATER MAIN REPAIR	**CUSTOMER SERVICE TURN ON/OFF**
Santa Ana, CA Total hours per main break repair: 98.6 (1987)	**Dunedin, FL** Turn on/offs per worker-hour: 3.5 (1992)
Duncanville, TX Average worker-hours per major leak repair: 20.39	**METERS READ PER READER**
Dunedin, FL Feet of water main replacement per worker-hour: 1.7 (1992)	**Fayetteville, AR** 49.8 per hour (1991)
MINOR LEAK REPAIR	**Santa Ana, CA** 363.2 per day (1987)
Duncanville, TX Average worker-hours per minor leak repair: 4.65	**Tallahassee, FL** 7,608 per month (1991)
VALVE MAINTENANCE	**Denton, TX** 306 per day
Duncanville, TX Average worker-hours per valve stack cleaned and raised: 1.33	**Raleigh, NC** Target: 300 per day (1993)
FIRE HYDRANT INSTALLATION/MAINTENANCE	**Shreveport, LA** 6,000 per month
Duncanville, TX Average worker-hours per fire hydrant installation: 12.0	**Winston-Salem, NC** 32.26 per worker-hour (0.031 worker-hours per meter) (1992)
Dunedin, FL Worker-hours per hydrant repair/replacement: 10.9 (1992)	**Dunedin, FL** 17.7 per hour (1992)
Santa Ana, CA Worker-hours per hydrant testing and repair: 2.25 (1987)	**Rochester, NY** 145 attempted reads per day; 87 reads completed (1991)[a]

a. Where meters are located inside buildings, the problem of "lock-outs" may restrict production rates.

Proper staffing for water system efficiency and effectiveness requires balancing the desire to have enough well-qualified employees to ensure sound operations and quick response to emergencies with the desire to avoid overstaffing. A good sense of workload, as well as of the human resources needed to match that workload, is necessary to strike the appropriate balance between those desires.

Charlotte reports its ratio of pipeline miles to water and sewer employees, a figure that has grown from 4.86 miles per employee in 1972 to 6.7 in 1995 (Figure 20.1). Overall rates from another community offer only a rough gauge for judging the adequacy of a system's employment level. More precise calculations of human resource requirements can be made using labor ratios for various water system tasks. The labor ratios reported by several municipalities are shown in Table 20.9.

TABLE 20.10 Allowable Pollution Concentrations

Pollutant	Monthly Average Concentration (milligrams/kilogram)[a]
Arsenic	41
Cadmium	39
Chromium	1,200
Copper	1,500
Lead	300
Mercury	17
Molybdenum	18
Nickel	420
Selenium	36
Zinc	2,800

SOURCE: Felix Sampayo, "EPA's Part 503 Regulations Will Increase POTW Standards," *American City & County, 109* (February 1994), p. 36. Reprinted by permission.
a. Dry weight basis.

Sewage Collection and Treatment

The quality of a community's wastewater collection and treatment system is of concern not only to local residents but also to state and federal environmental protection officials. Sludge from the treatment process is considered clean if various pollutants do not exceed allowable concentrations (Table 20.10).

Many cities report compliance with state and federal standards or regulations as evidence of sewage collection and treatment effectiveness. Some report the percentage of permit standards met during a given year. Some calculate the percentage of days in compliance with state and federal regulations. Others report the number of combined sewer overflows, the percentage of sewage bypassing treatment, the percentage of industrial samples in compliance with discharge limits, or the percentage of stream samples in compliance with ambient standards.

The two most common gauges of wastewater plant performance focus on the quality of treated water as it returns to the receiving stream. One measure pertains to biological oxygen demand (BOD) and the other to total suspended solids (TSS) in the water. Neither BOD nor TSS is desired; wastewater treatment is designed to remove as much of each as possible.

High levels of BOD and TSS can be harmful to lakes and streams. BOD is a measure of the oxygen required to decompose organic material. A high level of BOD in a pond or stream threatens the oxygen needs, and therefore the survival, of fish in that body of water. Suspended solids include organic and inorganic particles present in water. High levels of TSS restrict light, coat lake and stream beds, and thereby also threaten aquatic life.

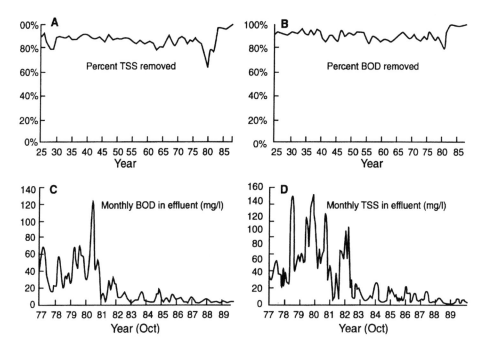

Figure 20.2. Indicators of Wastewater Treatment Effectiveness in Fort Worth, Texas
SOURCE: Fort Worth Water Department, cited in Samuel Nunn, "Organizational Improvement: The Case of Village Creek," *Public Productivity and Management Review, 16* (Winter 1992), p. 133. Used by permission.

Measures of BOD and TSS commonly are reported in one of two ways: (a) the percentage of BOD or TSS removed from the water or (b) the concentration of BOD or TSS remaining in the water following treatment. A high percentage of BOD and TSS removal is desirable, as is a low concentration of BOD and TSS remaining in the effluent following treatment. Charts showing the history of wastewater treatment performance in Fort Worth, Texas, reflecting both the percentage of TSS and BOD removed and the concentration of BOD and TSS remaining in the effluent, are provided in Figure 20.2.

Most cities that routinely report BOD and TSS removal indicate high percentages removed—often well above 90% (Table 20.11). Targeted concentrations in the treated effluent typically are 20 milligrams per liter or less.

The treatment of wastewater is only part of the job of sewer departments. Another major part is the operation and maintenance of the sewage collection system. Considerable variation in operating patterns and maintenance demand was evident among the cities reporting performance targets and operating statistics for wastewater collection systems (Table 20.12). Peoria, Arizona, for example, attempts to clean its system annually, compared with a 4-year cycle in Upper Arlington, Ohio.

Some cities report relatively few stoppages or backups per mile of sewer line, whereas others experience much heavier demand for maintenance services. Green-

TABLE 20.11 Indicators of Wastewater Treatment Effectiveness: Selected Cities

Biological Oxygen Demand (BOD)		Total Suspended Solids (TSS)	
Percentage of BOD Removed	*Remaining BOD in Effluent*	*Percentage of TSS Removed*	*TSS Remaining in Effluent*
Raleigh, NC 99.2% (1991)	**Rock Island, IL** Target: 20 mg/L or less Actual: 5.1 (plant 1), 0.6 (plant 2) (1990); 4.2 (plant 1), 0.6 (plant 2) (1991)	**Denton, TX** 99% (1991)	**Ocala, FL** Target for plant 1: Less than 5.0 mg/L Actual: 2.2 (1990); 1.5 (1991); 2.1 (1992)
Denton, TX 98% (1991)		**Corpus Christi, TX** 98% (1991)	
Oak Ridge, TN 98% (1991); 98% (1992)		**Houston, TX** Target: Remove at least 92% of suspended solids Actual: 95% (1992)	Target for plant 2: Annual average of 20 mg/L or less Actual: 20.8 (1991); 9.9 (1992)
Corpus Christi, TX 96% (1991)	**Ocala, FL** Target: Less than 20.0 mg/L Actual: 6.5 (plant 1) (1990); 2.0 (plant 1), 24.5 (plant 2) (1991); 2.2 (plant 1), 9.6 (plant 2) (1992)		
Grants Pass, OR 93.02% (YTD 12/92)			**Corpus Christi, TX** 6 mg/L (1991)
Fort Collins, CO Target: Maintain organic removal percentages above 90% Actual percent removal, organics: 95.0% (1991)	**Tallahassee, FL** Target: Maintain effluent BOD less than 20 mg/L Percentage of required BOD removed: 106% (1991)		**Tallahassee, FL** Target: Maintain effluent TSS less than 20 mg/L Percentage of required TSS removed: 105% (1991)
Portland, OR Target: 85% or greater Actual: 86% (1991); 89% (1992)	**Corpus Christi, TX** Effluent BOD: 8 mg/L (1991)		**Rock Island, IL** Target: 25 mg/L maximum Actual: 19.8 (plant 1), 0.6 (plant 2) (1990); 16.8 (plant 1), 1.0 (plant 2) (1991)
Sunnyvale, CA Target: Annual average of 60% BOD removal			

ville, South Carolina, for example, reported only 0.26 stoppages per mile—a figure about $\frac{1}{10}$ the rate reported by Savannah, Georgia. Such a result is only mildly surprising, for previous studies have shown remarkable variation among cities in the number of sewer breaks and backups (Table 20.13), attributed less to differences in the age of pipeline than to "problems of root infiltration, ground movement and poor bedding soils, faulty pipe installation, and inadequate protection from surface weights" (Peterson et al., 1984, p. 28).

As with staffing for other municipal operations, the proper staffing of wastewater collection and treatment operations requires a reasonable sense not only of the anticipated workload but also of the human resources required to handle assorted tasks. Production ratios for a variety of wastewater services in several cities are provided in Table 20.14.

Utilities Business Office

Field crews and plant operators are important to a utility's success, but so are the people in the office who receive the telephone calls, serve walk-up utility customers, prepare utility bills, process payments, and maintain accounting records. Unless office personnel perform their jobs well, customer relations—not to mention the utility's balance sheet—will suffer.

A 1991 survey of gas and electric utilities revealed a high level of administrative and clerical attention given both to customer relations and to the diligent collection of utility revenues (Table 20.15). The average utility customer had to wait only 37.5 seconds to speak to a company representative by telephone and 3 minutes in person. Among utility leaders, uncollectible expenses represented less than one half of 1% of total revenues. Among the 78 utilities providing information on the problem of delinquent accounts, one of every five reported at least 5% of its accounts to be more than 90 days delinquent. At the other end of the continuum, a similar portion reported less than 1% of the accounts to be delinquent for more than 90 days (Mercer Management Consulting, 1992, p. 7; used by permission).

Employees in a utilities business office typically deal with many customers throughout the day. Wichita, Kansas, for example, reported 16.28 inquiries handled per telephone representative per hour in 1991, compared with an average of approximately 12 per hour reported in a survey of 191 gas and electric utilities (Mercer Management Consulting, 1992, pp. 3-4; used by permission). A variety of indicators of business office effectiveness, ranging from error rates and collection rates to prompt customer service, have been reported by municipal utility offices and can serve as useful points of comparison for counterparts seeking benchmarks (Table 20.16).

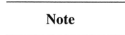

Note

1. Legislation affecting utilities often has pitted environmental and economic development proponents against one another. Local officials with direct managerial or general oversight responsibilities for public utilities are well advised to stay abreast of the latest developments and current standards.

TABLE 20.12 Routine Sewer Cleaning Cycles, Frequency of Stoppages, and Responsiveness to Emergencies and Complaints: Selected Cities

Sewer Cleaning Cycle	Frequency of Sewer Stoppages/Backups	Responsiveness	
		Emergencies	*General*
Peoria, AZ Targets: Clean entire system (lines under 12 inches) annually; clean all manholes and clean-outs within 18 months; clean larger mains as needed	**Boise, ID** Number of sewer backups per 100 miles of sewer line: 5.6 (.056/mi.) (1991)	**Denton, TX** Average response time: 15 minutes Average downtime: 30 minutes (1991)	**St. Petersburg, FL** Target: Respond to 95% within 3 hours Actual: 68.3% within 3 hours (1990)
	Corvallis, OR Target: No more than 10 sewer backups per 100 miles (.1/mi.) Actual: 37.4 (.374/mi.) (1992)	**Sunnyvale, CA** Target: Respond to emergency malfunction calls and hazardous discharge notifications within 20 minutes	**Orlando, FL** Target: Respond to citizen complaints within 4 hours
Fort Collins, CO Percentage of wastewater lines hydrocleaned per year: 78% (1991); 70% (projected, 1993)			**Chandler, AZ** Less than 24 hours (1991)
	Greenville, SC Stop-ups per mile of sewer line: 0.27 (1989); 0.26 (1990)	**Rock Island, IL** Average response to emergency calls: 0.8 hour (1990); 0.5 hour (1991)	**Shreveport, LA** Percentage of water and sewer complaints handled in 3 days or less: 90% (1991)
St. Charles, IL Target: Clean all sanitary sewer lines and manholes at least once every 2 years	**Anaheim, CA** Target: No more than one stoppage per 3 miles of sanitary sewer line (.33/mi.) Actual miles of sewer per stoppage: 6.51 (.15/mi.) (1991)	**Hurst, TX** Target: Respond to sewer main restriction complaints within 45 minutes Actual average response time to reports of sewer main stoppage: 34 minutes (1990)	**Fort Worth, TX** Target: Within 3 days Actual: Within 5 days (1991)
Upper Arlington, OH Target: Inspect and clean all sanitary sewer lines every 4 years			
	Winston-Salem, NC Main stoppages per mile of sewer main: 0.45 (1991); 0.48 (1992)		
	Fort Collins, CO Blockages per mile of line: 0.530 (1991)	**Lubbock, TX** Average time to respond and resolve customer request for service: 49 minutes (1991)	
	Savannah, GA Average stoppage rate per mile of sewer line: 2.05 (1988); 1.92 (1989); 2.69 (1990); 2.39 (1991); 2.70 (1992); 2.56 (1993)	**Portsmouth, VA** Target: Respond to stoppages in sewer mains and laterals within 1 hour	
		Reno, NV Target: Respond to sanitary sewer problems within 1 hour	

TABLE 20.12 *Continued*

Sewer Cleaning Cycle	Frequency of Sewer Stoppages/Backups	Responsiveness	
		Emergencies	*General*
		Oklahoma City, OK Target: Complete top priority jobs in average of 1.75 hours or less Actual: 1.5 (1991); 1.5 (1992)	
		Phoenix, AZ Target: 100% within 6 hours	
		Milwaukee, WI Same-day response to backwater flood control and other sewer trouble calls: 100% (1991); 100% (1992); 99% (1993)	
		Savannah, GA Target: Same day	
		New York City Average response time to backups: 8.7 hours (1991); 8.3 hours (1992) Percentage resolved within 24 hours: 96.9% (1991); 98.2% (1992)	

TABLE 20.13 Sewer Breaks and Backups: 1980

City Grouping	Average Number of Sewer Breaks per 1,000 Miles of Sanitary Sewers	Average Number of Backups per per 1,000 Miles of Sanitary Sewers
Region		
Northeast	39	1,927
North Central	97	483
South	201	1,239
West	163	215
Size		
Large	180	886
Small	85	892
Population Change, 1970-1980		
Growing	188	700
Declining	123	996
Distress		
High	116	1,626
Moderate	143	503
Not Distressed	179	741

SOURCE: Urban Institute survey, cited in George E. Peterson, Mary John Miller, Stephen R. Godwin, and Carol Shapiro, *Guide to Benchmarks of Urban Capital Condition* (Washington, DC: Urban Institute Press, 1984), pp. 30-31. Reprinted by permission.

TABLE 20.14 Production Ratios for Wastewater Services: Selected Cities

POLLUTION CONTROL

Reno, NV
  3 worker-hours per industrial user inspection;
  2 worker-hours per industrial user sample;
  4 worker-hours per treatment sample (1990)

SEWER LINE INSPECTION

Fayetteville, AR
  700 feet per 8 hours, using TV camera
  (3-person crew)

SEWER CLEANING

Duncanville, TX
  Average worker-hours per mile of sewer main
  cleaned: 15.91

Winston-Salem, NC
  Worker-hours per mile of sewer main cleaned:
  18.74 (1991); 16.02 (1992)

Santa Ana, CA
  Worker-hours per mile of sewer main cleaned:
  10.82 (1987)

Fayetteville, AR
  Sewer line cleaned per 8 hours (2-person crew):
  4,000 feet

Pasadena, CA
  Miles of sewer cleaned per person-day: 0.35
  mile (assumes 7-person crew with equipment)

Richmond, VA
  Linear feet of sewer lines "jet cleaned" per
  crew-day: 900 feet

Greenville, SC
  Miles of sewer line cleaned per worker per
  month: 1.5 (1987); 1.7 (1988); 2.3 (1989); 3.2
  (1990); 3.7 (1991)

Dunedin, FL
  Feet of sewer line cleaned per labor-hour: 68.9
  (1992)

MANHOLE REPAIR

Fayetteville, AR
  5.3 worker-hours per manhole repair

MAIN LINE REPAIR

Santa Ana, CA
  11.82 worker-hours per sewer main
  blockage/repair (1987)

Duncanville, TX
  16.25 worker-hours per repair

Fayetteville, AR
  20 worker-hours per repair

SERVICE LINE REPAIR

Duncanville, TX
  14.77 worker-hours per repair

SERVICE LINE STOPPAGE REMOVAL

Duncanville, TX
  2.56 worker-hours per stoppage

NEW SERVICE INSTALLATION

Duncanville, TX
  15.74 worker-hours per new service installation

TABLE 20.15 Selected Performance Marks of Gas and Electric Utility Leaders and Average Performers

Measurement	Leaders	Average Performers
Waiting time for customer to speak to company representative on telephone	< 20 seconds	37.5 seconds
Abandoned telephone calls	< 3%	7%
Waiting time for customer to speak to company representative in person	< 1 minute	3 minutes
Uncollectible expenses as a percentage of total revenue		
Electric	< 0.20%	0.42%
Gas	< 0.45%	1.24%
Combination	< 0.15%	0.44%

SOURCE: Excerpted from *Benchmarking the Quality of Utility Customer Services: 1991 Survey Results* (Lexington, MA: Mercer Management Consulting, Inc., 1992), p. iii. Reprinted by permission.
NOTE: N = 109 utilities.

TABLE 20.16 Indicators of Utilities Business Office Effectiveness: Selected Cities

MINIMAL CASHIER ERRORS

College Station, TX
Target: Error-free daily balance at least 95% of time

Sunnyvale, CA
Target: Daily cash no more than $1 out-of-balance at least 95% of time

ON-TIME MAILING OF BILLS

Sunnyvale, CA
Target: 100% of bills issued within 5 business days of schedule

Savannah, GA
Target: At least 99% of water bills mailed within 1 week of scheduled date
Actual: 98.5% (1988); 99.0% (1989); 98.5% (1990); 99.0% (1991)

Lubbock, TX
Percentage of accounts billed within 6 working days of meter reading: 99.88% (1991)
Percentage of accounts billed within 3 working days of meter reading: 99.68% (1991)

Decatur, IL
Target: 95% of billings ready for processing within 6 days (average) of meter readings

Winston-Salem, NC
Percentage of time billing period exceeded 70 days: 1% (1991); 5% (1992)

PROMPT PROCESSING OF RECEIPTS

Lubbock, TX
Percentage of payments posted on day of receipt: 98.9% (1991)

Raleigh, NC
Percentage of same-day deposits of utility bill payments: 79.8% (1991)

Hurst, TX
Percentage of payments processed within 1 day of receipt: 100% (1990)

Savannah, GA
Target: Process General Fund and Water/Sewer Fund receipts within 1 day at least 90% of time
Actual: Percentage of days 1-day processing target met: 90% (1988-1991)

COLLECTION RATE

Iowa City, IA
99.6% (1987); 99.7% (1988); 99.9% (1989); 100.0% (1990); 99.5% (1991)

Houston, TX
Collection rate (water and sewer): 99% (1992)

Savannah, GA
Target: Collect at least 98.5% of water, sewer, and refuse billings
Actual: 98.6% (1988); 98.5% (1989); 98.1% (1990); 98.7%; (1991); 98.3% (1992); 98.5% (1993)

Oklahoma City, OK
Percentage of billed water use collected: 95.2% (1991); 97.2% (1992)

Winston-Salem, NC
Percentage of billed water/sewer use collected: 99% (1991); 96% (1992)

New York City
Percentage of billed water use collected: 91.4% (1991); 87.2% (1992)

BAD DEBTS

Ft. Collins, CO
Bad debts written off as a percentage of total billed: 0.26% (1991)

College Station, TX
Target: 1% or less per year
Actual: 0.48% (1992)

Fayetteville, AR
Percentage of utility revenue dropped due to bad debt: 0.47% (1991); 0.49% (1992)
Percentage of final utility bills dropped due to bad debt: 17.2% (1991); 18.5% (1992)

Portsmouth, VA
Target: Less than 1%

LOBBY WAITING TIME

Denton, TX
5-minute average (1991)

Houston, TX
10-minute average (1992)

Long Beach, CA
Target: 10 minutes or less
Average wait during peak hours: 10 minutes (1991); 10 minutes (1992)

PROMPT UTILITY SERVICE

Lubbock, TX
Percentage of electric utility service orders worked day of receipt: 100% (1991)

College Station, TX
Target: Provide on date requested 99% of time

Houston, TX
Same-day new water service connection: 98% (1992)
Emergency turn-on: 3 hours (1992)

Chandler, AZ
Target: Response time of 24 hours or less to turn-on/turn-off requests
Actual: 100% (1991)

Savannah, GA
Target: Turn-on/turn-off service within 1 day of request

Charlotte, NC
Target: Complete 95% of customer-requested turn on/off orders within 1 work shift of request; 100% within 2 work shifts
Actual: 98% within 1 work shift of request; 100% within 2 work shifts (1987)

Peoria, AZ
Percentage of service requests handled within 48 hours: 80% (1991)

21

Public Works and Engineering

Typically, public works is one of the largest municipal departments in a community. The department is usually an amalgam of functions, often including engineering, streets, storm drainage, and a variety of other services directed toward the repair and maintenance of municipal equipment, properties, and infrastructure. Among the repair and maintenance functions often found in public works departments are six that will be reviewed in this chapter—street cleaning, street lighting, custodial services, building maintenance, radio maintenance, and vehicle and equipment maintenance. Other functions sometimes included in public works departments but treated in separate chapters of this volume are solid waste collection and disposal, traffic engineering and control, public utilities, and parks maintenance.

Sources of potential benchmarks for public works operations are varied. One logical source is the American Public Works Association, which, in fact, does endorse a set of "standards." Those standards, however, primarily are directed toward the adoption of "best practices" and the establishment of written policies and procedures covering various public works duties and issues. Although the adoption of modern methods and the establishment of written policies are undoubtedly worthwhile, process standards have limited value as performance benchmarks.

Fortunately, other options are available. In some cases, "engineered standards" have been developed in the private sector for operations that correspond closely to public works functions. In a few other cases, performance standards have been engineered specifically for public works operations. In still other instances, performance targets and performance records of respected municipalities can be used as the foundation for public works benchmarks.

Public Works Administration

The specific duties of public works administrators differ from one city to another. Many differences are attributable to variations in assigned functions—for instance, some public works directors are responsible for public utilities, refuse collection, and parks maintenance, whereas others are not. In all cases, however, public works administrators are responsible for coordination and control of assigned functions. Those in charge are expected to see that departmental objectives of efficiency and effectiveness are met and that operating units are responsive.

How quickly should public works departments respond to inquiries, complaints, or requests for services? The city of Fayetteville, Arkansas, reported that in one recent year its responses to 90% of citizen requests were completed within 10 days. Hurst, Texas, reported response to all "action line" calls within 5 working days. The Greenville, South Carolina, public works department responded to 70% of inquiries within 72 hours. Even quicker response was the norm in Corvallis, Oregon, where the target was response within 48 hours, and in Peoria, Arizona, where 95% of initial responses occurred within 24 hours.

Engineering

The skillful, conscientious, and prompt performance of engineering duties can produce major short-term and long-term benefits for a municipality. On the other hand, poor engineering performance can have major adverse ramifications for public relations, developer relations, and capital budgeting, as well as serious long-term consequences for community development. Unfortunately, quality engineering often goes unnoticed.

Several municipalities have established performance targets or reported experience on selected dimensions of engineering (Table 21.1). Some report impressive local standards of performance. Savannah, Georgia, for example, attempts to review all subdivision plats and site plans within 1 week of receipt. Orlando, Florida, reported processing 98% of final plats and development construction plans within 3 working days in 1991. Standards published by the LCC (1994) call for the checking of plans within 2 weeks by high-service engineering departments and within 4 weeks by medium-service departments (p. 39). Several cities have established ambitious plans for keeping engineering maps updated; in many cases, however, their ambitiousness outstrips reality. Hurst, Texas, for example, was only 40% successful in 1990 in meeting its extremely ambitious plan to update all relevant maps within 2 weeks.

Several municipalities focus on holding engineering costs at reasonable levels. Cincinnati, Ohio, for example, reported engineering costs for contract administration and quality control at 6.3% of construction costs for city projects in 1991. Reno, Nevada, reported design costs at 5% of capital construction project costs in 1990.

Engineers are often asked to provide cost estimates for capital projects. Accordingly, several cities track the accuracy of those projections. Although LCC (1994) standards suggest that the estimates of top engineering departments should be within 5% of the mark and those of medium-grade departments should be within 10% (p. 37), the average variance among cities reporting those statistics in the documents examined for this chapter ranged from 10% to 25%. Engineering involvement in major capital projects often extends beyond initial cost estimates to also include supervision of those projects. Project completion on time and within budget is an important objective of that supervision—an objective met 94% of the time by the city of Corvallis in 1992.

Streets

The bottom line for judging a municipal streets program is the condition of community roadways. Some cities attempting to gauge that condition use mechanical devices that record the degree of roughness of city streets. A more common technique, however, is the use of a simple scale called the present serviceability rating (PSR), ranging from 0 for *very poor* to 5 for *very good* (Table 21.2). A community may rate itself and track its progress in upgrading the condition of local roadways as measured by the change in average PSR or by change in the percentage of street miles with favorable ratings.

Some cities have devised their own street condition standards that differ somewhat from the PSR. The City of Savannah (1988), for example, uses a 4-point scale in which a rating of "1" is most desirable.

1. A street segment is rated *one* or *excellent* if it has no defects that are hazardous; no bumps, small cracks, or small holes which are larger than 2 inches in any dimension. The riding surface is equal to a newly paved street and provides a comfortable ride.
2. A street segment is rated *two* or *good* if it has two minor defects larger than 2 inches but less than 12 inches across, minor bumps less than 3½ inches high, or two major defects ½ inch or greater across, but has no safety hazards and provides a reasonably comfortable ride.
3. A street segment is rated *three* or *fair* when it has bumps more than 3½ inches high or two major defects 12 inches or greater across, but has no safety hazards and provides only a moderate degree of riding discomfort.

TABLE 21.1 Engineering: Performance Targets and Experience of Selected Cities

Prompt Completion of Engineering Work			Engineering Costs			
Review of Subdivision Plans	Review of Plans—General	Updating of Maps	Percentage of Construction Costs for Contract Administration, Quality Control, and Engineering	Percentage of Construction Costs for Design/Plans	Accurate Cost Projections for Capital Projects	Effective Supervision of Capital Projects
Savannah, GA Target: Review all subdivision plats and site plans within 1 week Actual: 96% (1988); 98% (1989); 100% (1990); 98% (1991)	**Orlando, FL** Percentage of sewer capacity applications and building plans reviewed within 5 working days: 98% (1991) Percentage of final plats and development construction plans processed within 3 working days: 98% (1991)	**Hurst, TX** Target: Update all relevant maps within 2 weeks after receiving "as-builts" Actual: 40% of relevant maps updated within 2 weeks (1990)	**Cincinnati, OH** Percentage for contract administration/quality control: 6.3% (1991)	**Reno, NV** Target: Keep design costs for capital construction projects to 5% of construction value Actual: 5% (1990)	**Peoria, AZ** Average variance between estimated and actual costs of city projects: 10% (1991)	**Corvallis, OR** Percentage of capital project design and construction schedules met: 85% (1991); 94% (1992)
Oak Ridge, TN Target: 1 week or less Actual: 85% (1991); 80% (1992)	**Irving, TX** Percentage of final plat construction plans reviewed within 5 working days: 95% (1991)	**Peoria, AZ** Percentage of subdivision plans entered on quarter section/street maps within 2 weeks of receipt: 3% (1991); 8% (projected, 1992)	**Reno, NV** Target: Not to exceed 7% of construction value Actual: 7% (1990)	**Corpus Christi, TX** Design cost as percentage of bid: 5.8% (1991)	**Winston-Salem, NC** Bid amount as percentage of estimated costs for major projects: 85% (1991); 84% (1992)	**Fayetteville, AR** Percentage of projects completed within budget and on time: 85% (1991)
Charlotte, NC Target: Review 95% of all subdivision and multifamily complex construction plans within 15 days Actual: 97% (1987); 100% (5 months, 1994)	**Greenville, SC** Percentage of drainage calculations reviewed within 5 days: 82% (1989); 75% (1990)	**Overland Park, KS** Target: Updated monthly to be at least 95% current with revisions to easements, rights-of-way, new plats, new construction	**Rock Island, IL** Target: In-house engineering costs not to exceed 15% of project construction costs Actual: 6.3% (1990); 5.1% (1991)	**Cincinnati, OH** Target: 6% or less for water main installation projects Actual: 6% (1991)	**Greenville, SC** Percentage of construction project estimates within 15% of actual bid: 71% (1989); 80% (1990)	**Savannah, GA** Percentage of capital projects completed on schedule: 77% (1988); 78% (1989); 80% (1990); 74% (1991); 82% (1992); 85% (1993) Percentage of capital projects completed within budget: 74% (1988); 80% (1989); 80% (1990); 78% (1991); 77% (1992); 80% (1993)
Winston-Salem, NC Percentage of subdivision and site plan reviews completed within 30 days: 86% (1991); 80% (1992)	**Peoria, AZ** Percentage of routine plats reviewed within 1 week: 100% (1991)	**Tucson, AZ** Target: Update water system map within 90 days of system upgrade/modification	**Winston-Salem, NC** Engineering costs including consultant, architect, testing, and in-house staff as percentage of project costs for building construction: 12.5% (1991); 14% (1992); for street construction: 20% (1991); 23% (1992); for utility construction: 16% (1991); 25% (1992)	**Rock Island, IL** Target: 7% or less Actual: 5.3% (1990); 3.9% (1991)	**Rock Island, IL** Target: Preliminary cost estimates within 20% of bid Actual: +3% (1990); +25.5% (1991)	**Charlotte, NC** Target: At least 90% of design projects completed within 30 workdays of estimate Actual: 90% (1987)
Jacksonville, FL Target: 42 days or less	**Milwaukee, WI** Percentage of driveway, street lighting, and utility plans and permits processed within 5 days: 100% (1991)				**Sunnyvale, CA** Target: At least 85% of projects within 25% of cost estimate (1994)	
	Hurst, TX Percentage of plans reviewed within 10 days: 95% (1990)					

Overland Park, KS

Target: Review of plans for developer-constructed streets, storm sewers, sidewalks within 2 weeks (inspection within 3 work hours. of request); review of right-of-way work permit applications within 3 workdays; review of building site plans within 2 weeks

Vancouver, WA

Target: Review developer requests for water/sewer connections or plan checks within 15 days

New York City

Sewer design reviews within 14 days: 95% (1991); 99% (1992)
Water and sewer site connection proposal reviews within 30 days: 80% (1991); 90% (1992)
Private drainage proposal reviews within 90 days: 78% (1991); 80% (1992)
Private sewer plan reviews within 30 days: 96% (1991); 96% (1992)

Long Beach, CA

Percentage of plans reviewed within 3 weeks: 95.0% (1991); 91.0% (1992)

Reno, NV

Targets: Review and process final maps (plats, plans, security, etc.) within 40 workdays, parcel maps within 30 days, and building permits within 14 days

Irving, TX

Percentage upgrading of water maps and sanitary sewer maps within 6 months of installation: 10% (1991)

Sunnyvale, CA

Target: Enter at least 95% of all physical facility modifications on 50-scale engineering maps and records within 1 year of modification (1994)

Greenville, SC

Percentage of construction projects completed within 30 days of projected completion date: 50% (1989); 95% (1990)

TABLE 21.2 Pavement Condition Rating: Federal Highway Administration

PSR[a]	Verbal Rating	Description
5	Very good	Only new (or nearly new) pavements are likely to be smooth enough and sufficiently free of cracks and patches to qualify for this category. All pavements constructed or resurfaced during the data year would normally be rated very good.
4	Good	Pavements in this category, although not quite as smooth as those described above, give a first-class ride and exhibit few, if any, visible signs of surface deterioration. Flexible pavements may be beginning to show evidence of rutting and fine random cracks. Rigid pavements may be beginning to show evidence of slight surface deterioration, such as minor cracks and spalling.
3	Fair	The riding qualities of pavements in this category are noticeably inferior to those of new pavements and may be barely tolerable for high-speed traffic. Surface defects of flexible pavements may include rutting, map cracking, and extensive patching. Rigid pavements in this group may have a few joint failures, faulting and cracking, and some pumping.
2	Poor	Pavements have deteriorated to such an extent that they affect the speed of free-flow traffic. Flexible pavement may have large potholes and deep cracks. Distress includes raveling, cracking, and rutting and occurs over 50% or more of the surface. Rigid pavement distress includes joint spalling, faulting, patching, cracking, and scaling and may include pumping and faulting.
1	Very poor	Pavements are in an extremely deteriorated condition. The facility is passable only at reduced speeds and with considerable ride discomfort. Large potholes and deep cracks exist. Distress occurs over 75% or more of the surface.

SOURCE: U.S. Department of Transportation, Federal Highway Administration, *Highway Performance Monitoring System Field Manual,* December 1987, Table IV-4.
a. Present Serviceability Rating (PSR) established by the American Association of State Highway Transportation Officials (AASHTO).

4. A street segment is rated *four* or *poor* when there are hazardous potholes or patches 12 inches or greater across; bumps are greater than $3\frac{1}{2}$ inches high; or the surface is considered rough or bumpy, resulting in an uncomfortable ride. (p. 24)

The condition of city streets is partly a result of construction standards and soil conditions and partly a product of maintenance practices. Many street construction factors, too numerous to be covered here, can influence the quality and durability of a roadway, but few elements are of greater significance than the roadway base. Although precise calculations of suitable base depths are well advised, a few rules of thumb may be helpful to the generalist administrator (Table 21.3). Base thick-

TABLE 21.3 Recommended Depth of Stabilized Roadway Base

Traffic Condition	Recommended Base Thickness
High volumes	8 inches (203.2 millimeters) or more
Moderate volumes	6 inches (152.4 millimeters)
Low volumes on a strong subgrade	5 inches (127 millimeters)
Very low volumes; parking areas	4 inches (101.6 millimeters)

SOURCE: William S. Foster, *Handbook of Municipal Administration and Engineering* (New York: McGraw-Hill, 1978), p. 1.19. Reprinted with permission of McGraw-Hill, Inc.
NOTE: The thicknesses prescribed above are rules of thumb that will serve most cases. Precise formulas providing more specific guidance are available. The depth of the base should depend on the anticipated amount of traffic, the supporting soil, and the use of the stabilized roadway.

nesses ranging from 4 to 8 inches normally are desired, depending on the nature and intensity of anticipated use of the roadway.

The program of continuing maintenance is also a major factor in street condition. Local officials and citizens who are concerned with the condition of city streets often focus on the number and depth of potholes, the promptness of pothole repair, plans for resurfacing of city streets (especially the choice of particular streets to be included or excluded from those plans), and the diligence of street crews in pursuit of the community's objectives.

When citizens express dismay over the productivity level of local street crews, they almost invariably mention what they perceive to be examples of deteriorated street conditions and instances of municipal laborers just "leaning on their shovels." Everyone who has ever wielded a shovel—even as a weekend amateur—knows that some shovel leaning provides a necessary breather in hard labor. But how much is too much? When does the frequency or duration of "breathers" begin to interfere with, rather than sustain, good performance? Knowing the answer requires a determination of what constitutes reasonable production in various tasks.

Among the performance documents examined, several cities reported production ratios and work crew configurations for various tasks (Table 21.4). Charlotte, North Carolina, for example, reported about 0.4 staff-hours per square yard of skin patching and 0.2 staff-hours per square yard for base failure repair. The city of Dunedin, Florida, reported 0.62 feet of curb repair per labor-hour and 0.73 feet of sidewalk repair per labor-hour. Additional sets of crew size and production standards are presented in Tables 21.5 and 21.6. Still further guidance on standard labor output may be found in a manual (updated annually) produced by R. S. Means Company for construction contractors.

Responsiveness is also an important aspect of street maintenance. High-quality street maintenance programs will not allow potholes and other road hazards to remain unrepaired for extended periods, for example. Among the examined performance reporting documents, several performance targets called for pothole repairs within 1 or 2 days of notification (Table 21.7). These targets are consistent with the standards established by the LCC (1994) that call for repair within 24

TABLE 21.4 Production Ratios for Street Maintenance Operations in Selected Cities

Pavement Repair		Curb and Sidewalk Repair	Miscellaneous Tasks
Asphalt	*Concrete*		

Charlotte, NC Targets: (a) 0.43 staff-hours per sq. yd. of skin patching; (b) 0.47 staff-hours per sq. yd. of base failure repair; (c) 2.5 staff-hours per sq. yd. of pothole repair Actual: (a) 0.44 staff-hours per sq. yd. of skin patching; (b) 0.20 staff-hours per sq. yd. of base failure repair; and (c) 1.86 staff-hours per sq. yd. of pothole repair (5 months, 1987) **Dunedin, FL** Surface repair: 15.2 sq. ft. per labor-hour (1.7 sq. yd. per labor-hour) (1992) Surface holes: 1.2 sq. ft. per labor-hour (1992) **Wichita, KS** Sq. yds. of asphalt pavement repairs (permanent) per labor-hour: 1.1 (1990); 1.6 (1991) Blocks of temporary pavement repairs per labor-hour: 2.9 (1990); 3.0 (1991) **ASPHALT REPAVING** **Pasadena, CA** Sq. ft. of asphalt repaving per person-hour: 30.8 (1986) Tons per person-hour: 0.176 (1986)	**Wichita, KS** Sq. yds. of concrete pavement repair (permanent) per labor-hour: 1.0 (1990); 1.3 (1991) **Little Rock, AR** Square yards of concrete replaced per day in concrete pavement repairs: 30 to 70 sq. yds. (7-person crew) **Lubbock, TX** Average worker-hours per paving cut and base failure repaired: 7.71 (1991)	**CURB** **Dunedin, FL** 0.62 ft. per labor-hour (1992) **Charlotte, NC** Target: 1.2 staff-hours or less per linear foot of curb and gutter replaced Actual: 1.58 staff-hours per linear foot (5 months, 1987) **SIDEWALK** **Dunedin, FL** 0.73 ft. per labor-hour (1992) **Santa Ana, CA** Sidewalk/curb forms built and cu. yds. of concrete poured and finished per direct worker-hour: 0.122 (1987)	**Little Rock, AR** Blocks of unimproved road prepared for chip sealing per day (i.e., ditches pulled, debris removed, roadway graded): 2 to 4 blocks (6-person crew) **Wichita, KS** Crack sealing of asphalt/concrete streets—lineal feet per labor-hour: 93 (1990); 65 (1991) **Little Rock, AR** Sq. yds. chip sealed: 10,000 to 16,000 sq. yds. per 10-hour day for 19-person crew **Little Rock, AR** Alley blocks graded and surface material placed per day: 3 to 5 blocks of alley (4-person crew) **Little Rock, AR** Guardrail installed: 25 to 75 linear feet (8 hours; 3-person crew) Guardrail repaired: 50 to 100 linear feet (8 hours; 3-person crew) Guardrail painted: 200 to 300 linear feet (8 hours; 3-person crew) **Little Rock, AR** Shoulder blocks of right-of-way mowed: 20 to 40 shoulder blocks per day (4-person crew) **Little Rock, AR** Alley blocks cleared of brush and debris: 1 to 3 alley blocks per day (4-person crew)

TABLE 21.5 Productions Standards for Selected Maintenance Operations: Fayetteville, Arkansas

Maintenance Operation	Crew Size	Average Daily Production
Right-of way mowing	4	20 lane miles (2 passes)
Pesticide spraying	1	75 to 100 gallons applied
Pothole patching (hot mix)	3	3 tons
Asphalt milling	4	1,000 square yards
Spot surface asphalt repair/replacement	8	450 square yards
Asphalt overlay	14	4,000 square yards
Asphalt surface treatment	4	5,000 square yards
Paving fabric application	5	4,000 square yards
Crack sealing	5	7,500 square yards
Street cut repairs	3	3 cuts repaired
Base repair	4	125 square yards
Shoulder repair	5	750 lineal feet
Curb and gutter installation/repair	3	25 lineal feet
Guardrail installation/repair	4	200 lineal feet
Street sweeping	1	30 curb miles
Clean catch basins and other drainage structures	4	15 structures
Wash drain tiles	4	150 lineal feet
Mechanical ditch cleaning	5	750 lineal feet
Install/replace French drain	4	300 feet
Install drain tile	4	100 feet
Concrete construction/repair	3	1.5 cubic yards

SOURCE: City of Fayetteville (Arkansas), *Public Works Maintenance Management System: System Manual,* January 1989.

hours by high-service departments and within 48 hours by medium-service departments (p. 43). Similarly, top operations establish ambitious performance targets for the repair of utility cuts and other general street and sidewalk maintenance.

Many street maintenance operations are responsible for restoring city streets to passable conditions during winter onslaughts of ice and snow. Denton, Texas, reports that its crews responded to inclement weather conditions within 2 hours of notification 90% of the time in 1991. Cincinnati's performance target is even more ambitious: mobilization within 45 minutes. Some cities target results rather than response. For example, Rock Island, Illinois, expects its crews to have snow removed from city streets within 48 hours after the end of a storm. In Sterling Heights, Michigan, the target is snow removal within 30 hours after snowfall ceases.

Fairly elaborate plans for snow removal are not unusual. The City of Overland Park, Kansas (1993), adopted a priority system for restoring passable conditions to its streets:

Priority I. Bare pavement on all heavily traveled thoroughfares . . . within $1\frac{1}{2}$ hours of snow or ice storm ending between hours of 7 a.m. and 9 p.m.

TABLE 21.6 Street Maintenance Standards

Title	Recommended Crew Size	Standard Labor Time	Hourly Crew Output
Joint and crack sealing	7	00.00182 man-hours/pound	862.0 (pounds)
Slurry seal	4	00.3024 man-hours/1000 lane feet	13.2 (1000 lane feet)
Premix patching	4	03.69 man-hours/ton	1.08 (tons)
Reshape shoulders	4	01.988 man-hours/1000 lane feet	2.08 (1000 lane feet)
Remove/install concrete pavement	7	07.56 man-hours/cubic yard	0.925 (cubic yards)
Remove/install bituminous pavement	8	00.1544 man-hours/ton	51.8 (tons)
Machine sweeping	1	00.240 man-hours/lane mile	4.16 (lane miles)
Machine flushing (streets)	1	00.275 man-hours/lane mile	3.64 (lane miles)
Other patching and paving	As required	—	—
Premix patching (miscellaneous)	2	01.044 man-hours/ton	1.92 (tons)
Regrade, gravel or dirt street	1	00.497 man-hours/1000 linear feet	2.01 (1000 linear feet)
Chip seal	6-8	00.529 man-hours/1000 linear feet	13.23 (1000 linear feet)
Oil surface	1	00.145 man-hours/1000 linear feet	6.89 (1000 linear feet)
Clean and reshape ditches	4	29.16 man-hours/1000 feet of ditch	0.137 (1000 feet of ditch)
Clean drainage structures	2-3	00.0764 man-hours/structure cleaned	26.18 (structures cleaned)
Manual policing	2	01.18 man-hours/cubic yard (of debris)	1.69 (cubic yards)
Machine mowing	1	00.046 man-hours/1000 square feet	21.74 (1000 square feet)
Stump removal	2	00.666 man-hours/stump	3.00 (stumps)
Other roadside and drainage	As required	—	—
Remove/install concrete curb	7	02.80 man-hours/cubic yard	2.5 (cubic yards)
Remove/install concrete sidewalk	5	06.09 man-hours/cubic yard	0.821 (cubic yards)
Remove/install miscellaneous concrete structures	7	—	—
Signal maintenance	2	00.73 man-hours/signal	2.74 (signals)
Pavement marking	2	02.53 man-hours/1000 linear feet	0.790 (1000 linear feet)
Sign fabrication	1	00.931 man-hours/sign	10.7 (signs)
Sign replacement	2	00.732 man-hours/sign	2.73 (signs)
Miscellaneous barricade, sign, and signal work	1	—	—
Spread sand or salt	2	00.16 man-hours/lane mile	12.5 (lane miles)
Control slick spots	2	—	—
Plow snow	1	01.61 man-hours/ lane mile (normal) 02.36 man-hours/ lane mile (heavy)	4.97 (lane miles— normal) 3.39 (lane miles— heavy)
Setup (instructions, tools)	—	0.220 hours/day/man	—
Travel and material handling	—	00.054 hours/mile/man	—
Cleanup	—	00.192 hours/day/man	—

SOURCE: Excerpted from Earl J. Ferguson, "Street Maintenance," p. 682, in George J. Washnis (Ed.), *Productivity Improvement Handbook for State & Local Government* (New York: John Wiley, 1980). Reprinted by permission.
NOTE: Indiscriminate use of these labor standards may result in significant errors. They are included as a guide only.

Priority II. Secondary thoroughfares on traffic routes to schools . . . will have been plowed, sanded, or salt and sand mix applied as conditions require within $1\frac{1}{2}$ hours of storm ending. Sand or salt/sand mix will be applied at a rate of 700 to 800 pounds per two-lane mile.

Priority III. Collector and industrial commercial streets . . . will have been plowed, sanded, or salt and sand mix applied within two hours of storm ending. Sand or salt/sand mix will be applied at a rate of 700 to 800 pounds per two-lane mile.

Priority IV. All other public streets . . . receiving 4- to 6-inch snowfall will be plowed and sand or salt/sand mix will be applied on steep grades and at intersections within 10 hours of snow ending. Application rate of abrasives, where required, will be 700 to 800 pounds per two-lane mile. (p. 5.151)

Sidewalks and curbs and gutters are also elements of the street maintenance program. The City of Sunnyvale, California (1993), uses a standard decision rule to determine whether to include particular sidewalks, curbs, and gutters in the upcoming year's repair schedule. Repair within 1 year is justified for a sidewalk when the differential between walking surfaces is greater than $\frac{3}{4}$ inch. For curbs and gutters, repair within 1 year is justified when displacement of the gutter lip is $\frac{3}{4}$ inch or greater and impounds water more than 10 feet up the gutter swale (p. 7).

Storm Drainage

As in the case of street maintenance, the quality of a municipal storm drainage program may be judged ultimately on its ability to achieve its objectives. Minimal flooding and rapid dissipation of standing water are signs of a successful program. In St. Petersburg, Florida, for example, the local performance standard calls for intersection drainage within 30 minutes following a normal rainfall. Observation of 336 intersections in 1990 revealed 333—or 99%—draining within the prescribed time (City of St. Petersburg, 1992, p. 243).

Diligent work crews, conscientious maintenance, and responsive services contribute to service proficiency and good public relations. Few cities report production ratios for storm drainage crews, but one that does is Little Rock, Arkansas (Table 21.8). More often reported are maintenance cycles and responsiveness of storm drainage operations to complaints and requests (Table 21.9).

Street Cleaning

The scope of street cleaning activities and the size of workforce needed to perform those activities are influenced by a variety of factors, including the frequency of street cleanings, miles of street requiring cleaning, terrain, climatic conditions, vehicle traffic, pedestrian traffic, parking conditions, building density, and the use of sand during the winter (Rubin, 1991, p. 55).[1]

TABLE 21.7 Prompt Road Repairs: Performance Targets and Experience of Selected Cities

Prompt Repair of Potholes and Road Hazards	*Prompt Patching of Utility Cuts*	*General Maintenance*
Dayton, Ohio Target: Repair road hazards within 24 hours of notification	**Tallahassee, FL** Utility service cut repair time: 1 day (1991)	**Reno, NV** Target: Routine pavement repairs within 2 weeks
Fort Collins, CO Target: Fill potholes within 1 business day of notification Actual: 98% (1991)	**Charlotte, NC** Target: Close 95% of utility cuts within 24 hours of completion by utility company Actual: 96% (1987)	**Orlando, FL** Target 1: Complete asphalt street patches within 15 workdays Actual: 100% (1991) Target 2: Complete brick street repair requests within 5 workdays Actual: 100% (1991) Target 3: Complete concrete repair requests within 15 workdays Actual: 100% (1991) Target 4: Repair broken/displaced sidewalks within 30 workdays of notification
Overland Park, KS Target 1: Chuckholes greater than 3 inches deep or with subbase exposed will be repaired within 24 hours Target 2: Deeper holes or overlay/surface failures will be repaired within 1 week	**Denton, TX** Percentage patched within 48 hours: 95% (1991)	
Raleigh, NC Target: Repair asphalt failures (e.g., potholes) within 24 hours of notification Actual: 79% (1991)	**Peoria, AZ** Target: Within 7 days	
Reno, NV Target: Repair potholes within 1 workday of notification	**Savannah, GA** Target: 7-14 days	
Sunnyvale, CA Target: Respond to all hazardous paving failures within 8 hours of report (1994)	**Rock Island, IL** Percentage repaired within 30 days: 89% (1990); 100% (1991)	
Lee's Summit, MO Target: Patch potholes within 2 workdays of notification		**Savannah, GA** Target: Priority III maintenance (street defect, tree root damage, curb defect, ROW mowing, overgrown lane, unpaved street grading, sidewalk repair, alligator cracking, shoulder repair) performed within 14-50 days of request Actual: 96% (1988); 82% (1989); 69% (1990); 100% (1991)
Orlando, FL Target: Repair reported potholes within 2 workdays Actual: 100% (1991)		
Denton, TX Target: Repair 85% of citizen-called-in potholes within 48 hours Actual: 90% (1991)		
Lubbock, TX Percentage of citizen-called-in potholes repaired within 48 hours: 81% (1991)		
Milwaukee, WI Percentage response (patching) within 4 days: 95% (1991); 100% (1992); 100% (1993)		**Milwaukee, WI** Percentage of sidewalk replacement requests serviced same year: 52% (1991); 85% (1992); 54% (1993)
Cincinnati, OH Target: Repair potholes within 5 days of service request		

TABLE 21.7 *Continued*

Prompt Repair of Potholes and Road Hazards	Prompt Patching of Utility Cuts	General Maintenance

Savannah, GA
  Target: Priority I (oil spill, washout, barricade
  request), 7 days; priority II (pothole, low
  patch, utility cut preparation, or surface
  restoration), 7-14 days
  Actual: priority I, 100% (1988-1991); priority
  II, 81% (1988), 62% (1989), 59% (1990),
  80% (1991)

New York City
  Average time between pothole/street cut
  identification and repair: 28.9 days (Bronx);
  14.0 days (Brooklyn); 15.2 days (Manhattan);
  27.9 days (Queens); 20.8 days (Staten Island);
  21.4 days (citywide) (1991)

Street sweeping cycles varied widely among the examined documents. Boston sweeps downtown streets "almost daily" and posted neighborhood streets twice monthly, whereas Milwaukee attempts at least four cleanings annually for each street (Table 21.10).

Factors such as terrain, traffic, and on-street parking that affect street sweeping activities generally also influence street sweeping production ratios reported by various communities. One authority has suggested that street sweepers should average 20 to 25 curb miles per shift (Foster, 1978, p. 6.5), but a study conducted by the City of Pasadena, California, Management Audit Team (1986), showed the operations in neighboring communities averaging 31.5 lineal street miles per sweeper-day and 3.9 street miles per sweeper-hour (p. 5.2). Worker-day averages among the examined documents ranged from 17 to 35 curb miles.

Street Lighting

Not only can street lighting affect a community's traffic safety; it can also influence the incidence of crime, the likelihood that offenders will be apprehended, and the general sense of security among residents. Proper illumination for traffic safety, however, differs from one roadway to another. Key factors are the luminance of a given road surface (Table 21.11), the type of road (e.g., expressway, collector street, etc.), and the commercial or residential nature of the area (Table 21.12). A community's choice among options for achieving desired illumination may have budgetary and aesthetic ramifications—different lighting sources yield different lighting efficiencies and intensities (Table 21.13).

TABLE 21.8 Production Targets for Storm Drain and Ditch Maintenance: Little Rock, Arkansas

---

Number of catch basins rebuilt per 8-hour day:
  0.5 to 1 (3-person crew)

Number of catch basins repaired per 8-hour day:
  2 to 5 (3-person crew)

Number of catch basins vacuumed per day:
  20 to 30 (3-person crew)

Linear feet of storm sewer removed and replaced per day:
  30 to 40 linear feet (6-person crew)

Linear feet of storm sewer and culverts cleaned per day, when cleaning by hand:
  100 to 150 linear feet (4-person crew)

Linear feet of street drainage ditch cleared by hand per day:
  750 to 1,000 (4-person crew)

Linear feet of off-street drainage ditch cleared by hand per day:
  50 to 100 (4-person crew)

Linear feet of drainage ditch dug out by backhoe per day:
  300 to 800 (5-person crew)

Linear feet of ditch cleared per day using heavy equipment:
  100 to 250 (6-person crew)

Square yards of riprap placed per day to stabilize ditch banks:
  30 to 50 square yards (6-person crew)

SOURCE: City of Little Rock (Arkansas), *Management Information Systems Manual*, 1990.

Some cities adopt streetlight policies with crime control in mind. The City of Savannah (1992), for example, prescribes a minimum lighting level of 0.25 foot-candles for high-crime areas, 0.04 foot-candles for high-density areas, and spacing of no greater than 500 feet between streetlights in low-density areas (p. 39).

Streetlight ratios among the examined documents ranged from one streetlight for every six residents in Oak Ridge, Tennessee, to one streetlight for every 53 residents of Waukesha, Wisconsin (Table 21.14). Newburgh, New York, reports 34.5 streetlights per mile. Several cities report prompt investigation of requests for new streetlights and prompt restoration of malfunctioning streetlights.

Custodial Services

Fairly detailed production standards have been developed for custodial services (Table 21.15). The application of such standards to a community's total custodial workload offers local officials a basis for judging the adequacy of current staffing.

A less precise gauge of custodial staffing adequacy may be obtained by comparing local staffing with other communities' ratios of custodians to square footage maintained (Table 21.16). A city with a ratio of 13,000 to 14,000 square feet per

TABLE 21.9 Storm Drainage Maintenance Cycles and Service Responsiveness: Selected Cities

Routine Maintenance Cycle	*Prompt Responses to/Resolution of Drainage Complaints/Requests*
Lubbock, TX Targets: Check/clean all storm sewer inlets at least once every 2 months; all "high-risk" inlets, once per week	**Lubbock, TX** Average response time on emergency calls: 15 minutes (1991)
Reno, NV Target: Clean catch basins twice per year Actual: 67% (1989-1990)	**Orlando, FL** Percentage of emergency responses within 2 hours: 100% (1991) Percentage of citizen inquiries/complaints investigated within 72 hours: 100% (1991)
Fort Collins, CO Target: Clean and/or check catch basins twice per year	**Overland Park, KS** Target: Curb inlet guards that are a hazard to children's safety will be repaired within 4 hours of notification
Orlando, FL Target: Maintain all drainage wells once each year, all open drainage facilities three times per year	**Denton, TX** Percentage of responses to drainage complaints/inquiries within 24 hours: 41% (1991)
Tallahassee, FL Targets: Enclosed system maintenance cycle, 4.7 years; detention facilities, 7.3 years; roadside ditches, 11 years; outfall ditch dredging, 9 years	**Cincinnati, OH** Target: Respond to complaints/requests within 48 hours Actual: 60% (1991)
Lee's Summit, MO Target: 6-year cycle	**Tallahassee, FL** Target: Respond to complaints/requests within 1 week
	Charlotte, NC Target: Resolve 95% of valid complaints within 30 days; 99% within 90 days Actual: 98% resolved within 30 days; 99% within 90 days (1987)

custodian would fit comfortably in the range occupied by Chesapeake, Virginia; Nashville-Davidson County, Tennessee; and Peoria, Arizona.

Building Maintenance

Many public works departments are responsible for the maintenance, repair, and occasional remodeling of city facilities. Some have developed systems for evaluating building conditions and the quality of their work. The City of Savannah (1992), for example, uses the following 4-point scale to rate the condition of municipal facilities:

I. All structural components of the building are in excellent condition—only preventive maintenance work is required.

II. All components are structurally sound; however, minor maintenance work is required.

TABLE 21.10 Street Sweeping Cycles and Production Ratios: Selected Cities

Street Sweeping Cycles	*Street Sweeping Production Ratios*
Boston, MA Targets: Posted neighborhood streets twice monthly; arterial streets weekly; downtown streets almost daily	**Peoria, AZ** Curb miles swept daily per sweeper operator: 29 (1991)
Reno, NV Target: Sweep downtown streets 5 times per week; all other streets once every 3 weeks	**Little Rock, AR** Curb miles per operator per 8-hour shift: 15-35
Richmond, VA Target: Commercial streets are swept 5 nights weekly; residential streets, every 4 to 6 weeks	**Iowa City, IA** Curb miles per unit per day (7 hours per day): 20 (1990); 20 (1991)
Orlando, FL Target: Sweep downtown area at least 3 times per week; residential areas 2.5 times per month	**Charlotte, NC** Target: At least 17 miles per worker-day Actual: 18.6 miles (5 months, 1994)
Peoria, AZ Target: 3 times per week downtown and roadside parking areas; 17-day cycle for arterial and residential streets	**Irving, TX** Lane miles swept per crew-hour: 5.0 (1991)
Fort Collins, CO Targets: Sweep arterials, collectors, and bike lanes twice per month, downtown eight times per month, and residential streets twice per year; additional sweeping after snowstorms	**Santa Ana, CA** Curb miles machine swept per direct worker-hour: 3.97 (1987)
Lee's Summit, MO Target: Sweep downtown streets weekly; other curb-and-guttered streets once per year	**Study of Los Angeles-area cities, 1984**[a] Curb miles cleaned per hour: 3.09 (sample mean); 3.46 (contractor); 2.71 (city employees)
St. Charles, IL Target: Average of once per week during nonwinter months	**Oak Ridge, TN** Curb miles cleaned mechanically per staff-hour: 4.47 (1991); 2.82 (1992)
Sunnyvale, CA Target: Average of 26 sweepings per street per year	**Winston-Salem, NC** Worker-hours per curb mile of street cleaned: 0.83 (1987)
Wilmington, DE Target: At least twice per month	
Overland Park, KS Targets: All streets having curb and gutter will be swept at least 3 times per year; downtown and high-priority streets, every 2 weeks; sand and debris swept from first-, second-, and third-priority streets within 1 month following snow/ice storm	
Lubbock, TX Percentage of downtown swept monthly: 75% (1991) Percentage of thoroughfares swept monthly: 61% (1991) Percentage of residential streets swept once per year: 56% (1991)	
College Park, MD Target: 8 per year	
Hurst, TX Target: Maintain a 45-day street cleaning cycle	
Tallahassee, FL Target: Sweep every street having curb and gutter 8 times per year	
Milwaukee, WI Target: At least 4 cleanings annually for each street	

a. Barbara J. Stevens (Ed.), *Delivering Municipal Services Efficiently: A Comparison of Municipal and Private Service Delivery* (Prepared by Ecodata, Inc., for the U.S. Department of Housing and Urban Development, June 1984), p. 52.

TABLE 21.11 Luminance of Various Roadway Surfaces

Road Surface Class	Luminance Coefficient (Q_o)	Description	Mode of Reflectance
R1	0.10	Portland cement concrete road surface. Asphalt road surface with a minimum of 15% of the aggregates composed of artificial brightener (e.g., Synopal) aggregates (e.g., labradorite, quartzite)	Mostly diffuse
R2	0.07	Asphalt road surface with an aggregate composed of a minimum 60 percent gravel (size greater than 10 millimeters)	Mixed (diffuse and specular)
R3	0.07	Asphalt road surface (regular and carpet seal) with dark aggregates (e.g., trap rock, blast furnace slag); rough texture after some months of use (typical highways)	Slightly specular
R4	0.08	Asphalt road surface with very smooth texture	Mostly specular

SOURCE: *American National Standard Practice for Roadway Lighting,* ANSI/IESNA RP-8-83 (1983). Published by the Illuminating Engineering Society of North America, 120 Wall Street, 17th Floor, New York, NY 10005. Reprinted by permission.

TABLE 21.12 Recommended Roadway and Walkway Illumination[a]

Roadway and Walkway Classification		Average Illuminance Pavement Classification[b]						Uniformity Avg./Min.
		R1		R2 & R3		R4		
		Foot-Candles	Lux	Foot-Candles	Lux	Foot-Candles	Lux	
Expressway[c,d]	Commercial	0.9	10	1.3	14	1.2	13	
	Intermediate	0.7	8	1.1	12	0.9	10	3:1
	Residential	0.6	6	0.8	9	0.7	8	
Major[d]	Commercial	1.1	12	1.6	17	1.4	15	
	Intermediate	0.8	9	1.2	13	1.0	11	3:1
	Residential	0.6	6	0.8	9	0.7	8	
Collector[d]	Commercial	0.7	8	1.1	12	0.9	10	
	Intermediate	0.6	6	0.8	9	0.7	8	4:1
	Residential	0.4	4	0.6	6	0.5	5	
Local[d]	Commercial	0.6	6	0.8	9	0.7	8	
	Intermediate	0.5	5	0.7	7	0.6	6	6:1
	Residential	0.3	3	0.4	4	0.4	4	
Alleys	Commercial	0.4	4	0.6	6	0.5	5	
	Intermediate	0.3	3	0.4	4	0.4	4	6:1
	Residential	0.2	2	0.3	3	0.3	3	
Sidewalks	Commercial	0.9	10	1.3	14	1.2	13	3:1
	Intermediate	0.6	6	0.8	9	0.7	8	4:1
	Residential	0.3	3	0.4	4	0.4	4	6:1
Pedestrian Ways and Bicycle Lanes[e]		1.4	15	2.0	22	1.8	19	3:1

SOURCE: From *An Informational Guide for Roadway Lighting,* Copyright © 1984 by the American Association of State Highway and Transportation Officials, Washington, D.C. Used by permission.
a. Average illuminance on the traveled way or on the pavement area between curb lines of curbed roadways.
b. See Table 21.11 for descriptions of various road surface classifications.
c. Both mainline and ramps. Expressways with full control of access are covered separately as freeways.
d. Adapted from *American National Standard Practice for Roadway Lighting,* ANSI/IESNA RP-8-83 (1983). Published by the Illuminating Engineering Society of North America, 120 Wall Street, 17th Floor, New York, NY 10005. Used here by permission.
e. This assumes a separate facility. Facilities adjacent to a vehicular roadway should use the illuminance or luminance levels for that roadway.

TABLE 21.13 Average Life and Lumens per Watt for Typical Light Sources

Source	Average Life (hours)	Initial Lumens/ Watt Including Ballast
Incandescent (100-1,500 W)	1,000-3,000	17-23 (No Ballast)
Fluorescent (30-215 W)	20,000	62-63
Mercury (100-1,000 W)	24,000	35-59
Metal Halide (175-1,000 W)	15,000	67-116
High Pressure Sodium (70-1,000 W)	20,000	64-130
Low Pressure Sodium (35-180 W)	18,000	80-150

SOURCE: Center for Design Planning, *Streetscape Equipment Sourcebook 2* (Washington, DC: Urban Land Institute, 1979), p. 22. Reprinted by permission.

III.   Structure shows indications of slight deterioration in several component parts of the building. Maintenance work is required within 6 months.

IV.   Structure shows severe deterioration. Immediate major maintenance is required. (p. 129)

The building maintenance operation in Savannah has as its objective ensuring that at least 95% of all city facilities meet condition II or better.

Several cities have established performance targets for prompt response to maintenance requests (Table 21.17). The trick is to achieve a reasonably high level of responsiveness without doing so through overstaffing.

A major concern of local governments is compliance with provisions of the federal Americans With Disabilities Act of 1990. Many building maintenance operations have become involved in retrofitting existing properties for compliance. A sample of building accessibility standards, in this case for public libraries in Illinois, is provided in Table 21.18.

Another concern of most building maintenance operations is the unfortunate need to repair the work of vandals. Some cities have even established performance standards for the prompt removal of graffiti from city facilities. Oakland, California, strives to remove graffiti within 2 workdays; Peoria, Arizona, and Oklahoma City attempt to have it removed within 24 hours.

Radio Maintenance

Many municipal departments—especially those in the public safety field—depend heavily on two-way radio communications. Radio malfunctions can severely damage service effectiveness. Extended downtime is intolerable.

Radio maintenance, whether performed by contract or in-house, must be prompt and reliable. Several cities have established ambitious performance targets and have achieved admirable performance records in the prompt repair of radio equipment (Table 21.19).

TABLE 21.14 Streetlight Ratios and Service Responsiveness: Selected Cities

Streetlight Ratios		Service Responsiveness	
Population per Streetlight	Streetlights per Mile	Prompt Response to Streetlight Requests	Prompt Response to Streetlight Malfunctions
Oak Ridge, TN 6.0 (1993)	**Newburgh, NY** 34.5 (1991)	**Orlando, FL** Percentage of streetlight requests investigated within 1 week: 100% (1991)	**Lubbock, TX** Percentage of critical streetlight orders resulting from damages cleared within 1 hour of order receipt: 97% (1991) Percentage of noncritical streetlight orders cleared within 24 hours of order receipt: 89% (1991)
Greenville, SC 9.6 (1991)	**Greenville, SC** 27.2 (1991)	**Winston-Salem, NC** Percentage of streetlight requests processed within 10 days: 95% (1992)	
Boston, MA 9.73 (1992)[a]	**Tallahassee, FL** 26.3 (1991)		
Tallahassee, FL 10.2 (1991)	**Vancouver, WA** 24.3 (1991)	**Charlotte, NC** Targets: Investigate and respond to (a) 95% of individual requests within 30 days; (b) 95% of neighborhood requests within 60 days Actual: (a) 95%; (b) 80% (1987)	**Sunnyvale, CA** Targets: (a) Replace at least 95% of all burned-out streetlights within 24 hours of notification; (b) correct at least 95% of all minor electrical problems within 24 hours; (c) correct at least 90% of all major electrical problems within 72 hours (1994)
Vancouver, WA 10.2 (1991)	**Oak Ridge, TN** 22.4 (1993)		
Rock Hill, SC 10.4 (1992)	**Denton, TX** 17 (1991)		
Longview, TX 10.7 (1992)	**Peoria, AZ** 16.0 (1991)		
Sioux Falls, SD 12.2 (1991)	**Longview, TX** 14.6 (1991)	**Milwaukee, WI** Percentage responded to within 30 days: 83% (1991); 90% (1992); 87% (1993)	**Milwaukee, WI** Percentage of circuit troubles repaired within 24 hours: 98% (1991); 98.7% (1992); 99% (1993) Percentage of unit troubles repaired within 30 days: 85% (1991); 91.5% (1992); 92% (1993)
Peoria, AZ 14.0 (1991)	**Phoenix, AZ** 13.6 (1992)		
Denton, TX 14.7 (1991)	**San Luis Obispo, CA** 13.6 (1992)		
Phoenix, AZ 19.6 (1992)	**Iowa City, IA** 10.8 (1991)		
Irving, TX 23.5 (1992)	**Southfield, MI** 6.9 (1991)		**Tallahassee, FL** Malfunctioning streetlights repaired within average of 2 days from time reported (1991)
Iowa City, IA 25.8 (1991)	**Blacksburg, VA** 6.3 (1994)		
Blacksburg, VA 27.6 (1994)	**Waukesha, WI** 5.4 (1992)		**Charlotte, NC** Service restored in average of 2.2 days (1987)
San Luis Obispo, CA 27.8 (1992)			**Boston, MA** Percentage of citizen complaints addressed within 5 days: 84% (1992)
Southfield, MI 45.5 (1991)			**San Jose, CA** Percentage of streetlight outages repaired within 2 weeks: 97% (1988); 66% (1989)
Waukesha, WI 52.8 (1992)			

a. Includes city-owned and utility-owned streetlights in Boston. For city-owned streetlights only, the ratio was 17.95 persons per streetlight.

TABLE 21.15 Production Rate Recommendations for Custodial Services

Custodial Service	Time Standard
Basic Service	
Trash removal	2.50 minutes per 1,000 square feet
Clean ashtrays	1.85 minutes per 1,000 square feet
Dust/spot clean horizontally exposed surfaces	2.00 minutes per 1,000 square feet
Spot clean vertical surfaces (e.g., walls, light switches, doors, door frames)	1.10 minutes per 1,000 square feet
Clean glass door (both sides)	2.75 minutes per unit
Clean drinking fountain	1.15 minutes per unit
Clean sink, faucets, and adjacent surface	2.00 minutes per unit
Wash 4′ × 4′ lunchroom table	3.00 minutes per unit
Dust venetian blinds	1.50 minutes per unit
Polish wood desk	5.50 minutes per unit
Restroom Service	
Machine scrub/disinfect restroom floors (150-250 square foot average area)	15 minutes per room
Wash restroom partitions	5.00 minutes per stall
Thoroughly clean and restock	2.50 minutes per fixture (sinks, toilets, urinals)
Floor Service (Carpets)	
Spot vacuum (upright)	3.00 minutes per 1,000 square feet
Traffic lane area vacuum (upright)	5.00 minutes per 1,000 square feet
Detail vacuum (corners, edges, knee wells, etc.)	8.00 minutes per 1,000 square feet
Full vacuum—all exposed carpet (upright)	10.5 minutes per 1,000 square feet
Clean carpet (rotary)	60 minutes per 1,000 square feet
Floor Service (Hard and Resilient)	
Dust mop	
Unobstructed	5.00 minutes per 1,000 square feet
Obstructed	10 minutes per 1,000 square feet
Broom sweep (24" push broom)	7.00 minutes per 1,000 square feet
Damp mop	
Unobstructed	12.65 minutes per 1,000 square feet
Obstructed	14.50 minutes per 1,000 square feet
Stair Service	
Sweep stairs and clean handrails	6.00 minutes per floor
Vacuum carpeted stairs and clean handrails	11.5 minutes per floor

SOURCE: Excerpted from *BSCAI Production Rate Recommendations,* pp. 1-3. Copyright © 1992 Building Service Contractors Association International. All rights reserved. Reprinted with permission of the Building Service Contractors Association International, 10201 Lee Highway, Suite 225, Fairfax, Virginia 22030.

TABLE 21.16 Production Ratios for Custodial Services: Selected Cities

City	Square Footage Maintained per Custodian
New York City	26,800 (1991); 32,600 (1992)
Nashville-Davidson County, TN	14,618—library (1992)
Chesapeake, VA	13,646 (1991)
Peoria, AZ	13,500 (1991)
San Antonio, TX	11,000—office (1991)

Vehicle and Equipment Maintenance

The repair and maintenance of city vehicles and equipment often is the responsibility of a central garage. The proficiency of such an operation may be judged by various combinations of indicators.

TABLE 21.17 Prompt Response to Building Maintenance Requests: Selected Cities

City	Performance Target	Actual
Fayetteville, AR		Average response time of 2 hours to general service requests (1991)
Orlando, FL	Respond to service calls within 8 hours	90% within 8 hours (1991)
Alexandria, VA		Average number of days to complete a work order: 2.0 (1990); 2.0 (1991)
Oakland, CA	Respond to service requests within 2 days of receipt	100% within 2 days (1992)
Winston-Salem, NC	Respond to emergency requests within 1 hour; minor maintenance requests at recreation centers within 48 hours	Responded to 68% of emergency requests within 1 hour; 95% of recreation center requests within 48 hours (1992)
Reno, NV	Respond to calls for emergency building repairs within 1 hour; to calls for routine repairs within 3 working days	
Sunnyvale, CA	Complete 90% of emergency repairs within 1 day of notification; complete 80% of nonemergency repairs within 1 week	
Phoenix, AZ	Respond within 6 days	
Hurst, TX	Complete 65% of work orders within 7 days of receipt	Completed within 7 days: 0% (1990); 50% (estimate, 1991)
Fort Collins, CO	Complete more than 90% of service calls within 10 days	88% completed within 10 days (1991)

TABLE 21.18 Selected Building Accessibility Standards for Persons With Disabilities: Public Libraries in Illinois

Automobile parking

Minimum 96-inch-wide parking space, plus adjacent access aisle minimum of 96-inch width for aggregate of 192 inches (16 feet). Two spaces may use the same access aisle.

Accessible route to circulation desk and the materials catalog

Minimum door openings of 32 inches and minimum 36-inch-wide passageways

Drinking fountain and public telephone

Wheelchair accessible

Toilet facility

Wheelchair accessible

Patron work area

At least 5% or minimum of one of each item: fixed seating, table or study carrel. Area needed: clear floor space of 30 inches × 48 inches; knee clearance at least 27 inches high, 30 inches wide, and 19 inches deep. Top of work surfaces up to a maximum of 36 inches from floor.

Shelf heights

Shelf height for the materials catalog, reference, and current periodicals within the accessible reach area: 48-inch maximum high forward reach and 54-inch maximum high side reach of a person in a wheelchair; minimum forward reach not less than 15 inches above floor and not less than 9 inches above floor minimum side reach.

Carpet

Height of carpet pile: $1/2$ inch maximum height.

SOURCE: An excerpt from *Avenues to Excellence II: Standards for Public Libraries in Illinois* (Chicago: Illinois Library Association, 1989), pp. 33-34. Used by permission.
NOTE: The Illinois Library Association encourages persons responsible for public library facilities to check the latest standards for accessibility for persons who are disabled and applicable federal, state, and local regulations.

TABLE 21.19 Prompt Repair of Radios: Selected Cities

Repair of Public Safety Radios	*General Radio Repair*
Portland, OR 100% of police, water, and maintenance mobile radios returned to service within 1 hour; 95% of fire mobile radios returned to service within 2 hours (1991)	**Rock Island, IL** Average downtime: 2.16 hours (1990); 1.23 hours (1991)
	Raleigh, NC 90.3% of repairs completed in 4 hours or less (1991)
Milwaukee, WI Percentage restored within 24 hours: 100% (1991); 100% (1992); 99% (1993)	**Portland, OR** Percentage of time portable radios are returned to service within 8 hours: 98% (1991)
Savannah, GA Percentage restored within 48 hours: 99.9% (1988); 100% (1989); 100% (1990); 100% (1991); 100% (1992)	**Charlotte, NC** Target: Repair 90% of malfunctions within 8 hours Actual: 99% (1987)
	Milwaukee, WI Percentage restored within 24 hours: 98% (1991); 98.2% (1992); 98% (1993)
Cincinnati, OH 80% of police/fire portables repaired within 5 days (1991); 90% of other public safety equipment repaired within 10 days (estimate, 1992)	**Lubbock, TX** Percentage of communication equipment returned to user within 24 hours: 90% (1991)
	Fort Worth, TX Target: Complete 90% of repairs within 24 hours
	Savannah, GA Target: Complete at least 98% of all radio and electronic equipment repairs within 48 hours Actual: 99%+ (1988-1994)
	Cincinnati, OH 80% of non-public safety equipment repaired within 10 days (1992 estimate)

A fairly sophisticated approach to assessing the performance of the municipal garage draws on the engineered standards developed for private shops. Commercially available rate manuals identify performance standards for mechanics, declaring that a competent mechanic should be able to perform a given repair job on a specified type of vehicle in a particular amount of time.[2] A city garage that can usually meet or beat these or other engineered standards may justifiably declare itself to be an efficient shop.

Several of the performance documents examined for this chapter declared rates of mechanic efficiency, presumably tied to engineered standards. Fayetteville, Arkansas, reported average mechanic efficiency of 95% per repair in 1991. Denton, Texas, reported that 79% of its repairs in 1992 were completed within the flat rate time, whereas Lubbock, Texas, reported 85% completed in accordance with engineered time standards in 1991. The most ambitious target among the documents examined belonged to Rock Island, Illinois, in which the city garage strives for quality maintenance and repairs at rates lower than the private sector. In fact, the Rock Island municipal garage attempts to exceed industry productivity standards by at least 12%—and reports having exceeded those standards by 29% in 1990 and by 34% in 1991 (City of Rock Island, 1993, pp. 08.252, 08.254).

Other measures of the quality of vehicle and equipment maintenance services include the promptness of repairs, the extent to which initial repairs are done properly and need no rework, and the availability rate of municipal vehicles and equipment (Table 21.20). Some cities target 24-hour turnaround for vehicle repair and report high percentages of achievement. Several have established a performance target of having less than 3% of repairs returned for subsequent rework. Most of the reporting cities expected more than 90% of their vehicles and equipment to be available for use at any given time.

The proper staffing of a vehicle and equipment maintenance operation depends on a variety of factors, not the least of which are the types of vehicles and equipment to be maintained, the nature of their use, and their age. A study conducted in Washington, D.C., in the early 1980s drew input from a group of government fleet maintenance managers and offered the following rules of thumb for staffing ratios: one mechanic for every 50 automobiles and light trucks; one mechanic for every 40 heavy trucks; one mechanic for every 30 pieces of heavy equipment; and one mechanic for every 120 mowers and rollers (City of Washington, D.C., n.d., p. 48). The application of such rules of thumb, however, should be tempered by consideration of the age of the equipment being maintained.

Many city governments establish age and mileage guidelines for the replacement of various types of vehicles. Most, however, tend to exceed their own guidelines (Figure 21.1). Although extending the life of a passenger car, for example, from 80,000 to 90,000 miles may be wise from a capital expenditure standpoint and may be a credit to the excellent equipment maintenance that allows such an extension, local officials should recognize that the mechanic workforce needed for maintaining an aging fleet is likely to be greater than that needed for an inventory of newer vehicles and equipment.

TABLE 21.20 Promptness and Quality of Vehicle and Equipment Maintenance Services: Selected Cities

Prompt Repairs	Returns for Rework	Equipment Availability Rate		
		Police Emergency Vehicles	Fire Emergency Vehicles	Overall Rate
Knoxville, TN Percentage of equipment repairs completed within 24 hours: 90%—heavy equipment; 98%—light equipment (1991) **Long Beach, CA** Target: 24-hour turnaround for at least 80% of vehicles in for repair Actual: 80% (1991); 80% (1992) **Charlotte, NC** Target: 77% of all scheduled and unscheduled maintenance completed within 1 workday; 87% within 3 workdays Actual: 85.6% within 1 day; 93.4% within 3 workdays (1987) **Reno, NV** Percentage of service and minor repair jobs completed within 2 working days: 95% (1990) **Fort Worth, TX** Percentage completed within 3 working days: 72% (1991)	**Greenville, SC** "Reworks" as percentage of total work orders: 0.24% (1989); 0.30% (1990) **Rock Island, IL** Target: Recall work of less than 1% **Lubbock, TX** Percentage of repairs returned for corrective action: 1.25% (estimate, 1992) **Oklahoma City, OK** Target: Fewer than 2% of repairs returned for correction Actual: 0.2% (1992) **Charlotte, NC** Target: No more than 2% of all repairs returned for rework Actual: 1% (1987); 0.2% (5 months, 1994) **Peoria, AZ** Target: 2% rejection rate or less Actual: 5% (1991) **Hurst, TX** Target: Repeat repair rate of 3% or less **Fort Worth, TX** Target: Repeat repair rate of 4% or less Actual: 4% (1991) **Denton, TX** Target: Repeat repair rate of 5% or less Actual: 10% (estimated, 1992)	**San Jose, CA** Police patrol car fleet availability: 93% (1988) **Boston, MA** Percentage of marked vehicles available each day: 92% (1992) **Houston, TX** Target: At least 92% availability **New York City** Percentage of police vehicles out of service: 9.8% (1991); 9.6% (1992) **Tucson, AZ** Target: At least 85% of fleet available	**Raleigh, NC** 99.23% (1991) **Winston-Salem, NC** 97% (1992) **San Jose, CA** Target: Downtime of less than 3% Actual: 4% (1989) **Shreveport, LA** 95% (1991) **Charlotte, NC** Target: Unscheduled downtime of less than 10% Actual: 5% (1987) **Tallahassee, FL** Percentage of vehicle operational time: 96.4% (1991) **Boston, MA** Percentage of vehicle uptime (average): 93.9% (1992) **Tacoma, WA** Apparatus availability: 91.6% (1991) **Corpus Christi, TX** 90% (1991) **New York City** Percentage of hours unavailable: engines, 18%; ladders, 18%; support vehicles, 5% (1992)	**Addison, TX** 97% **Orlando, FL** Fleet availability: 97% (1991) **Richardson, TX** 97% **Sunnyvale, CA** Target: 97% **Little Rock, AR** 96.7% (1991) **Carrollton, TX** 96% **Wichita, KS** Vehicles: 97% (1990); 95% (1991) **Rock Island, IL** Target: 95% or greater Actual: 96.6% (1990); 96.6% (1991) **College Station, TX** Target: 95% **Fort Worth, TX** Target: 95% Actual: 95% (1991) **Plano, TX** 95% **Greenville, SC** Fleet availability average: 93.5% (1991) **San Jose, CA** 93% (1988) **Coral Gables, FL** Target: 92% Actual: 95% (1992) **Oakland, CA** Target: Fleet availability of at least 92% Average downtime: 10% (1991); 7.5% (1992) **Chandler, AZ** Vehicles: 92%; equipment: 85.4% (1991) **Portland, OR** Target: 90% of fleet **Tucson, AZ** Target: 90%

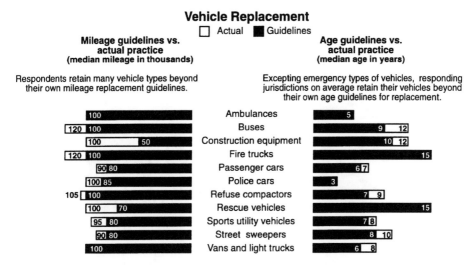

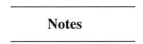

Figure 21.1. Vehicle Replacement Guidelines Versus Actual Practice
SOURCE: *City & State* and Stone and Webster Management Consultants, Inc. Adapted from original graphic by Jerry Parks appearing in Julie Bennett, "Budget Woes Seize Fleet Buying Cycles," *City & State, 10* (August 2, 1993), p. 16. Reprinted by permission of *Governing,* successor to *City & State,* and Stone and Webster Management Consultants, Inc.

Notes

1. These nine factors were identified by Marc A. Rubin (1991, p. 54) for their potential to influence service efforts and accomplishments indicators. Output indicators recommended by Rubin are number of street miles cleaned, percentage of street miles receiving regular street sweeping, and tons of refuse collected. Recommended outcome indicators are percentage of street sweepings not completed on schedule, average customer satisfaction rating, and percentage of streets rated acceptably clean. Recommended efficiency indicators are cost per mile of street cleaned and cost per ton of refuse collected.

2. Two of the most popular of the rate manuals are known as the *Chilton Guide* (Chilton Book Company) and the *Mitchell Manual* (Mitchell International). Each is revised annually.

22 Purchasing and Warehousing

Centralized purchasing and warehousing are the strategies adopted by many municipalities to avoid duplication of purchasing and storage efforts across the various departments that acquire common supplies, to secure economies of scale through bulk purchases, and to derive the benefits of purchasing expertise developed through specialization. In these cities, a department needing a particular item may find it already on hand in the central warehouse or may request its acquisition through the purchasing department, which will then contact vendors to secure the most favorable price.

Typically, a requisition prepared by the requesting department initiates the purchasing process. The purchasing department may seek telephone quotes from suppliers, written quotations, or formal bids, depending in most cases on the anticipated price of the item. Telephone quotations may be obtained quickly and often are deemed sufficient for low-cost items. Securing formal bids takes much longer but is the more appropriate option for high-priced items.

In Vancouver, Washington, for example, the purchasing department secures telephone quotes for materials, supplies, and equipment costing less than $2,500. Written quotations are requested for items costing between $2,500 and $10,000. Formal bids, kept sealed until the announced opening time and date, are solicited for items costing more than $10,000. The formal bidding process in Vancouver usually requires 3 to 4 weeks (City of Vancouver, 1991).

264

Rules of Thumb for Purchasing

To derive maximum value from a centralized purchasing operation, the purchase of equipment and supplies needed by different departments should be well coordinated for quantity discounts; multiple vendors should compete on major transactions; and the cost of administering the centralized purchasing program should be relatively small, compared either with the dollar value of all transactions or with documented savings. An authority in the field has offered these rules of thumb:

- Dividing the total dollar value of all transactions by the total number of all transactions [yields the] average value of each purchase. If the average value is less than $1,000, [the] overall process of buying [should be reviewed].
- [The] total of purchase orders issued should not exceed 25% of the total transactions. Purchase orders are expensive, and should be used only when nothing else will do. Change orders and cancellations are indications of not having things together at the outset.
- A large number of bids indicates a lack of scheduling, and a large number of re-bids indicates a lack of initial preparation.
- Sole source buying can reach as high as 5 percent of total purchases, but should seldom exceed 1 percent. A high percentage of this type buying, even when buying for libraries, points to insufficient bid list development, and will lock [the municipality] into the initial purchases of major systems.
- Local purchasing or bid awards should always be done without local preference consideration. If local merchants are partners in pre-bid conferences, they will win 75 to 80% of all bids anyway simply because they will know more about what [is desired]. If [the city] is small [it] may not have local access to all [of its] needs, but in medium to large cities 75 to 80% should be average. Local preference should never be allowed; competition is still the best way for everyone to get what they need.
- There is no rule, or average, on how many employees make up a good purchasing department, but if the total budget exceeds 1% of the total dollar value of all transactions, [the department] may not be working at [its] most efficient. (Butler, 1985, p. 43; reprinted with permission of *Texas Town & City,* the official publication of the Texas Municipal League)

Normally, an aggressive centralized purchasing operation can outperform decentralized arrangements in securing competitive prices for goods and services. Even when consolidation of orders is impractical and the transaction value does not justify formal bids, telephone quotes may be secured from several vendors. Armed with a comprehensive list of potential suppliers, an aggressive purchasing office normally can claim a high percentage of competitive purchases—regardless of the mix of formal bids and informal quotes.

How well do various cities do in securing competitive purchases, while holding down the costs of centralized purchasing? Boston reported an average of three bidders per contract in 1992. Fort Collins, Colorado, reported 4.7 qualified bidders per bid in 1991, with purchasing expenses equaling 0.43% of the cost of goods and services bought. Decatur, Illinois, reported that competitively bid purchases constituted 50% of total dollar volume in 1991, a figure still well short of its target of 70%.

TABLE 22.1 Prompt Purchase: Performance Targets and Actual Experience of Selected Cities

City	Issuance of Purchase Order
Chandler, AZ	Target 1: Process routine requests of less than $2,000 within 5 workdays Actual: 66.2% (1991) Target 2: Process priority requests for less than $2,000 within 8 work hours Actual: 98.9% (1991)
Lubbock, TX	Processing time for requisitions: 3 days (1991); 1 day (projected, 1992)
Fayetteville, AR	Average processing time when sealed bids are required: 45 days (1991) Average processing time when sealed bids are not required: 2 days (1991)
Oak Ridge, TN	Average time required to process purchase order: 2 days (1991)
Denton, TX	Targets: 3 days for standard requests; 21 days for requests requiring formal bids
Shreveport, LA	Percentage of requisitions processed within 3 days: 95% (1991)
Hurst, TX	Percentage of purchase orders processed within 4 days: 98% (1990)
Portland, OR	Target 1: At least 70% of requisitions of less than $5,000 processed within 4 days Actual: 65% (1991) Target 2: At least 85% of requisitions greater than $5,000 processed within 15 days Actual: 78% (1991)
Long Beach, CA	Target 1: Award at least 92% of formal bids within 90 days Actual: 99% (1991); 90% (1992) Target 2: Award at least 92% of informal bids within 30 days Actual: 92% (1991); 90% (1992) Target 3: Process at least 90% of requisitions within 5 days Actual: 85% (1991); 87% (1992)
Orlando, FL	Percentage of sealed bids awarded within 60 days: 92% (1991) Percentage of nonbid purchase requests awarded within 5 workdays: 100% (1991)
Reno, NV	Percentage of formal bids opened within 6 weeks: 96% (1990) Percentage of informal quotes secured within 3 weeks: 97% (1990) Percentage of requisitions processed to purchase orders within 5 days: 93% (1990)
Tallahassee, FL	Percentage of requisitions greater than $500 processed within 5 days: 79.9% (1991)
Richmond, VA	Processing time for requisitions requiring formal bid: 30+ days Processing time for requisitions requiring informal bid/quotes: 10 days
Fort Collins, CO	Elapsed time from requisition to purchase order: 64.6 days when sealed bids (1991); 11 days without sealed bids (estimate, 1992)
Largo, FL	Turnaround time for bids: 18 workdays (1990); 22 workdays (1991) Turnaround time for quotes: 12 workdays (1990); 14 workdays (1991)
Irving, TX	Target: Issue at least 96% of purchase orders within 15 days when sealed bid is not required
Sunnyvale, CA	Targets: At least 90% of orders requiring formal bids issued within 9 weeks; at least 90% of orders requiring informal quotes issued within 3 weeks (1994)
Cincinnati, OH	Average time required: 23 days (1991)
Tucson, AZ	Targets: Issuance of purchase order within 100 days if bid required; within 25 days, if less than $5,000
Oakland, CA	Target: Process at least 95% of purchase requests within 34 days
Milwaukee, WI	Average time to place order under formal bid process: 88.2 days (1991); 123.5 days (1992); 107.4 days (1993) Average time to place order using informal bids: 27 days (1991); 43.9 days (1992); 35 days (1994)
Savannah, GA	Target 1: Formal bids awarded within 8 weeks Actual: 90% (1988); 90% (1989); 95% (1990); 95% (1991) Target 2: Informal quotes awarded within 4 weeks Actual: 90% (1988); 90% (1989); 95% (1990); 98% (1991) Target 3: Issue purchase order within 14 days of bid award Actual: 100% (1988-1991)

TABLE 22.2 Prompt Award of Contracts: National Survey of Cities

	Average Time From Receipt of Requisition to Award of Contract				
	51 Calendar Days or More	*31-50 Calendar Days*	*26-30 Calendar Days*	*20-25 Calendar Days*	*Less Than 20 Calendar Days*
Percentage of responding municipalities (*N* = 138)	5%	16%	15%	18%	46%

SOURCE: *1993 Procurement Practices Survey,* National Institute of Governmental Purchasing, Inc., Reston, VA. Used by permission. (See National Institute of Governmental Purchasing, 1994.)

Benchmarks

How long does it take a proficient purchasing department to move from receipt of a requisition to the issuance of a purchase order? Generally, the quickest departments can advance to that stage in a few days for items not requiring formal bids. Although many cities take less than a week to process routine purchase orders, the requirement of formal bids for some items slows the process substantially, often requiring a month or more (Table 22.1). Such delays can be frustrating to operating departments anxious to receive the needed items. Steps to streamline the process should be taken where practical, but formal bidding for major purchases safeguards the municipality's integrity and promotes competition. A 1993 survey of 138 municipalities conducted by the National Institute of Governmental Purchasing (1994) indicated that almost half the responding cities were able to move from requisition to award of contract in less than 20 days and that relatively few required more than 30 days (Table 22.2).

Many cities with centralized purchasing operations also maintain a warehouse where items can be secured and held in advance of need. A carefully managed warehouse can be a money saver for a municipality, but one that is poorly managed can be costly.

A good warehouse operation will stock supplies needed on a recurring basis but will not laden its shelves with seldom used items. Cincinnati, for example, estimated that 75% of its requests in 1993 could be filled from its warehouse stock. Little Rock's inventory reportedly met 82% of its need for automotive parts in 1991.

The decision to place a particular item in the warehouse inventory should be based on the anticipated need for that item in the future. Once such a decision is made, proper inventory management calls for periodic restocking to ensure availability. Out-of-stock ratios of less than 10% are common among leading municipalities (Table 22.3).

A city wishing to avoid high out-of-stock ratios could simply overstock all items to minimize the chance of running out, but such a strategy would be poor inventory management because it would unnecessarily tie up municipal resources in idle inventory and would commit warehouse space unnecessarily. Much like well-

TABLE 22.3 Adequacy of Warehouse Inventory: Selected Cities

City	Adequacy of Inventory
Tallahassee, FL	Percentage of items in stock when requested: 92.5%, vehicle repair parts; 97%, other (1991)
New York City	Stock outs: 4.7% (1991); 3.8% (1992)
Lubbock, TX	Orders filled on demand: 96.1% (1991) Office supplies requested that are in stock: 95% (1991) Required vehicle repair parts not in stock: 5% (1991)
St. Petersburg, FL	Target: Out-of-stock ratio of less than 4%
Sunnyvale, CA	Target: Outage of 4% or less in inventoried items
Tucson, AZ	Target: 96% of requisitioned items filled from stock
Hurst, TX	Target: Inventory stock sufficient to meet 95% of needs Actual: 95% (1990)
Peoria, AZ	Stock outs: 5% (1991)
Oakland, CA	Target: Outage of less than 6% in inventoried items
Orlando, FL	Percentage of requested fleet parts filled from stock: 92% (1991)
Winston-Salem, NC	Percentage of orders filled on demand for items advertised as available: 90% (1992)

managed libraries, well-managed warehouses experience high inventory turnover. They have carefully chosen stock that is "popular," and they manage their supply for maximum advantage. Among cities reporting such statistics, inventory turnover rates ranged from once per year to more than three times annually—that is, the value of items issued from the warehouse during 1 year was at least as great as the average value of the inventory and in some cases was more than three times its average value (Table 22.4).

Other aspects of warehouse management include prompt processing of newly received warehouse items and proper inventory control both for accuracy and for

TABLE 22.4 Warehouse Inventory Turnover Rate: Selected Cities

City	Inventory Turnover Rate
Savannah, GA	Target: 3.00 Actual: 3.06 (1988); 2.90 (1989); 3.52 (1990); 3.60 (1991); 3.93 (1992)
Largo, FL	Actual: 3.1, central stores; 3.8, central automotive parts (1991)
Amarillo, TX	Actual: 3.1 (1991)
Wichita, KS	Estimate: 2.47 (1992)
Milwaukee, WI	Estimate: 2 (1993)
Winston-Salem, NC	Actual: 1.79 (1992)
Lubbock, TX	Actual: 0.97 (1991) Estimate: 1.75 (1992)
Little Rock, AR	Actual: 1.15, automotive parts (1991)
St. Petersburg, FL	Target: 1 to 2

TABLE 22.5 Prompt Acquisition/Issuance of Requested Items: Selected Cities

In-Stock Items (office supplies, etc.)	*Special Order Items*
Reno, NV Percentage of items requested from central stores that are delivered within 36 hours: 98% (1990)	**Orlando, FL** Percentage of special order parts for city fleet that are filled within 24 hours: 97% (1991)
Peoria, AZ Target: At least 95% within 2 days	**Tallahassee, FL** Percentage of vehicle repair parts available within 48 hours (including warehouse and special order): 95.1% (1991)
Chandler, AZ Target: Within 2 workdays Actual: 81% (1991)	**Cincinnati, OH** Target 1: Special order items within 168 hours Actual: 80% (1991) Target 2: Emergency items within 24 hours Actual: 100% (1991)
Portland, OR Target: At least 67% within 48 hours Actual: 69% (1991)	
Sunnyvale, CA Target: Within 2 workdays	**Chandler, AZ** Target: Fill requests for noninventory materials within 7 workdays Actual: 84% (1991)
Cincinnati, OH Target: Within 72 hours Estimate: 90% (1992)	

minimum loss or theft. If warehouse bins are empty because inventory records incorrectly report that supplies are still available or because newly received items remain unpacked, the projects of operating departments can be delayed. Some cities report high rates of inventory accuracy. Corpus Christi, Texas, for example, reported 98% accuracy of its warehouse inventory in 1991, and Tallahassee, Florida, reported a stock item variance of only 0.5%, excluding vehicle repair parts, in 1991. St. Petersburg, Florida, expects its physical inventory to be within 2% of book value. Wichita, Kansas, conducts monthly spot checks and reported its item location accuracy at 95% and item count accuracy at 85% in 1991. Getting newly received items on the shelves is also important. Sunnyvale, California, expects at least 95% of such items to be available within 2 working days of receipt.

Warehouse losses can be a serious matter to municipalities and also of concern to auditors. In some cases, such losses are accounting problems—that is, items may have been removed from the warehouse, used in legitimate city projects, but not properly recorded. Such a problem is serious and should be corrected, but it is less serious than the problem of warehouse theft. Inadequate security measures can result in significant losses.

No amount of losses should be considered acceptable, but what level of shrinkage is realistic? Even with minimal loss due to theft, some bookkeeping errors and failures to properly check out emergency items are virtually inevitable. Cincinnati, Ohio, estimated inventory shrinkage in 1993 equal to 4% of inventory value. Oakland, California, has an inventory shrinkage target of 2% or less. Savannah, Georgia's target for warehouse shrinkage is 1.5% of inventory or less.

The best purchasing and warehousing operations get the proper items in the hands of requesting departments quickly. If needed items are maintained in the warehouse inventory, they often can be delivered to the requesting department

TABLE 22.6 Prompt Authorization for Payment: Selected Cities

City	Prompt Authorization for Payment of Invoices
Lubbock, TX	Percentage of payment authorizations processed in 1 day: 98% (1991)
Oak Ridge, TN	Target: Process at least 95% of payment authorizations within 1 day after receipt of invoice Actual: 75%
Milwaukee, WI	Average number of days until payment authorization: 7.8 (1991)
Shreveport, LA	Percentage of invoices processed within 8 days: 90% (1991)
Savannah, GA	Target: Take advantage of at least 70% of all prompt-payment discounts
Boston, MA	Percentage of invoices processed within 20 days: 57% (1992)

within a couple of days (Table 22.5). If special order arrangements are necessary, delays are likely to result.

Once an item has been ordered and received, the purchasing department typically is involved in authorizing payment of the vendor. Prompt authorization by the purchasing department following notification of receipt from the operating department facilitates effective cash management and increases the likelihood of securing prompt-payment discounts offered by many vendors (Table 22.6).

23

Solid Waste Collection

In most American communities, the municipal role in the removal of household refuse has grown considerably since the 18th and early 19th centuries, when that task, according to one authority, was left to "unreliable contractors, scavengers, and the pigs" (Yates, 1977, p. 70). Although refuse collection is a popular candidate for intergovernmental and private contracting, even cities choosing one of those options remain heavily involved. They still specify service scope and quality, monitor contract compliance, and retain ultimate responsibility for services.[1] In either of the two predominant modes of municipal involvement—direct service delivery or contracting—the municipality has need for a system of performance monitoring.

Performance Measures

Recommended performance measures for solid waste collection emphasize the amount of refuse collected, the efficiency with which it is collected (e.g., unit costs), collection reliability, community cleanliness, and citizen satisfaction (Rubin, 1991). Measures recommended for the related functions of solid waste recycling and disposal typically address the percentage of households participating in a recycling program, the amount of waste recycled (either as a percentage of all solid waste or as an average quantity per household), unit costs, net recycling income, tons of waste disposed in landfills or by other means, cubic yards of landfill used, unit costs for disposal, and the degree of local compliance with state and federal environmental standards governing solid waste disposal.

Refuse Collection Benchmarks

Understandably, officials in communities in which municipal crews collect the garbage often are interested in production ratios for sanitation employees. They want assurances that their own crews are performing a reasonable amount of work. Although some guidance in that regard may be provided by production ratios reported by other municipalities (Table 23.1), such ratios should not be applied indiscriminately. Different types of refuse collection equipment, modes of operation, work rules, collection frequency (e.g., once a week or twice a week), collection location (e.g., curbside, alley, or backdoor), terrain, distance between stops, and even the prevalence of on-street parking may affect production ratios. A municipal official whose local production ratio is deemed unacceptable relative to others should consider those factors and possible changes in mode of operation, as well as modifications in equipment size and type, before leaping to conclusions regarding the diligence of refuse collection employees.

A common concern among city officials who are disturbed by unfavorable production ratios is the problem of employee absenteeism among sanitation personnel. A benchmark for assessing that concern was provided by a federally sponsored study of service delivery among Los Angeles-area cities in the mid-1980s that revealed an average absenteeism rate of 10.6% for sanitation employees (Stevens, 1984, p. 173). The rate was 13.4% in communities using city crews for refuse collection and 7.9% for cities using contractors. Among cities examined for this chapter, New York City reported a sick leave rate of 3.85% and a line-of-duty injury absence rate of 1.8% for a total paid absence rate among uniformed sanitation employees of 5.65% in 1992. Cincinnati, Ohio, reported an employee absenteeism rate of 3.7% in 1991, well within its target of 5% of total hours or less. Savannah, Georgia, reported annual sick leave use per sanitation employee ranging from a high of 6.9 days to a low of 5.4 days from 1988 through 1991.

Municipal officials attempting to sort out the relevance of equipment size for refuse collection productivity will be interested in the production ratios specified in a City of Philadelphia study prepared in 1991. That report called for the collection of an average of 12.66 tons per 8-hour shift when using a 32-cubic-yard compactor, 11.84 tons per 8-hour shift when using a 20-cubic-yard compactor, and an average of 9.21 tons per 8-hour shift when using a 9-cubic-yard compactor (p. 4).

The quality of refuse collection services often is measured by collection reliability relative to announced schedules, the extent to which collection stops are

TABLE 23.1 Production Ratios for Sanitation Employees: Selected Cities

STOPS PER COLLECTION EMPLOYEE	LABOR RATIOS
Shreveport, LA 1,176 per week (1991)	**College Park, MD** Target: 1,000 pounds per labor-hour
Corpus Christi, TX 58,777 per year (1992)	**Dunedin, FL** 845 pounds per labor-hour (1992)
	Charlotte, NC Tonnage per sanitation staff member: 784 (1989); 660 (1990); 618 (1991); 651 (1992); 625 (1993)
	Peoria, AZ 230.45 tons per month per operator (1991)

occasionally missed, and the satisfaction of residents with collection services, often gauged by the number of complaints received. Citizens expect their garbage to be picked up according to the announced collection schedule, and some cities have an extremely low tolerance for deviation. Sunnyvale, California, for example, targets 99.9% compliance or better.

Similarly, sanitation officials attempt to keep missed collections to a minimum because each such instance imposes costs in equipment and labor to return to the site to correct the oversight, as well as taking a toll in public relations. Some "missed collections" are bum raps, the result of a resident depositing garbage at the collection location after the crew has already passed. Even in such instances, the sanitation department has little to gain by arguing the point. Most cities reporting statistics on missed collections experienced 10 to 25 misses per 10,000 scheduled stops, with a low of 3 per 10,000 stops reported by Corpus Christi (Table 23.2).

Statistics on refuse collection complaints are reported in a variety of ways. The Los Angeles-area study noted earlier (Stevens, 1984), for example, reported approximately 21 complaints per year for each 1,000 households (Table 23.3). Alternatively, some municipalities report complaints per 10,000 persons or per 10,000 collections. Charlotte, North Carolina, sets its performance target at minimizing complaints per refuse collection crew, with each crew striving to receive no more than 5 valid complaints per month.

TABLE 23.2 Missed Collections: Selected Cities

City	Missed Collections per 10,000 Scheduled Stops
Corpus Christi, TX	3 (estimate, 1993)
Oak Ridge, TN	10 (1992)
Shreveport, LA	10 (1991)
Winston-Salem, NC	25 (1991); 28 (1992)
Fayetteville, AR	36.7 (1991)

TABLE 23.3 Refuse Collection Complaints: Selected Cities

COMPLAINTS PER YEAR PER 1,000 HOUSEHOLDS	COMPLAINTS PER 10,000 COLLECTIONS
Study of Los Angeles-area cities, 1984[a]	**Alexandria, VA**
20.7 among cities with municipal crews	0.2 (1989); 0.4 (1990); 0.2 (1991)
21.2 among cities with contractors	**Corpus Christi, TX**
COMPLAINTS PER 10,000 POPULATION	1.5 (estimate, 1992)
Greenville, SC	**Raleigh, NC**
10.3 per year (1989); 6.3 per year (1990)	1.78 (1991)
Denton, TX	**OTHER**
31.2 per year (1991)	**Charlotte, NC**
Oklahoma City, OK	Target: No more than 25% of crews receiving
0.49 per day (1991); 0.25 per day (1992)	more than 5 valid complaints per month, and no
	more than 25% of supervisors having more than
	one crew exceeding the 5-complaint limit

a. Barbara J. Stevens (Ed.), *Delivering Municipal Services Efficiently: A Comparison of Municipal and Private Service Delivery* (Prepared by Ecodata, Inc., for the U.S. Department of Housing and Urban Development, June 1984), p. 161.

Note

1. Although most municipalities either provide refuse collection services directly through city crews or contract with another government or a private firm for that service, some cities assume a lesser role—either regulating the private market through franchise arrangements or refraining from any involvement whatsoever.

24 Traffic Engineering and Control

Pity the poor traffic engineer. Sometimes it seems that everyone who has ever driven a car thinks he or she has the answers to the community's traffic problems. Too bad those answers are rarely the same as the ones offered by the city's official traffic engineer!

A stop sign here, a traffic signal there, and a left-turn arrow with a turning lane over there should do the job nicely. With three simple moves, the problems plaguing one particular motorist could be solved. The speeding that occurs on the neighborhood's streets could be reduced, lengthy delays in making a difficult left-turning maneuver from the neighborhood onto a major thoroughfare could be cut, and an equally difficult left turn across traffic into the office complex could be simplified. Any traffic engineer who cannot grasp these simple solutions must not be trying to understand! Or could it be that the traffic engineer has more than a single motorist in mind?

Like so many other public servants, the traffic engineer is invisible when things are running smoothly. But let a few traffic snarls develop, inconvenience a few drivers for the sake of traffic flow, or have the community experience a rash of accidents—perhaps a tragic fatality—and the traffic engineer suddenly may be thrust into the eye of a storm.

Standards

Traffic engineers strive to design and regulate transportation arteries to allow safe movement with minimum delay. Professional guidance is available in the form of standards on the use of uniform traffic control markings and guidelines identifying the conditions that call for the use of traffic control devices. Often, the "solutions" that seem so obvious to amateur traffic engineers impose a host of problems that are apparent to their professional counterparts—for example, the stop sign or traffic signal that eases conditions for a few motorists may cause travel delays for many others. Add enough stops along a given route and traffic flow is seriously impaired.

The *Manual on Uniform Traffic Control Devices for Streets and Highways* (U.S. Department of Transportation, 1988) provides standards that govern the size of traffic control signs, their shape and color, the size of their letters or figures, their placement relative to roadways and intersections, and their height. The manual also provides guidance in identifying the circumstances that warrant various traffic control devices. The installation of a traffic signal, for instance, may be warranted by one or a combination of 11 factors:

- *Warrant 1—Minimum vehicular volume.* This warrant is reached when the hourly traffic volume, sustained for eight hours of an average day, reaches 500 vehicles traveling either direction on a major street and 150 vehicles traveling in one direction on a minor, intersecting street, if each is a two-lane street with one lane of traffic moving in each direction.

- *Warrant 2—Interruption of continuous traffic.* This warrant is met, for example, when 75 vehicles per hour traveling a single direction on a minor street (one lane each direction) attempt to enter or cross a major street (two lanes each direction) carrying 900 vehicles per hour.

- *Warrant 3—Minimum pedestrian volume.* This warrant is met when 190 or more pedestrians attempt to cross a major street during any single hour or when 100 or more cross during each of four hours.

- *Warrant 4—School crossing.* This warrant is met when traffic engineering studies indicate inadequate gaps in the traffic stream.

- *Warrant 5—Progressive movement.* This warrant is met, for example, when the nearest signals are too far apart to effect vehicle platooning and speed control.

- *Warrant 6—Accident experience.* This warrant is met when five or more accidents of a type that could be avoided with a traffic signal occur within a 12-month period, the location experiences substantial vehicular and pedestrian volume (Warrants 1, 2, and 3), other remedies have proven ineffective, and signal installation will not seriously disrupt traffic flow.

- *Warrant 7—Systems warrant.* In some cases, traffic signal installation may be warranted to encourage concentration and organization of traffic flow networks.

- *Warrant 8—Combination of warrants.* Even in cases where no single warrant is met, meeting Warrants 1 and 2 at the 80% level may be sufficient to justify signal installation.

- *Warrant 9—Four-hour volumes.* High volumes of vehicles through an intersection, in various major street-minor street combinations during each of four hours of an average day, may justify signal installation.
- *Warrant 10—Peak hour delay.* This warrant is met when extreme delays are experienced by drivers on a minor street attempting to enter or cross a major street during any single hour of an average weekday.
- *Warrant 11—Peak hour volume.* Like Warrant 10, this warrant addresses the difficulties experienced by drivers at a high-volume intersection during rush-hour traffic. (pp. 4C.3-4C.8)

The degree to which these conditions are required prior to signal installations differs from community to community. Emotion and politics—as well as professional expertise—often weigh heavily in such decisions.

Other Possible Benchmarks

Responsible traffic engineers strive for traffic flow and safety. In that regard, some cities declare precise traffic flow targets. For example, Savannah, Georgia, has identified first and second priority transportation corridors and has established a traffic flow target of maintaining overall average speed of at least 15 miles per hour during peak traffic hours. That target was met 91% of the time in 1993. Winston-Salem, North Carolina, reported average travel time within coordinated signal sections of 2.42 minutes per mile during peak times and 2.34 minutes per mile during off-peak times in 1992.

Traffic safety can be reported in several ways. Portland, Oregon, defines as "high-accident intersections" those with 20 or more accidents in a 4-year period and reported 5.6 such intersections in 1992. More common methods are to report accidents per 1,000 population or accidents per 1 million vehicle miles (Table 24.1).

Professional conduct for traffic engineers requires the conscientious application of professional norms to local traffic considerations, but it also implies responsiveness and diligence in the conduct of all traffic engineering duties. The traffic engineering department in Orlando, Florida, for example, reported providing complete responses to most citizen requests within 1 week and the completion of most major studies within 30 days.

Good traffic flow depends not only on sound engineering but also on the installation and maintenance of appropriate traffic control devices. Traffic control signs, for instance, should conform to uniform standards in appearance and placement. Pavement markings should be properly designed and maintained. The LCC (1994) considers the repainting of centerlines, curb markings, crosswalks, and legends on a 9- to 18-month cycle to be the mark of high-level service; an 18- to 30-month cycle to reflect medium service; and less frequent repainting to indicate a low-service level (p. 43). Many cities examined for this chapter attempt to repaint pavement markings at least annually (Table 24.2).

TABLE 24.1 Traffic Engineering: Performance Targets and Actual Experience of Selected Cities

Prompt Investigation of Traffic Problems	Prompt Completion of Traffic Studies	Prompt Review of Plans	Accident Rate
Orlando, FL Target: Provide complete response to citizen requests for information/assistance within 1 week Actual: 90% (1991)	**Long Beach, CA** Percentage of traffic engineering studies completed within 3 weeks: 60.0% (1991); 65.6% (1992)	**Lubbock, TX** Percentage of subdivision plats checked within 1 workday: 100% (1991)	**REPORTED ACCIDENTS PER 1,000 POPULATION** **Iowa City, IA** 40.7 (1990)
Rock Island, IL 11.7 days (1990); 26.2 days (1991)	**Cincinnati, OH** Target: Within 30 days	**Raleigh, NC** 1.6 days for plot plans (1991); 1.475 days for construction plans (1991)	**Tallahassee, FL** 65.3 (1991)
Reno, NV Target: Within 14 days Actual: 65% (1990)	**Orlando, FL** Percentage of major studies completed within 30 days: 95% (1991)	**Reno, NV** Percentage of development applications reviewed within 10 days: 95% (1990)	**ACCIDENTS PER MILLION VEHICLE MILES** **Sunnyvale, CA** Targets: (a) 2.97 accidents per 1 million miles traveled; (b) 0.14 bicycle accidents per 1 million miles traveled; (c) 0.06 pedestrian accidents per 1 million miles traveled (1994)
Lubbock, TX Percentage of citizen requests resolved within 15 workdays: 66% (1991)		**Long Beach, CA** Percentage reviewed within 3 weeks: 95.0% (1991); 91.0% (1992)	
Tacoma, WA 29 days (1990); 30 days (1991)		**Cincinnati, OH** Target: Within 30 days for zoning changes, building permits, PUD[a] projects, streetscape projects	**Winston-Salem, NC** 3.08 accidents per 1 million vehicle miles; 1.61 accident injuries per 1 million vehicle miles; 0.01 accident deaths per 1 million vehicle miles (1992)
Raleigh, NC Traffic signal requests: 35 days; traffic sign requests: 4 days; individual streetlight requests: 0.94 days (1991)			**Charlotte, NC** Target: Not to exceed 6.44 traffic accidents per million miles traveled Actual: 4.75 (5 months, 1994)
Corvallis, OR Target: Evaluate and respond to neighborhood traffic/parking concerns within 60 days Actual: 90% (1991); 100% (1992)			**Savannah, GA** 16.6 accidents per 1 million vehicle miles

a. PUD = planned unit development.

When stop or yield signs are damaged or when traffic signals malfunction, prompt response is important. Many cities report a response time target of 4 hours or less for the repair or replacement of damaged stop or yield signs (Table 24.3).

TABLE 24.2 Pavement Marking Cycles: Selected Cities

City	Pavement Marking Cycle
Fort Collins, CO	Target: Repaint most crosswalks and traffic lane symbols once per year, more frequently on arterials and collectors; repaint traffic lanes on arterials four times per year, on collectors three times, and residentials once per year
Lubbock, TX	Target: Repaint thoroughfares twice per year; marked collector streets and crosswalks, annually
Overland Park, KS	Target: Twice per year (unless thermoplastic)
Phoenix, AZ	Target: Repaint major streets and intersections twice per year
Milwaukee, WI	Average marking frequency per year for lanes and centerlines: 1.3 (1991); 1.5 (1992); 1.2 (1993) Average marking frequency per year for crosswalks and other markings: 0.6 (1991); 0.9 (1992)
Cincinnati, OH	Target: Repaint centerlines, lane lines, and edge lines throughout the city at least once a year
Shreveport, LA	Target: Annually
Sunnyvale, CA	Target: 12-month maintenance cycle for markings on arterials; 24-month cycle off arterials
Rock Island, IL	Target: Repaint center and lane lines annually; curbs on a 2-year cycle; stop bars, crosswalks, and lane use markings at intersections on a 2-year cycle
Tallahassee, FL	Target: 2-year cycle

Because of their customary placement at hazardous, high-volume intersections, targeted response times for malfunctioning traffic signals were even quicker.

Proper staffing for the maintenance of traffic control devices balances the need for quick response with the desire to avoid overstaffing. The prudent administrator, as well as the wily budget officer, is likely to want statistics on the incidence of traffic signal problems, the normal rate of traffic control sign repair and replacement, and relevant production ratios for maintenance personnel. Such statistics are collected by many municipalities. Tallahassee, Florida, for example, had 9.34 trouble reports per signal during one recent year, compared with 8.8 and 15 trouble calls per signalized intersection in Nashville-Davidson County, Tennessee, and Raleigh, North Carolina, respectively. Nashville-Davidson County had 73 signalized intersections per full-time signal technician in 1991.

Several cities report production ratios for a variety of traffic control activities (Table 24.4). Although a given city's own production ratios may not match those reported here precisely, substantial variation may suggest the need for review.

TABLE 24.3 Prompt Response to Damaged Traffic Control Signs and Malfunctioning
Traffic Signals: Selected Cities

Response to Damaged Traffic Control Signs		Response to Malfunctioning
High-Priority Signs	General	Traffic Signals

Lubbock, TX
Service calls on stop/yield signs responded to within 1 hour: 99% (1991)

Reno, NV
Target: Respond to calls regarding damaged stop or yield signs within 1 hour

Denton, TX
Responded with 1 hour to 80% of service calls (1991)

Orlando, FL
Target: Within 2 hours at hazardous intersections
Actual: 100% (1991)

Overland Park, KS
Target: Respond within 2 hours

Tucson, AZ
Target: Within 2 hours

Charlotte, NC
Percentage of damaged stop or yield signs repaired/replaced within 3 hours of notification: 100% (5 months, 1994)

Dayton, Ohio
Target: Repair or replace within 4 hours of notification

Sunnyvale, CA
Target: Respond within 4 hours of receipt of report

Boston, MA
Percentage of public safety signs (high priority) replaced or repaired within 24 hours: 100% (1992)

Winston-Salem, NC
Percentage of stop or yield signs replaced or repaired within 24 hours of complaint: 100% (1992)

Lubbock, TX
Service calls on traffic control signs responded to within 1 workday: 99% (1991)

Orlando, FL
Target 1: Within 2 hours at hazardous intersections
Actual: 100% (1991)
Target 2: Within 24 hours if not creating a traffic hazard (excludes street name signs)
Actual: 98% (1991)

Reno, NV
Targets: Respond to calls regarding damaged stop or yield signs within 1 hour; other traffic control signs within 1 workday; street name signs within 2 weeks

Fort Collins, CO
Average time for repair/replacement of signs: 2 workdays (1991)

Tucson, AZ
Target: High priority within 2 hours of notification; lower priority, within 1 week

Overland Park, KS
Target: At least 90% of damaged, lower-priority signs corrected within 10 workdays (e.g., street name signs)

Boston, MA
Target: Repair or replace non-public safety signs (i.e., lower priority) within 15 days
Actual: 50% (1992)

Sunnyvale, CA
Target: Respond within 30 days on damaged, lower-priority signs

Lubbock, TX
Percentage of emergency service calls responded to within 30 minutes: 98% (1991)

Shreveport, LA
Target: Respond to traffic signal trouble calls within 30 minutes
Actual: 97% (1991)

St. Petersburg, FL
Target: Respond to traffic signal trouble calls within 30 minutes
Actual: 96.2% compliance (1990)

Raleigh, NC
Average response and repair time per complaint: 58 minutes (1991)

Study of LA-area cities, 1984[a]
Average time taken to reach repair site: 32.5 min. (city crews); 43 min. (contractor)
Average time to complete repair work after arrival: 77 min. (city crews); 37 min. (contractor)

Corvallis, OR
Target: Respond to signal malfunctions within 1 hour

Charlotte, NC
Target: Respond to 90% of requests for emergency service on traffic control equipment within 1 hour; 100% within 2 hours
Actual: 92% within 1 hour; 100% within 2 hours (1987)

Greenville, SC
Percentage of emergency service requests responded to within 1 hour: 90% (1990)

Sunnyvale, CA
Target: Commence at least 90% of emergency repairs to traffic signals within 1 hour of notification (1994)

Milwaukee, WI
Percentage responded to within 1 hour: 88% (1991); 88.3% (1992); 85% (1993)
Percentage restored within 1 workday: 99% (1991); 97.7% (1992); 97% (1993)

Denton, TX
Responded within 1 hour to 85% of service calls (1991)

TABLE 24.3 *Continued*

Response to Damaged Traffic Control Signs		Response to Malfunctioning Traffic Signals
High-Priority Signs	*General*	
		Boston, MA
		Percentage of calls for traffic signal repair responded to within 2 hours: 95% (1992)
		Percentage of calls for traffic signal repair resolved within 2 hours: 75% (1992)
		Percentage of malfunctioning pedestrian signal indicators and traffic signals replaced or repaired within 24 hours: 100% (1992)
		Overland Park, KS
		Target: Correct or alleviate every priority malfunction within 2 hours working time
		Dayton, OH
		Target: Restore minimum service within 4 hours of notification
		Cincinnati, OH
		Target: Make 98% of problems safe within 8 hours
		Actual: 98% (1991)
		Orlando, FL
		Target: Replace failed signal lamps within 2 hours of notification; restore failed signals to minimum requirements within 8 hours of notification
		Actual: 100% (1991)
		Reno, NV
		Percentage of traffic signal problems resolved within 1 day: 90% (1990)
		Tallahassee, FL
		Target: Same-day response

a. Barbara J. Stevens (Ed.), *Delivering Municipal Services Efficiently: A Comparison of Municipal and Private Service Delivery* (Prepared by Ecodata, Inc., for U.S. Department of Housing and Urban Development, June 1984), p. 279.

TABLE 24.4 Production Ratios: Traffic Control Operations in Selected Cities

STREET NAME SIGN INSTALLATION

Little Rock, AR

Street name markers installed per day: 5 to 8 if new posts required (2-person crew); 15 to 20 if mounted on existing posts (2-person crew)

STREET NAME SIGN REPAIR

Little Rock, AR

Street name markers repaired/replaced per day: 30 to 50 (2-person crew)

LANE MARKINGS

Raleigh, NC

Staff-hours per mile of lane markings: 4.2

Wichita, KS

Lineal feet of pavement marking per worker-hour: 832 ft. (1990); 608 ft. (1991)

Santa Ana, CA

Lineal feet of pavement marking per direct worker-hour: 648 ft. (1987)

LEGEND/ARROW INSTALLATION (THERMOPLASTIC)

Little Rock, AR

Legends/arrows installed in traffic lanes per day: 4 to 6 locations (2-person crew)

LEGEND/ARROW PAINTING

Little Rock, AR

Legends/arrows painted in traffic lanes per day: 10 to 12 locations (2-person crew)

RAISED PAVEMENT MARKING

Little Rock, AR

Number of raised pavement markings per day: 30 to 50 (2-person crew)

LANE LINE INSTALLATION (THERMOPLASTIC)

Little Rock, AR

Linear feet installed per day: 3,600 feet (4-person crew)

LANE LINE PAINTING

Cincinnati, OH

Linear feet of lane lines painted per day: 100,000 (2-person crew)

Little Rock, AR

Linear feet of lane lines painted per day: 20,000 to 60,000 (4.2-person crew)

Winston-Salem, NC

Worker-hours per gallon of traffic paint used: 0.84 (1992)

MAINTENANCE OF TRAFFIC CONTROL SIGNS

Cincinnati, OH

Number of traffic control signs repaired (straightened, replaced, etc.) per day: 40 (2-person crew)

Little Rock, AR

Number of traffic control signs and posts installed per day: 14 to 18 (2-person crew) Number of traffic control signs and posts removed per day: 30 to 50 (2-person crew) Number of traffic control signs repaired (straightened, replaced, etc.) per day: 30 to 50 (2-person crew)

Dunedin, FL

Signs repaired per worker-hour: 1.3 (1992) Signs replaced per worker-hour: 1.3 (1992) Posts replaced per worker-hour: 1.1 (1992) Posts maintained per worker-hour: 1.1 (1992) Posts removed per worker-hour: 1.8 (1992)

Santa Ana, CA

Signs fabricated, installed, replaced, or repaired per direct worker-hour: 0.965 (1987)

Winston-Salem, NC

Worker-hours per sign maintenance activity: 1.03 (1992)

MOUNTING OVERHEAD SIGNS

Little Rock, AR

Number of overhead signs mounted on traffic signal equipment per day: 4 locations (2-person crew)

TRAFFIC SIGNAL CONSTRUCTION

Fort Collins, CO

Total work-hours needed to construct traffic signal: 312 (1991)

TRAFFIC SIGNAL REPAIR/ MAINTENANCE

Winston-Salem, NC

Worker-hours per signal failure correction: 0.45 (1992)

Wichita, KS

Average worker-hours per traffic signal repair: 1.29 (1990); 1.08 (1991)

Duncanville, TX

Average worker-hours to maintain traffic signal (field): 2.7

Little Rock, AR

Traffic signals repaired per day: 2 locations (2-person crew)

25

Performance Milestones

A city may be judged on several different dimensions. Is it a good place to live? Is it a decent place to raise a family? How are the schools? Is the climate pleasant? What entertainment opportunities are available? Does the community support the arts? Is the local economy healthy? Is it growing? Is the community clean—physically and politically? Are municipal services good?

A city deemed to be outstanding in one dimension might be decidedly *un*remarkable—banished from the list of top cities—in another. Being number one is more than sheer luck, but it helps if the criteria used to establish the ranking coincide with the community's strengths and overlook points of weakness. On some lists, it also helps to have strong neighbors.

A city's reputation—favorable or unfavorable—may belong to an entire region. Reference to *city* often means *community* in a fairly broad sense. For instance, travelers who say they are from Dallas and report favorably on their hometown and its local government services actually may reside in suburban Mesquite or Plano and, without intending any deceit, may be reporting on living conditions and services in one of those cities. In such cases, the entire metropolitan area takes on a common identity symbolized by the central city. Some lists are no more exact than that.

Among local government officials, the definition of *city* is much more precise, and the relevance of *city rankings* is much less ambiguous. To them, city refers to the corporate municipal entity and its geographic territory. In this sense, it may be all right to judge a *community* on its climate, but a *city* should be judged only on political, economic, and social characteristics or dimensions of performance that local government officials can influence.

Popular Ratings

Community ratings are popular. People like to know how their town stacks up against others. Some rankings are consistent with the narrower definition of city, but the authors of many lists have little concern with precise municipal boundaries and adopt a broader community view. The popular ranking found in the *Places Rated Almanac* (Boyer & Savageau, 1989; Savageau & Boyer, 1993) is like that. Although city names are used, the authors carefully explain that they are rating metro areas. The top "places" to live—Seattle and Cincinnati headed the lists in 1989 and 1993, respectively—are selected on the basis of climate, five types of facilities (health care, education, recreation, transportation, and the arts), and four economic and social indicators (crime, costs of living, housing, and jobs).

Similarly, recognition as an outstanding locale for business, retirement living, or affordable housing typically is an honor intended for the named city and its environs rather than strictly for that city alone. Although such lists are no doubt of interest to municipal officials—and especially to their chamber of commerce brethren—such ratings are of dubious value as scorecards for municipal government.

Several national programs do honor municipal boundaries, recognize the subtleties of local identity, and concentrate on civic matters, but even these often focus on other aspects of municipal government rather than on the effective delivery of traditional services. Citizens and local government officials desiring models of citizen participation, for example, may consult the honor roll of recipients of the National Civic League's (NCL) All-America Cities awards. If they wish to assess their own community's civic infrastructure, they may do so by applying NCL's Civic Index, which addresses 10 important aspects of civic life—only one element of which focuses on government performance (Table 25.1).

Other popular, third-party assessments of municipal performance often extol innovation—especially innovation in a particular local government service. The Management Innovation Awards of the International City/County Management Association (ICMA), the Technology Achievement Awards sponsored by Public Technology, Inc., the Innovations in State and Local Government program of the Ford Foundation and the John F. Kennedy School of Government at Harvard University, and the Exemplary State and Local (EXSL) Awards Program sponsored by the National Center for Public Productivity all belong to this genre. Each provides important recognition to recipient governments and encouragement to others, but none of these esteemed programs purports to serve as an assessment device for judging municipal services in general.

TABLE 25.1 Ten Elements of the National Civic League's "Civic Index"

1. Citizen participation
2. Community leadership
3. Government performance
4. Volunteerism and philanthropy
5. Intergroup relations
6. Civic education
7. Community information sharing
8. Capacity for cooperation and consensus building
9. Community vision and pride
10. Intercommunity cooperation

Standards for Various Services

Standards and guidelines for several municipal functions, as described in previous chapters, have been offered by professional associations and others having a specific interest in those services. Although often developed by specialists with important insights regarding service delivery problems and possibilities, these guidelines are sometimes vulnerable to self-serving motives and, if adopted, should be applied with that caveat in mind.

Typically, single-service standards or guidelines emphasize process over product, or resource input over performance output. Even where standards for a particular specialty focus on performance objectives or prescribe reasonable performance expectations, they do so out of context, with little regard for the full mosaic of municipal services.

Multiservice Performance Guidelines

Even before Clarence Ridley and Herbert Simon (1943) raised the performance management profile of local government by encouraging comprehensive measurement of municipal activities, a series of halting steps signaled the fits and starts of a never-quite-ready-for-prime-time effort to compare municipal performance meaningfully.[1] The rationale has always been that the performance of all units could benefit by the effective exchange of relevant performance data, but securing sufficient commitment and resources has proved through the years to be a recurring problem. Current efforts by ICMA and The Innovations Groups once again offer renewed hope for that strategy.[2]

Even in the absence of a comprehensive database of relevant performance statistics, a few promising initiatives focusing on good management and the delivery of basic municipal services have undergirded a growing interest in performance benchmarks. They have done so by turning attention toward the need for standards, heightening the profile of multicity comparisons, and renewing interest among municipal officials in the possibility of being recognized as a competent service provider or perhaps even as "one of the best."[3] For example, local government practitioners have demonstrated a growing desire to know how their operations stack up against others and a willingness to be held accountable to reasonable standards. Evidence of that desire comes in several forms, not the least of which is the recent publication of workbooks and guidelines on standards and service levels for local governments in California, Washington, and Pennsylvania (LCC, 1994; *Level of Service Standards,* 1994; SPRPC, 1990).

Benchmarking

Benchmarking is a practice that has grown rapidly in popularity since the 1980s. Although variations are common, true benchmarking involves the identification of best-in-class performers, the comparison of local performance outputs and results with those of top performers, the analysis of practices that account for any performance gaps, and the development and implementation of strategies to adjust the performance gap in one's favor.

This volume's collection of recommended standards, guidelines, performance records, and performance targets of individual cities offers a head start. These municipal benchmarks provide a pool of standards and performance data that allows initial assessments and sets the stage for cities wishing to undertake a more extensive benchmarking project.

Although leading municipal performers in several service activities have been identified, others remain untapped. The careful choice of suitable benchmark partners, chosen for their performance excellence, general comparability on relevant characteristics, and willingness to cooperate, is vital to the success of a full-scale benchmarking project.[4]

A municipality can benefit from simply comparing its current performance with relevant benchmarks. In doing so, important strengths and weaknesses may be revealed and local officials may discover aspects of the operation that deserve detailed analysis. They may even decide to rearrange some priorities. To get full value from a benchmarking project, however, a municipality must take the next steps—supplementing the measures found here with others it collects itself, analyzing operations, and developing performance improvement strategies.

Performance Milestones

Resting on one's laurels is hazardous in city government. *Below average* performance is rarely acceptable; *good* performance can be made better; and those judged *the best* can hope to retain that status only through continuous improvement.

The purpose of identifying municipal benchmarks is to provide guideposts to city governments as they attempt to improve operations. Managers who desire easy answers and quick closure will find neither in benchmarking. Instead, they will find a means by which they can gauge the status of their performance and plot changes to enhance it. By comparing their own performance marks with those of other respected cities or with relevant standards, local officials can decide where improvements are needed and may be able to identify models that will prove helpful as they design those improvements.

City officials are encouraged to consult the full array of possible benchmarks for each service area. From that array, a subset of performance indicators has been assembled and labeled *performance milestones.* These are not benchmarks—the achievements of best-in-class performers. Benchmarks may be found in the individual chapters on municipal specialties. Instead, performance milestones denote standing just below best-in-class and serve as guideposts for cities on their way toward the upper echelon. In a sense, they are *near-benchmarks.* They represent good performance and worthy objectives for most municipalities (Table 25.2).

The performance targets identified in Table 25.2 are not destinations. They are milestones on the journey to optimum performance. Although a journey to an optimum place really has no end, some routes are more satisfying than others. Cities plotting their course by focusing on the performance milestones are traversing such a route. Those that have already passed the milestones must travel on, scouting out new guideposts on the basis of their own experience. As seasoned travelers know, there are few rest stops along this route.

Notes

1. For example, Downs and Larkey (1986, p. 22) note that in perhaps the earliest of such initiatives, the Census Bureau collected municipal performance statistics for a few years in the 1890s, until controversy over data interpretation and budget problems brought that effort to an end.

2. ICMA's Comparative Performance Measurement Consortium and The Innovation Groups' Standard Performance Reporting for City/County Services project are designed to compile and share performance information among local governments.

3. The international Bertelsmann Prize, awarded in 1993 to Phoenix, Arizona, as one of the two best-managed cities in the world, is an example of the type of recognition that focuses on the municipality as a multiservice entity.

4. Several excellent books on benchmarking have been published in recent years. Some of the best are Boxwell (1994), Camp (1989), Fitz-enz (1993), Karlof and Ostblom (1993), Leibfried and McNair (1992), and Spendolini (1992). For more specific application to the public sector, see Bruder and Gray (1994).

TABLE 25.2 Performance Milestones

Department/ Function	Performance Variable	Performance Milestone
Animal control	Percentage of emergency responses within 25 minutes, routine responses within 2 hours	90%
	Daily hours of shelter availability or other arrangement for receiving sick/injured animals	24 hours
City attorney	Percentage of oral opinions on routine matters within 24 hours; written opinions on routine matters within 10 days, written opinions on nonroutine matters within 30 days	90%
City clerk	Percentage of city council meeting minutes prepared within 10 days of the meeting	80%
	Percentage of requests for municipal records or information receiving same-day service	95%
Courts	Number of cases heard and processed per employee-year	400
	Collection rate for traffic fines	80%
Data processing	System availability during users' work hours	98%
	Percentage of responses to problems within 1 hour; resolution within 24 hours	90%
Development Administration	Average review time for subdivision plats and rezoning requests	30 days
	Percentage of commercial building permits processed within 10 days; residential, 3 days	95%
	Percentage of building inspections conducted within 1 day of request	95%
Emergency Communications	Percentage of 911 calls answered within 10 seconds	90%
	Percentage of emergency calls dispatched within 2 minutes	75%
EMS	Average response time from call to arrival	6 min.
Finance	Receipt of GFOA certificates for financial reporting and budgeting	yes
	Percentage of monthly reports issued by 10th of following month	80%
	Amount by which return on investments exceeds average 3-month Treasury bill rate	10%
Fire	ISO public protection classification (fire insurance rate)	5
	Percentage of emergency responses within 6 minutes	95%
	Percentage of fires contained within room(s) involved on arrival	85%
Library	Circulation per capita	7
	Percentage of not-immediately-available material provided within 30 days	80%
Parks and recreation	Developed/total park acres per 1,000 population	5 acres/ 7 acres
	Tennis courts per 10,000 population	2
Personnel Administration	Percentage of positions filled within 90 days of vacancy	80%
	Percentage of position audits completed within 30 days of request	85%
Police	Crime rate as percentage of "projected norm" for population cluster and state or region[a]	90%
	Percentage of emergency responses within 6 minutes	95%

TABLE 25.2 *Continued*

Department/ Function	Performance Variable	Performance Milestone
Property appraisal	Average assessment-to-sales ratio	90%-100%
	Percentage of appraisals upheld at Board of Equalization appeal	50%
Public health	Local public health statistics as percentage of selected national targets[b]	100%
	Annual inspections per food service establishment	3
Public transit	Percentage on-time performance—peak/nonpeak periods	90%/95%
	Fares as a percentage of operating expenditures	35%[c]
Public utilities	Maximum line loss—gas/electric/water	2%/6%/14%
	Compliance with AWWA standards of water quality[d]	100%
	Percentage of BOD/TSS removed from wastewater	90%
Public works and engineering	Percentage of potholes and road hazards repaired within 2 days of notification	90%
	Percentage of low bids for major projects within 15% of engineer's estimate	80%
Purchasing and warehousing	Average processing time from requisition to purchase order— sealed bids/other	60 days/ 10 days
	Out-of-stock ratio (inventoried items in warehouse)	6% or less
Solid waste collection	Missed collections per 10,000 scheduled stops	20 or less
	Stops per collection employee per week	1,100
Traffic control	Percentage of centerlines, lane lines, and symbols on major streets repainted during year	75%
	Average response time to damaged stop and yield signs; to malfunctioning traffic signals	3 hours; 1 hour

a. See Table 16.9.
b. See Table 18.1.
c. 20%, if serving a population of less than 200,000.
d. See Table 20.5.

References

Allan, I. J. (1990, November). *Unreserved fund balance and local government finance* (Research Bulletin series). Chicago: Government Finance Officers Association.

Almy, R. R., Gloudemans, R. J., & Thimgan, G. E. (1991). *Assessment practices: Self-evaluation guide.* Chicago: International Association of Assessing Officers.

American Association of State Highway and Transportation Officials (AASHTO). (1984). *An informational guide for roadway lighting.* Washington, DC: Author.

American Library Association. (1962). *Interim standards for small public libraries.* Chicago: Author.

American Library Association. (1967). *Minimum standards for public library systems.* Chicago: Author.

American national standard practice for roadway lighting (ANSI/IESNA RP-8-83). (1983). New York: Illuminating Engineering Society of North America.

American Public Health Association. (1981). *Public swimming pools: Recommended regulations for design and construction, operation and maintenance.* Washington, DC: Author.

American Public Health Association. (1991). *Healthy communities 2000: Model standards* (3rd ed.). Washington, DC: Author.

American Public Power Association. (1993, March). *Selected financial and operating ratios of public power systems: 1991.* Washington, DC: Author.

American Red Cross. (1995). *Lifeguarding today.* St. Louis, MO: Mosby Lifeline.

American Water Works Association. (1992). *1992-1993 officers and committee directory.* Denver, CO: Author.

Americans With Disabilities Act of 1990, Pub. L. No. 101-336, 42 U.S.C. § 12100 et seq.

Ammons, D. N. (1994, Spring). The role of professional associations in establishing and promoting performance standards for local government. *Public Productivity and Management Review, 17,* 281-298.

Ammons, D. N. (1995, January/February). Overcoming the inadequacies of performance measurement in local government: The case of libraries and leisure services. *Public Administration Review, 55,* 37-47.

Ammons, D. N. (1996a). Local government standards via professional associations: How useful are they for gauging performance? In A. Halachmi & G. Bouckaert (Eds.), *Organizational performance and measurement in the public sector* (Chap. 10, pp. 201-222). Westport, CN: Quorum.

Ammons, D. N. (1996b). *Local policies on municipal fund balances.* Unpublished manuscript, University of Georgia, Carl Vinson Institute of Government, Athens.

Attanucci, J. P., Jaeger, L., & Becker, J. (1979, April). *Bus service evaluation procedures: A review: Short range transit planning* (Special Studies in Transportation Planning). Washington, DC: U.S. Department of Transportation.

Baker, S. L., & Lancaster, F. W. (1991). *The measurement and evaluation of library services* (2nd ed.). Arlington, VA: Information Resources Press.

Bannon, J. J. (1976). *Leisure resources: Its comprehensive planning.* Englewood Cliffs, NJ: Prentice Hall.

Barrett, K., & Greene, R. (1994, February 1). The state of the cities: Managing for results. *Financial World, 163,* 40-49.

Bennett, J. (1993, August 2). Budget woes seize fleet buying cycles. *City & State, 10,* 9, 16.

Bens, C. K. (1986, November). Strategies for implementing performance measurement. *Management Information Service Report, 18,* 1-14. (Available from International City/County Management Association, Washington, DC)

Boxwell, R. J., Jr. (1994). *Benchmarking for competitive advantage.* New York: McGraw-Hill.

Boyer, R., & Savageau, D. (1989). *Places rated almanac: Your guide to finding the best places to live in America.* New York: Simon & Schuster.

Brown, Z., & Webster, C. (1913). *Buying list of books for small libraries.* Chicago: American Library Association.

Bruder, K. A., Jr., & Gray, E. M. (1994, September). Public-sector benchmarking: A practical approach. *Public Management, 76,* S9-S14.

Building Service Contractors Association International. (1992). *BSCAI production rate recommendations.* Fairfax, VA: Author.

Butler, R. F. (1985, March). Measuring productivity in local government purchasing. *Texas Town & City, 72,* 41-45.

Camp, R. C. (1989). *Benchmarking: The search for industry best practices that lead to superior performance.* Milwaukee, WI: American Society for Quality Control Press.

Carpenter, V. L., Ruchala, L., & Waller, J. B., Jr. (1991). *Public health: Service efforts and accomplishments reporting—Its time has come.* Norwalk, CT: Governmental Accounting Standards Board.

Carter, L., & Vogt, A. J. (1989, Winter). Fund balance in local government budgeting and finance. *Popular Government, 54,* 33-41.

Center for Design Planning. (1979). *Streetscape equipment sourcebook 2.* Washington, DC: Urban Land Institute.

Chilton Book Company. (annual). *Chilton's labor guide and parts manual.* Radnor, PA: Author.

Chute, A. (1992, June). *Public libraries in the U.S.: 1990* (NCES 92-028). Washington, DC: U.S. Department of Education, Office of Educational Research and Improvement.

City of Alexandria (VA). *Fiscal year 1992-1993 approved budget.*

City of Chandler (AZ). *Annual budget 1992-1993.*

City of Charlotte (NC). *FY88-FY89 objectives.*

City of Charlotte (NC). *FY95 operating plan.*

City of Dallas (TX). (1989, August). *Dallas police department's recommendations to improve response time to citizens' calls for police service.*

City of Dallas (TX) Fire Department. (1992, April 20). *Comparative analysis with index cities.*

City of Dallas (TX) Parks and Recreation Department. (1992, April). *City comparison study.*

City of Dayton (OH) Office of Management and Budget. (1991). *1991 program strategies.*

City of Fayetteville (AR). (1989, January). *Public works maintenance management system: System manual.*

City of Fort Collins (CO). *1993 annual budget: Volume I. Budget overview.*

City of Greenville (SC). *1991 year-end performance report.*

City of Largo (FL). *Annual budget: Fiscal year 1993.*

City of Little Rock (AR). (1990). *Management information systems manual.*

City of Overland Park (KS). *1993 budget.*

City of Pasadena (CA) Management Audit Team. (1986, October). *Opportunities for improving efficiency and effectiveness: Department of public works.*

City of Philadelphia (PA) Office of the City Controller. (1991, January). *Performance audit: Streets department/sanitation division.*

City of Phoenix (AZ). *The Phoenix budget 1993-94.*

City of Phoenix (AZ) Auditor Department. (1991, December 31). *Fire department performance indicators.*

City of Portland (OR) Office of the City Auditor. (1992, May). *An evaluation of city financial trends: 1980-92.*

City of Portland (OR) Office of the City Auditor. (1993, January). *Service efforts and accomplishments: 1991-92* (Rep. No. 176).

City of Rock Island (IL). (1993). *Budget for fiscal year 1993-1994.*

City of Santa Ana (CA). *Annual budget 1988-89.*

City of Savannah (GA) Community, Housing, and Economic Development Department. (1988, October). *Responsive public services programs.*

City of Savannah (GA). (1992). *1992 program of work.*

City of St. Petersburg (FL). (1992). *Approved 1992 program budget.*

City of Sunnyvale (CA). (1993, June). *Performance indicators.*

City of Tucson (AZ). (1989, December). *City of Tucson parks and recreation master plan 2000: Planning our recreational future.*

City of Vancouver (WA). (1991). *1991-1992 biennial budget.*

City of Washington, D.C. (1989, March). *Improving ambulance operations in Washington, D.C.: A blueprint for change.* Washington, DC: Office of the City Administrator, Productivity Management Services.

City of Washington, D.C. (n.d.). *Improving vehicle maintenance productivity in the D.C. Department of Transportation* (Productivity Rep. No. 4). Washington, DC: Department of Transportation and Office of the City Administrator, Productivity Management Services.

Clawson, M. (1963). *Land and water for recreation.* Chicago: Rand McNally.

Commission on Accreditation for Law Enforcement Agencies (CALEA). (1993). *Standards for law enforcement agencies: The standards manual of the law enforcement agency accreditation program.* Fairfax, VA: Author. (Updated periodically)

Crowell, A., & Sokol, S. (1993, May). Playing in the gray: Quality of life and the municipal bond rating game. *Public Management, 75,* 2-6.

Decker, L. L., & Manion, P. (1987, March-April). Performance measurement in Phoenix: Trends and a new direction. *National Civic Review, 76,* 119-129.

Donahue, R. (Ed.). (1986). *Park maintenance standards.* Alexandria, VA: National Recreation and Park Association.

Downs, G. W., & Larkey, P. D. (1986). *The search for government efficiency: From hubris to helplessness.* New York: Random House.

Drebin, A., & Brannon, M. (1992). *Police department programs: Service efforts and accomplishments reporting—Its time has come.* Norwalk, CT: Governmental Accounting Standards Board.

Drucker, P. F. (1975). *Management: Tasks, responsibilities and practices.* New York: Harper & Row.

Erwin, J. (1993, January). Firefighters see red: Crew size and deaths or injuries are debated. *Public Management, 75,* 2-5.

Fast facts. (1992, October). *Public Management, 74,* 21.

Fast facts. (1995, January). *Public Management, 77,* 28.

Ferguson, E. J. (1980). Street maintenance. In G. J. Washnis (Ed.), *Productivity improvement handbook for state & local government.* New York: John Wiley.

Fiction catalog. New York: H. W. Wilson. (Published every few years, with supplements)

Fitz-enz, J. (1993). *Benchmarking staff performance.* San Francisco: Jossey-Bass.

Foster, W. S. (1978). *Handbook of municipal administration and engineering.* New York: McGraw-Hill.

Fowler, P. (1992, February). The establishment of health indicators in municipal health departments. *Texas Town & City, 79,* 20-24.

Fox, S. F. (1993, Spring). Professional norms and actual practices in local personnel administration: A status report. *Review of Public Personnel Administration, 13,* 5-28.

Glover, M. (1993, October 7). *Performance measurement: Local government efforts and successes.* Paper presented at the annual meeting of the Southeastern Conference on Public Administration, Cocoa Beach, FL.

Gold, S. M. (1973). *Urban recreation planning.* Philadelphia: Lea & Febiger.

Gordon, B. B., Drozda, W., & Stacey, G. S. (1969, December 31). *Cost-effectiveness in fire protection* (Working paper). Columbus, OH: Urban Studies Center, Battelle Memorial Institute.

Government Finance Officers Association (GFOA). (1993). *Distinguished budget presentation awards program: Awards criteria.* Chicago: Author.

Governmental Accounting Standards Board (GASB). (1992, December 18). *Preliminary views of the Governmental Accounting Standards Board on concepts related to service efforts and accomplishments reporting.* Norwalk, CT: Author.

Granito, J. (1991). Evaluation and planning of public fire protection. In A. E. Cote & J. L. Linville (Eds.), *Fire protection handbook* (17th ed., Sec. 10, Chap. 4). Quincy, MA: National Fire Protection Association.

Grizzle, G. A. (1987, Spring). Linking performance to funding decisions: What is the budgeter's role? *Public Productivity Review, 10,* 33-44.

Groves, S. M. (1980). *Financial trend monitoring system: A practitioner's handbook: Handbook 2. Evaluating financial condition.* Washington, DC: International City Management Association.

Groves, S. M., & Godsey, W. M. (1980). *Evaluating financial condition.* Washington, DC: International City Management Association.

Groves, S. M., Godsey, W. M., & Shulman, M. A. (1981, Summer). Financial indicators for local government. *Public Budgeting & Finance, 1,* 5-19.

Groves, S. M., & Valente, M. G. (1986). *Evaluating financial condition: A handbook for local government.* Washington, DC: International City Management Association.

Guidelines for public libraries (IFLA Pub. No. 6). (1986). Munich, Germany: K. G. Saur for the International Federation of Library Associations and Institutions.

Haley, A. J. (1985). Municipal recreation and park standards in the United States: Central cities and suburbs, 1975-1980. *Leisure Sciences, 7*(2), 175-188.

Hall, J. R., Jr., & Cote, A. E. (1991). America's fire problem and fire protection. In A. E. Cote & J. L. Linville (Eds.), *Fire protection handbook* (17th ed., Sec. 1, Chap. 1). Quincy, MA: National Fire Protection Association.

Hall, J. R., Jr., Koss, M., Schainblatt, A. H., Karter, M. J., Jr., & McNerney, T. C. (1979). *Fire code inspections and fire prevention: What methods lead to success?* Boston: National Fire Protection Association.

Handy, G. L. (1993, September). Local animal control management. *Management Information Service Report, 15,* 1-20. (Available from International City/County Management Association, Washington, DC)

Hatry, H. P. (1978, January/February). The status of productivity measurement in the public sector. *Public Administration Review, 38,* 28-33.

Hatry, H. P. (1980a, May-June). Local government uses for performance measurement. *Intergovernmental Personnel Notes,* 13-15. (Available from U.S. Office of Personnel Management, Washington, DC)

Hatry, H. P. (1980b, December). Performance measurement principles and techniques: An overview for local government. *Public Productivity Review, 4,* 312-339.

Hatry, H. P., Blair, L. H., Fisk, D. M., Greiner, J. M., Hall, J. R., Jr., & Schaenman, P. S. (1992). *How effective are your community services? Procedures for measuring their quality* (2d ed). Washington, DC: Urban Institute and International City/County Management Association.

Hatry, H. P., Fountain, J. R., Jr., & Sullivan, J. M. (1990). Overview. In H. P. Hatry, J. R. Fountain, J. M. Sullivan, & L. Kremer (Eds.), *Service efforts and accomplishments reporting: Its time has come—An overview.* Norwalk, CT: Governmental Accounting Standards Board.

Humane Society of the United States. (1982, April). *HSUS guidelines for animal shelter policies.* Washington, DC: Author.

Humane Society of the United States. (1986). *Responsible animal regulation.* Washington, DC: Author.

Humane Society of the United States. (1988, November). *HSUS shelter guidelines pertaining to potentially dangerous dogs.* Washington, DC: Author.

Humane Society of the United States. (1990, September). *HSUS guidelines for the operation of an animal shelter.* Washington, DC: Author.

Humane Society of the United States. (n.d.). *HSUS guidelines for responsible pet adoptions.* Washington, DC: Author.

Illinois Library Association. (1989). *Avenues to excellence II: Standards for public library service in Illinois.* Chicago: Author.

Insurance Services Office, Inc. (1980). *Fire suppression rating schedule* (6-80 ed.). New York: Author.

International Association of Fire Fighters. (1993). *Safe fire fighter staffing: Critical considerations.* Washington, DC: Author.

Karlof, B., & Ostblom, S. (1993). *Benchmarking: A signpost to excellence in quality and productivity.* New York: John Wiley.

Karter, M. J., Jr. (1992, October). *U.S. fire department profile through 1991.* Quincy, MA: National Fire Protection Association.

Karter, M. J., Jr. (1993, July/August). NFPA surveys U.S. fire departments. *NFPA Journal, 87,* 59-62.

Keillor, G. (1985). *Lake Wobegon days.* New York: Penguin.

Keillor, G. (1987). *Leaving home.* New York: Viking.

Kraus, R. G., & Curtis, J. E. (1982). *Creative management in recreation and parks.* St. Louis, MO: C. V. Mosby.

Krohe, J., Jr. (1990, December). Park standards are up in the air. *Planning, 56,* 10-13.

Lancaster, R. A. (Ed.). (1983). *Recreation, park and open space standards and guidelines.* Alexandria, VA: National Recreation and Park Association.

League of California Cities (LCC). (1994). *A "how to" guide for assessing effective service levels in California cities.* Sacramento: Author.

LeGrotte, R. B. (1987). *Performance measurement in major U.S. cities.* Unpublished master's thesis, North Texas State University, Denton.

Leibfried, K., & McNair, C. (1992). *Benchmarking: A tool for continuous improvement.* New York: HarperCollins.

Level of service standards: Measures for maintaining the quality of community life. (1994, September). Kirkland: Municipal Research and Services Center of Washington.

MacManus, S. A. (1984, Autumn). Coping with retrenchment: Why local governments need to restructure their budget document formats. *Public Budgeting and Finance, 4,* 58-66.

Massey, J., & Tyer, C. (1990, April-June). Local government fund balances: How much is enough? *The South Carolina Forum, 1,* 40-46.

McGowan, R. P., & Poister, T. H. (1985, February). Impact of productivity measurement systems on municipal performance. *Policy Studies Review, 4,* 532-540.

Mercer Management Consulting, Inc. (1992). *Benchmarking the quality of utility customer services: 1991 survey results.* Lexington, MA: Author.

Mitchell International. (annual). *Mechanical labor estimating guide.* San Diego, CA: Author.

Moody's Public Finance Department. (1988). *1988 medians: Selected indicators of municipal performance.* New York: Moody's Investors Service, Inc.

Moulton, C., Wright, P., & Rindy, K. (1991, April 1). The role of animal shelters in controlling pet overpopulation. *Journal of the American Veterinary Medical Association, 198,* 1172-1176.

National Arbor Day Foundation. (n.d.). *Tree City USA application.* Nebraska City, NE: Author.

National Arborist Association. (1987). *Standard for fertilizing shade and ornamental trees.* Amherst, NH: Author.

National Arborist Association. (1988). *Pruning standards for shade trees.* Amherst, NH: Author.

National Civic League. (1993). *Civic index.* Denver, CO: Author.

National Commission on Accreditation for Local Park and Recreation Agencies. (1992, November). *Quick-start manual: Local park and recreation agency accreditation standards.*

National Commission on the State and Local Public Service. (1993). *Hard truths/tough choices: An agenda for state and local reform.* Albany, NY: Nelson A. Rockefeller Institute of Government.

National Fire Protection Association. (1991). *NFPA 1911: Service tests of pumps on fire department apparatus.* Quincy, MA: Author.

National Fire Protection Association. (1992). *NFPA 1500: Standard on fire department occupational safety and health program.* Quincy, MA: Author.

National Institute of Governmental Purchasing. (1994). *Results of the 1993 survey of procurement practices.* Reston, VA: Author.

Nunn, S. (1992, Winter). Organizational improvement: The case of Village Creek. *Public Productivity and Management Review, 16,* 117-136.

O'Connell, G. B. (1989, June). Rate your city: Here's how! *Public Management, 71,* 7-10.

Olson, L. R., Shuck, D. C., Vinyon, M. J., & Lilja, J. B. (1980, March). The need for adequate fund balance levels. *Minnesota Cities, 65,* 9-12, 21.

Osborne, D., & Gaebler, T. (1992). *Reinventing government: How the entrepreneurial spirit is transforming the public sector.* Reading, MA: Addison-Wesley.

O'Toole, D. E., & Marshall, J. (1987, October). Budgeting practices in local government: The state of the art. *Government Finance Review, 3,* 11-16.

O'Toole, D. E., & Stipak, B. (1988, Fall). Budgeting and productivity revisited: The local government picture. *Public Productivity Review, 12,* 1-12.

Pacific Northwest collection assessment manual (3rd ed.). (1990). Salem: Oregon State Library Foundation.

Parry, R. W., Jr., Sharp, F. C., Vreeland, J., & Wallace, W. A. (1991). *Fire department programs: Service efforts and accomplishments reporting—Its time has come.* Norwalk, CT: Governmental Accounting Standards Board.

Peters, T. J., & Waterman, R. H., Jr. (1982). *In search of excellence: Lessons from America's best-run companies.* New York: Harper & Row.

Peterson, G. E., Miller, M. J., Godwin, S. R., & Shapiro, C. (1984). *Guide to benchmarks of urban capital condition.* Washington, DC: Urban Institute.

Peterson, W. (1991). Fire department occupational safety and health. In A. E. Cote & J. L. Linville (Eds.), *Fire protection handbook* (17th ed., pp. 9.4-9.13). Quincy, MA: National Fire Protection Association.

Poister, T. H. (1992). Productivity monitoring: Systems, indicators, and analysis. In M. Holzer (Ed.), *Public productivity handbook* (pp. 195-211). New York: Marcel Dekker.

Poister, T. H., & McGowan, R. P. (1984). Municipal management capacity: Productivity improvement and strategies for handling fiscal stress. In *The Municipal Year Book 1984* (pp. 206-214). Washington, DC: International City Management Association.

Public Library Association. (1992). *Statistical report '92: Public library data service.* Chicago: American Library Association.

Public library catalog. New York: H. W. Wilson. (Published every few years, with supplements)

R. S. Means Company. (annual). *Means site work cost data.* Kingston, MA: Author.

Rider, K. L. (1979, May). The economics of the distribution of municipal fire protection services. *Review of Economics and Statistics, 61,* 249-258.

Ridley, C. E., & Simon, H. A. (1943). *Measuring municipal activities: A survey of suggested criteria for appraising administration.* Chicago: International City Managers' Association.

Roberts, E. G. (1991). Fire department organization. In A. E. Cote & J. L. Linville (Eds.), *Fire protection handbook* (17th ed., Sec. 9, Chap. 4). Quincy, MA: National Fire Protection Association.

Rosenhan, A. K. (1991). Fire department apparatus, equipment, and protective clothing. In A. E. Cote & J. L. Linville (Eds.), *Fire protection handbook* (17th ed.). Quincy, MA: National Fire Protection Association.

Rubin, M. A. (1991). *Sanitation collection and disposal: Service efforts and accomplishments reporting—Its time has come.* Norwalk, CT: Governmental Accounting Standards Board.

Sampayo, F. (1994, February). EPA's Part 503 regulations will increase POTW standards. *American City & County, 109,* 36.

Savageau, D., & Boyer, R. (1993). *Places rated almanac: Your guide to finding the best places to live in North America.* New York: Simon & Schuster.

Schultz, G. (1991). Water distribution systems. In A. E. Cote & J. L. Linville (Eds.), *Fire protection handbook* (17th ed.). Quincy, MA: National Fire Protection Association.

Shivers, J. S., & Hjelte, G. (1971). *Planning recreational places.* Cranbury, NJ: Fairleigh Dickinson University Press.

Smith, G. N. (1992, February 18). Mayoral measures. *Financial World, 161,* 8.

Southwestern Pennsylvania Regional Planning Commission (SPRPC). (1990, Spring). *Standards for effective local government: A workbook for performance assessment.* Pittsburgh: Author.

Spendolini, M. J. (1992). *The benchmarking book.* New York: AMACOM.

Standard & Poor's. (1983). *Credit overview: Municipal ratings.* New York: Author.

Standard & Poor's. (1995). *Municipal finance criteria: 1995.* New York: Author.

Standards for North Carolina public libraries. (1988). North Carolina Library Association/Public Library Section and North Carolina Public Library Directors Association. Raleigh: Author.

Standards for public library service in Ohio: 1992 revision and survey of compliance. (1995). Columbus: Ohio Library Council.

Stauffer, E. (1991). Fire department emergency communication systems. In A. E. Cote & J. L. Linville (Eds.), *Fire protection handbook* (17th ed., Sec. 4, Chap. 1). Quincy, MA: National Fire Protection Association.

Stevens, B. J. (Ed.). (1984, June). *Delivering municipal services efficiently: A comparison of municipal and private service delivery.* New York: Ecodata, Inc. (Tech. rep. prepared for the U.S. Department of Housing and Urban Development, Washington, DC: Government Printing Office)

Terry, J. (1991, June 10). *Research project into performance budget measures of typical public service agencies used by select cities nationwide* [Memorandum]. Boston: City of Boston.

Thompson, L. H. (1991, May 8). *Service to the public: How effective and responsive is the government?* (Statement before the U.S. House of Representatives Committee on Ways and Means, GAO/T-HRD-91-26). Washington, DC: U.S. General Accounting Office.

U.S. Department of Justice, Bureau of Justice Statistics. (1992). *Law enforcement management and administrative statistics, 1990: Data for individual state and local agencies with 100 or more officers.* Washington, DC: Government Printing Office.

U.S. Department of Justice, Federal Bureau of Investigation. (1994). *Crime in the United States 1993: Uniform crime reports.* Washington, DC: Government Printing Office.

U.S. Department of Transportation, Federal Highway Administration. (1987, December). *Highway performance monitoring system field manual.* Washington, DC: Government Printing Office.

U.S. Department of Transportation, Federal Highway Administration. (1988). *Manual on uniform traffic control devices for streets and highways.* Washington, DC: Government Printing Office.

U.S. Department of Transportation, Federal Transit Administration. (1994a, December). *Transit profiles: Agencies in urbanized areas exceeding 200,000 population.* Washington, DC: Author.

U.S. Department of Transportation, Federal Transit Administration. (1994b, December). *Transit profiles: Agencies in urbanized areas with a population of less than 200,000.* Washington, DC: Author.

U.S. Department of Transportation, Federal Transit Administration. (1994c, December). *Transit profiles: The thirty largest agencies.* Washington, DC: Author.

U.S. General Accounting Office. (1980, November). *Additional federal aid for urban water distribution systems should wait until needs are clearly established* [Rep. to Cong.]. Washington, DC: Author.

U.S. General Accounting Office. (1993, February). *Performance budgeting: State experiences and implications for the federal government* (GAO/AFMD-93-41). Washington, DC: Author.

Usher, C. L., & Cornia, G. C. (1981, March/April). Goal setting and performance assessment in municipal budgeting. *Public Administration Review, 41,* 229-235.

Van House, N. A., Lynch, M. J., McClure, C. R., Zweizig, D. L., & Rodger, E. J. (1987). *Output measures for public libraries: A manual of standardized procedures* (2nd ed.). Chicago: American Library Association.

Walker, R. A. (1989). *Parks division quality standards manual.* Sunnyvale, CA: City of Sunnyvale.

Wallace, W. A. (1991). *Mass transit: Service efforts and accomplishments reporting—Its time has come.* Norwalk, CT: Governmental Accounting Standards Board.

Weech, T. L. (1988, Summer). Small public libraries and public library standards. *Public Libraries, 27,* 72-74.

Withers, F. N. (1974). *Standards for library service: An international survey.* Paris: UNESCO Press.

Wright, P., Cassidy, B. A., & Finney, M. (1986, July). Local animal control management. *Management Information Service Report, 18,* 1-20. (Available from International City/County Management Association, Washington, DC)

Yates, D. (1977). *The ungovernable city: The politics of urban problems and policy making.* Cambridge: MIT Press.

Municipal Documents

City of Alexandria (VA). *Fiscal year 1992-1993 approved budget.*
City of Alexandria (VA). *Fiscal year 1994-1995 approved budget.*

City of Amarillo (TX). *Annual budget: 1992-1993.*

City of Anaheim (CA). *Resource allocation plan: 1992/93.*

City of Auburn (AL). *FY93 annual budget.*

City of Bakersfield (CA). *Annual budget, 1989-90.*

City of Boise (ID). *Annual budget 1992-93.*

City of Boston (MA). *The mayor's management report: FY92: Volume I.*

City of Boston (MA). *The mayor's management report: FY92: Volume II.*

City of Burbank (CA) Fire Department. *Management reporting system: Year end report 1988-89.*

City of Burbank (CA). *1992-1994 biennial budget.*

City of Carrollton (TX). *Comprehensive annual financial report: Fiscal year ended September 30, 1993.*

City of Chandler (AZ). *Annual budget 1992-1993.*

City of Charlotte (NC). *FY88-FY89 objectives.*

City of Charlotte (NC) Budget and Evaluation Department. (1994, January). *Achievement of objectives: FY94 mid-year status report.*

City of Charlotte (NC). *FY95 operating plan.*

City of Chesapeake (VA). *Approved operating budget: 1992-93.*

City of Cincinnati (OH) Research, Evaluation and Budget. (1992, May 1). *Annual performance report: 1991 measures of success.*

City of Cincinnati (OH). *1993/1994 budget: Program information.*

City of College Park (MD). (1992, May 26). *Adopted operating and capital budget for the fiscal year ending June 30, 1993 and five year capital improvement plan.*

City of College Station (TX). *1992-1993 annual budget.*

City of Columbus (OH). *Proposed 1993 budget.*

City of Coral Gables (FL). *1992-93 budget.*

City of Corpus Christi (TX). *Annual budget: Fiscal year 1992-1993.*

City of Corvallis (OR). *Quarterly operating report: Fourth quarter ending June 30, 1992.*

City of Corvallis (OR). *Fiscal year 1992-93 annual budget.*

City of Dallas (TX). (1989, August). *Dallas police department's recommendations to improve response time to citizens' calls for police service.*

City of Dallas (TX). (1990, February). *Emergency medical service.*

City of Dallas (TX) Fire Department. (1992, April 20). *Comparative analysis with index cities.*

City of Dallas (TX) Parks and Recreation Department. (1992, April). *City comparison study.*

City of Dayton (OH) Office of Management and Budget. (1991). *1991 program strategies.*

City of Decatur (IL). *Annual performance budget, fiscal year 1992-1993: Volume I.*

City of Denton (TX). *Annual program of services: 1992-93.*

City of Duncanville (TX). *Annual budget: Fiscal year October 1, 1991-September 30, 1992.*

City of Dunedin (FL). *Fiscal year 1992/1993 adopted annual operating & capital budgets.*

City of Fairfield (CA). *Proposed 1992/93 budget and ten-year financial plan.*

City of Fayetteville (AR). (1989, January). *Public works maintenance management system: System manual.*

City of Fayetteville (AR). *Annual budget and work program: 1993.*

City of Fayetteville (AR). *Annual budget and work program: 1994.*

City of Fort Collins (CO). *1993 Annual budget: Volume I. Budget overview.*

City of Fort Collins (CO). *1993 Annual budget: Volume II. Program performance budget.*

City of Fort Lauderdale (FL). *Annual operating budget: Fiscal year 1992-1993.*

City of Fort Smith (AR). *Fiscal year 1993 budget.*

City of Fort Worth (TX). *1992-1993 annual budget and program objectives.*

City of Fullerton (CA). *Annual budget 1990-91.*

City of Grand Junction (CO). *Comprehensive annual financial report for the fiscal year ended December 31, 1991.*

City of Grand Junction (CO). *1992-1993 municipal budgets.*

City of Grand Prairie (TX). *Monthly performance report: Month ending December 31, 1992.*

City of Grants Pass (OR). *Adopted operating/capital budget: Fiscal year 1992-1993.*

City of Grants Pass (OR). (1992, December). *Manager's monthly report: Statistics on operations of city of Grants Pass.*

City of Greenbelt (MD). *Adopted budget for fiscal year 1992-1993.*

City of Greenville (SC). *1991 year-end performance report.*

City of Greenville (SC). *1992 adopted budget.*

City of Gresham (OR). *Adopted operating/capital budget document: Fiscal year 1992-93.*

City of Houston (TX). *Adopted budget for fiscal year ending June 30, 1993: Volume one.*

City of Houston (TX). *Adopted budget for fiscal year ending June 30, 1993: Volume two.*

City of Houston (TX) Finance and Administration Department. *FY '93 monthly status report for the period ended December 31, 1992.*

City of Huntington Beach (CA). *Annual budget 1989-90.*

City of Hurst (TX). *Annual operating budget for fiscal year of 1991-92.*

City of Indianapolis (IN). *1993 annual operating budget.*

City of Iowa City (IA). *Comprehensive annual financial report for the fiscal year ended June 30, 1992.*

City of Iowa City (IA). *Financial plan for FY 93-95.*

City of Irving (TX). *Annual operating budget: Municipal service programs 1992-1993.*

City of Jacksonville (FL). *Annual budget for the fiscal year ending September 30, 1993.*

City of Knoxville (TN). *FY92-93 annual operating budget.*

City of Largo (FL). *Annual budget: Fiscal year 1993.*

City of League City (TX). *Annual budget for fiscal year 1992-93.*

City of Lee's Summit (MO). *Operating budget: Year beginning July 1, 1992.*

City of Lewiston (ID). *Annual budget: Fiscal year 1992-1993.*

City of Liberty (MO). (1992, November). *Service demand issue paper.* (Prepared by P. Gentrup, assistant to the city administrator)

City of Little Rock (AR). (1990). *Management information systems manual.*

City of Little Rock (AR). *1993 annual operating budget.*

City of Long Beach (CA). *Adopted resource allocation plan: Fiscal year 1992-93/operating detail.*

City of Longview (TX). *Annual budget 1992-1993.*

City of Longview (TX). *Annual budget 1993-1994.*

City of Loveland (CO). *1993 budget.*

City of Lubbock (TX). *Goals and objectives: 1992-93.*

City of Milwaukee (WI). *1993 budget.*

City of Milwaukee (WI). *Special ADAP supplement to the proposed executive budget: 1993.*

City of Milwaukee (WI). *1994 budget.*

City of Milwaukee (WI). *1995 budget.*

City of New York (NY). (1992, September 17). *The mayor's management report.*

City of Newburgh (NY). *The fiscal budget for the year 1993.*

City of Newnan (GA). *Annual budget 1993.*

City of Oak Ridge (TN). *Goals & objectives program: Annual report fiscal year 1992.*

City of Oak Ridge (TN). *1993 budget.*

City of Oak Ridge (TN). *1994 budget.*

City of Oakland (CA). *1992-93 adopted policy budget* (Vol. 2).

City of Ocala (FL). *1992-1993 budget.*

City of Ogden (UT). *Budget: Fiscal year 1992-93.*

City of Oklahoma City (OK). *Annual budget: Fiscal year 1992/93.*

City of Orlando (FL). *Annual budget: 1992-1993.*

City of Overland Park (KS). *1993 budget.*

City of Pasadena (CA) Management Audit Team. (1986, October). *Opportunities for improving efficiency and effectiveness: Department of public works.*

City of Peoria (AZ). *1991-92 approved budget and financial plan.*

City of Philadelphia (PA) Office of the City Controller. (1991, January). *Performance audit: Streets department/sanitation division.*

City of Phoenix (AZ). *The Phoenix budget 1992-93.*

City of Phoenix (AZ). *The Phoenix budget 1993-94.*

City of Phoenix (AZ) Auditor Department. (1991, December 31). *Fire department performance indicators.*

City of Phoenix (AZ) Auditor Department. (1992, May 6). *Parks, recreation and library department performance indicators.*

City of Phoenix (AZ) Auditor Department. (1992, June 12). *Police department performance indicators.*

City of Portland (OR) Office of the City Auditor. (1992, May). *An evaluation of city financial trends: 1980-91.*

City of Portland (OR). *FY 1992-93 adopted budget: Volume I.*

City of Portland (OR) Office of the City Auditor. (1993, January). *Service efforts and accomplishments: 1991-92* (Rep. No. 176).

City of Portsmouth (VA). *Adopted annual operating budget for the fiscal year ending June 30, 1995.*

City of Raleigh (NC). *Description of operating programs: FY 1992-93.*

City of Reno (NV). *Program and financial plan: Fiscal year 1991-1992.*

City of Richmond (VA). *General fund budget 1993-1994: Proposed.*

City of Rochester (NY). *1992-93 budget.*

City of Rochester (NY). *1992-93 second quarter report: Program performance.*

City of Rock Hill (SC). *Budget and work program 1993.*

City of Rock Island (IL). (1993). *Budget for fiscal year 1993-1994.*

City of Sacramento (CA). (1992, September-November). *Budget workshop materials.*

City of Saginaw (TX). *Annual budget: Fiscal year 1992-1993.*

City of San Antonio (TX). *Adopted annual budget 1992-1993: Detail information.*

City of San Jose (CA). *Adopted operating budget 1989-90.*

City of San Luis Obispo (CA). *1991-93 financial plan & approved 1991-1992 budget.*

City of San Luis Obispo (CA). (1992, July). *1991-93 financial plan supplement and approved 1992-93 budget.*

City of Sanford (NC). *Annual operating budget: Fiscal year 1992-93.*

City of Santa Ana (CA). *Annual budget 1988-89.*

City of Santa Fe (NM). *Annual operating budget: Fiscal year 1992-93.*

City of Savannah (GA). (1992). *1992 program of work.*

City of Savannah (GA). *1994 service program & budget.*

City of Savannah (GA). Community, Housing and Economic Development Department. (1988, October). *Responsive public services programs.*

City of Shreveport (LA). *1993 annual operating budget.*

City of Sierra Vista (AZ). *Annual budget: Fiscal year 1992/1993.*

City of Sioux Falls (SD). *Comprehensive annual financial report: Fiscal year ended December 31, 1991.*

City of Smyrna (GA). *Annual financial plan for the fiscal year ending June 30, 1993.*

City of Southfield (MI). *Municipal budget: July 1992-June 1993.*

City of St. Charles (IL). *Budget plan 1992-1993.*

City of St. Petersburg (FL). (1992). *Approved 1992 program budget.*

City of Sterling Heights (MI). *1992/93 annual budget.*

City of Sunnyvale (CA). (1993, June). *Performance indicators.*

City of Tacoma (WA). *Preliminary program biennial budget: 1993/1994.*

City of Tallahassee (FL). *Fiscal year 1993 approved detail budget.*

City of Tucson (AZ). (1989, December). *City of Tucson parks and recreation master plan 2000: Planning our recreational future.*

City of Tucson (AZ). *Cost report summary: Fiscal year 1991-92.* (Prepared by S. Postil, Department of Budget and Research)

City of Tucson (AZ). *Fiscal year 1992-1993 budget.*

City of Upper Arlington (OH). *Annual budget: Fiscal year 1992.*

City of Vancouver (WA). (1991). *1991-1992 biennial budget.*

City of Washington, D.C. (1989, June). *Improving emergency medical services.* Office of the City Administrator, Productivity Management Services.

City of Washington, D.C. (1989, March). *Improving ambulance operations in Washington, D.C.: A blueprint for change.* Washington, DC: Office of the City Administrator, Productivity Management Services.

City of Washington, D.C. (n.d.). *Improving vehicle maintenance productivity in the D.C. Department of Transportation* (Productivity Rep. No. 4). Washington, DC: Department of Transportation and Office of the City Administrator, Productivity Management Services.

City of Waukesha (WI). *Annual budget management plan: 1993.*

City of Wichita (KS). *1993/94 annual budget performance measures.*

City of Wilmington (DE). *Annual budget: Fiscal year 1993.*

City of Wilmington (DE) Office of Management and Budget. *MAPP: Management's administrative planning process, FY 1994.*

City of Winston-Salem (NC). *Annual budget program 1986-87: Year-end report.*

City of Winston-Salem (NC). *Annual budget program for FY92-93.*

City of Winston-Salem (NC). (1993, January). *Year-end reports for FY91-92.*

Metropolitan Government of Nashville and Davidson County (TN). *Fiscal year 1993 operating budget.*

Sedgwick County (KS). *1991 budget.*

Town of Addison (TX). *Annual budget: Fiscal year 1992-93.*

Town of Avon (CT). *Annual operating budget 1992-1993.*

Town of Blacksburg (VA). *Town manager's recommended budget: Fiscal year 1995-1996.*

Village of Hanover Park (IL). *Comprehensive annual financial report: Fiscal year ended April 30, 1992.*

Village of Hanover Park (IL). *1992-1993 annual budget.*

Index of Municipalities

Subject and Author Index